NEUROMETHODS

For further volumes:
http://www.springer.com/series/7657

Neural Tracing Methods

Tracing Neurons and Their Connections

Edited by

Benjamin R. Arenkiel

Baylor College of Medicine, Houston, TX, USA

Editor
Benjamin R. Arenkiel
Baylor College of Medicine
Houston, TX, USA

ISSN 0893-2336 ISSN 1940-6045 (electronic)
ISBN 978-1-4939-1962-8 ISBN 978-1-4939-1963-5 (eBook)
DOI 10.1007/978-1-4939-1963-5
Springer New York Heidelberg Dordrecht London

Library of Congress Control Number: 2014953902

Cover illustration: A confocal image of mitral cells in a fixed olfactory bulb slice that were retrogradely labeled using engineered rabies virus injected into the piriform cortex.

Printed on acid-free paper

Humana Press is a brand of Springer
Springer is part of Springer Science+Business Media (www.springer.com)

Series Preface

Experimental life sciences have two basic foundations: concepts and tools. The *Neuromethods* series focuses on the tools and techniques unique to the investigation of the nervous system and excitable cells. It will not, however, shortchange the concept side of things as care has been taken to integrate these tools within the context of the concepts and questions under investigation. In this way, the series is unique in that it not only collects protocols but also includes theoretical background information and critiques which led to the methods and their development. Thus it gives the reader a better understanding of the origin of the techniques and their potential future development. The *Neuromethods* publishing program strikes a balance between recent and exciting developments like those concerning new animal models of disease, imaging, *in vivo* methods, and more established techniques, including, for example, immunocytochemistry and electrophysiological technologies. New trainees in neurosciences still need a sound footing in these older methods in order to apply a critical approach to their results.

Under the guidance of its founders, Alan Boulton and Glen Baker, the *Neuromethods* series has been a success since its first volume published through Humana Press in 1985. The series continues to flourish through many changes over the years. It is now published under the umbrella of Springer Protocols. While methods involving brain research have changed a lot since the series started, the publishing environment and technology have changed even more radically. Neuromethods has the distinct layout and style of the Springer Protocols program, designed specifically for readability and ease of reference in a laboratory setting.

The careful application of methods is potentially the most important step in the process of scientific inquiry. In the past, new methodologies led the way in developing new disciplines in the biological and medical sciences. For example, Physiology emerged out of Anatomy in the nineteenth century by harnessing new methods based on the newly discovered phenomenon of electricity. Nowadays, the relationships between disciplines and methods are more complex. Methods are now widely shared between disciplines and research areas. New developments in electronic publishing make it possible for scientists that encounter new methods to quickly find sources of information electronically. The design of individual volumes and chapters in this series takes this new access technology into account. Springer Protocols makes it possible to download single protocols separately. In addition, Springer makes its print-on-demand technology available globally. A print copy can therefore be acquired quickly and for a competitive price anywhere in the world.

Wolfgang Walz

Preface

The nervous system is built upon an elaborate collection of cells that form precise patterns of synaptic connections underlying complex behaviors and physiologies. Proper nervous system function emerges from a remarkably robust set of genetic blueprints that specify evolutionarily conserved patterns of circuit organization and connectivity between morphologically and functionally diverse sets of neurons.

Although it is well established that individual neurons represent the elemental building blocks of the brain, understanding how neurons "wire up" through synapses and ultimately contribute to complex neural circuit function remains one of biology's major challenges. Our current understanding of how interconnected neuronal populations produce perception, memory, and behavior remains nascent. In most cases, it is far from clear how brain processing emerges from the multitude of cells that make up the nervous system. To begin to unravel the details of complex nervous system function, we must consider not only the morphological and physiological properties of individual neurons, but also the nature and structure of connections formed between different cell types, and how electrical activity propagates through intact neural circuits. Thus, tracing neurons and their patterns of connectivity in the brain remains a cornerstone to both basic and clinical neuroscience.

Much of what we currently know about neuronal architecture and cellular diversity has stemmed from classical histological staining, immunohistochemical, and electron microscopy methods. Recent advances in molecular genetics, viral engineering, and imaging technologies have begun to reveal previously unattainable details about the cellular morphologies and subcellular structures that are unique to the different types of neurons that make up the brain. Moreover, new approaches to genetically mark and manipulate neuronal activity now allow investigators to probe patterns of functional circuit connectivity with electrophysiological and temporal precision. This ever expanding molecular and genetic "tool box" for tracing neurons and their synaptic connections heralds a new and exciting era of functional neuroanatomy.

Importantly, understanding the fundamentals of brain wiring requires detailed investigation of many cell types, different neural circuits, and these details within multiple species. It is imperative to acknowledge that not all technologies or approaches are equally useful or informative. Thus, comprehensive experimentation and thorough analysis of neuroanatomical architecture and function is required.

In the accompanying volume, *Neural Tracing Methods: Tracing Neurons and Their Connections*, part of Springer's *Neuromethods* series, we have set out to compile a comprehensive collection of chapters authored by inventors and expert users, describing state-of-the-art neuronal tracing and functional analysis methods. From classical lipophilic dye and conjugated lectin tracing techniques, to electrophysiological, in vivo imaging, viral tract tracing, and emerging genetic methods to mark, manipulate, and monitor neural circuits, this volume includes reference to an arsenal of tools and technologies currently being implemented in model systems ranging from flies to mice.

The main goal of this volume is to familiarize readers to the diverse range of technologies and approaches currently available for probing neuron and circuit architecture, and, when possible, attach detailed protocols to help guide readers toward practical application. Although this is a rapidly moving field, the methods described here are establishing the foundation for a wealth of information we are continually gathering about the form and function of the nervous system. The near future will undoubtedly see a rapid expansion in our understanding of brain processing.

In moving forward, it is important to first mention that in all cases the authors have made every attempt to include the primary references to the origins and applications of a given method. Also, I would like to offer my sincere appreciation to all of the authors who have graciously agreed to make contributions to this volume. We are currently at a point in the field of neuroscience where it is enormously beneficial to share our evolving knowledge and insights with each other; knowing what is possible enables new and even brighter ideas to flourish.

Houston, TX, USA ***Benjamin R. Arenkiel***

Contents

Series Preface ... *v*
Preface ... *vii*
Contributors ... *xi*

1 A Survey of Current Neuroanatomical Tracing Techniques ... 1
Floris G. Wouterlood

2 Wheat Germ Agglutinin (WGA) Tracing: A Classic Approach for Unraveling Neural Circuitry ... 51
Sabrina L. Levy, Joshua J. White, and Roy V. Sillitoe

3 Retrograde Tract-Tracing "Plus": Adding Extra Value to Retrogradely Traced Neurons ... 67
José L. Lanciego

4 Simultaneous Collection of In Vivo Functional and Anatomical Data from Individual Neurons in Awake Mice ... 85
Brittany N. Cazakoff and Stephen D. Shea

5 Single-Cell Electroporation for In Vivo Imaging of Neuronal Morphology and Growth Dynamics ... 101
Sharmin Hossain, Kaspar Podgorski, and Kurt Haas

6 Practical Methods for In Vivo Cortical Physiology with 2-Photon Microscopy and Bulk Loading of Fluorescent Calcium Indicator Dyes ... 117
Stephen D. Van Hooser, Elizabeth N. Johnson, Ye Li, Mark Mazurek, Julie H. Culp, Arani Roy, Rishabh Kasliwal, and Kelly Flavahan

7 Optogenetic Dissection of Neural Circuit Function in Behaving Animals... 143
Carolina Gutierrez Herrera, Antoine Adamantidis, Feng Zhang, Karl Deisseroth, and Luis de Lecea

8 Remote Control of Neural Activity Using Chemical Genetics ... 161
Andrew J. Murray and Peer Wulff

9 Generation of BAC Transgenic Mice for Functional Analysis of Neural Circuits ... 177
Jonathan T. Ting and Guoping Feng

10 Engineered Rabies Virus for Transsynaptic Circuit Tracing ... 217
Jennifer Selever and Benjamin R. Arenkiel

11 Genetic Labeling of Synapses ... 231
Carlos Lois and Wolfgang Kelsch

12 Genetic Pathways to Circuit Understanding in Drosophila ... 249
Jennifer J. Esch, Yvette E. Fisher, Jonathan C.S. Leong, and Thomas R. Clandinin

Index ... *275*

Contributors

ANTOINE ADAMANTIDIS • *Department of Psychiatry, Douglas Mental Health University Institute, McGill University, Montreal, QC, Canada*
BENJAMIN R. ARENKIEL • *Department of Molecular & Human Genetics, and Department of Neuroscience, Baylor College of Medicine, Houston, TX, USA; Jan and Dan Duncan Neurological Research Institute, Texas Children's Hospital, Houston, TX, USA*
BRITTANY N. CAZAKOFF • *Cold Spring Harbor Laboratory, Cold Spring Harbor, NY, USA*
THOMAS R. CLANDININ • *Department of Neurobiology, Stanford University, Stanford, CA, USA*
JULIE H. CULP • *Center for Neuroscience, University of California, Davis, CA, USA*
KARL DEISSEROTH • *Department of Bioengineering, Stanford University, Stanford, CA, USA*
JENNIFER J. ESCH • *Department of Neurobiology, Stanford University, Stanford, CA, USA*
GUOPING FENG • *Department of Brain and Cognitive Sciences, McGovern Institute for Brain Research, Massachusetts Institute of Technology, Cambridge, MA, USA*
YVETTE E. FISHER • *Department of Neurobiology, Stanford University, Stanford, CA, USA*
KELLY FLAVAHAN • *Departments of Biology, Brandeis University, Waltham, MA, USA*
KURT HAAS • *Department of Cellular and Physiological Sciences, University of British Columbia, Vancouver, BC, Canada; Brain Research CentreUniversity of British Columbia, Vancouver, BC, Canada*
CAROLINA GUTIERREZ HERRERA • *Department of Psychiatry, Douglas Mental Health University Institute, McGill University, Montreal, QC, Canada*
STEPHEN D. VAN HOOSER • *Departments of Biology, Brandeis University, Waltham, MA, USA; Department of Neurobiology, Duck University Medical Center, Durham, NC, USA*
SHARMIN HOSSAIN • *Department of Cellular and Physiological Sciences, University of British Columbia, Vancouver, BC, Canada; Brain Research CentreUniversity of British Columbia, Vancouver, BC, Canada*
ELIZABETH N. JOHNSON • *Department of Neurobiology, Duke University Medical Center, Durham, NC, USA; Institute for Brain ScienceDuke University Medical Center, Durham, NC, USA*
RISHABH KASLIWAL • *Department of Neurobiology, Duke University Medical Center, Durham, NC, USA; Institute for Brain ScienceDuke University Medical Center, Durham, NC, USA*
WOLFGANG KELSCH • *Central Institute of Mental Health, Medical Faculty Mannheim of Heidelberg University, Mannheim, Germany*
JOSÉ L. LANCIEGO • *Laboratory of Basal Ganglia Neuromorphology, Department of Neurosciences, Center for Applied Medical Research (CIMA and CIBERNED), University of Navarra, Pamplona, Spain*
LUIS DE LECEA • *Department of Psychiatry and Behavioral Sciences, Stanford University, Stanford, CA, USA*

JONATHAN C.S. LEONG • *Department of Neurobiology, Stanford University, Stanford, CA, USA*

SABRINA L. LEVY • *Department of Pathology & Immunology, and Department of Neuroscience, Baylor College of Medicine, Houston, TX, USA; Jan and Dan Duncan Neurological Research Institute, Texas Children's Hospital, Houston, TX, USA*

YE LI • *Howard Hughes Medical Institute, University of California, Berkeley, CA, USA*

CARLOS LOIS • *Department of Neurobiology, University of Massachusetts Medical School, Worcester, MA, USA*

MARK MAZUREK • *Department of Biology, Metropolitan State University of Denver, Denver, CO, USA*

ANDREW J. MURRAY • *Department of Biochemistry and Molecular Biophysics, Columbia University, New York, NY, USA*

KASPAR PODGORSKI • *Department of Cellular and Physiological Sciences, University of British Columbia, Vancouver, BC, Canada; Brain Research Centre, University of British Columbia, Vancouver, BC, Canada*

ARANI ROY • *Departments of Biology, Brandeis University, Waltham, MA, USA*

JENNIFER SELEVER • *Department of Molecular & Human Genetics, Baylor College of Medicine, Houston, TX, USA; Jan and Dan Duncan Neurological Research Institute, Texas Children's Hospital, Houston, TX, USA*

STEPHEN D. SHEA • *Cold Spring Harbor Laboratory, Cold Spring Harbor, NY, USA*

ROY V. SILLITOE • *Department of Pathology & Immunology, and Department of Neuroscience, Baylor College of Medicine, Houston, TX, USA; Jan and Dan Duncan Neurological Research Institute, Texas Children's Hospital, Houston, TX, USA*

JONATHAN T. TING • *Department of Brain and Cognitive Sciences, McGovern Institute for Brain Research, Massachusetts Institute of Technology, Cambridge, MA, USA*

JOSHUA J. WHITE • *Department of Pathology & Immunology, and Department of Neuroscience, Baylor College of Medicine, Houston, TX, USA; Jan and Dan Duncan Neurological Research Institute, Texas Children's Hospital, Houston, TX, USA*

FLORIS G. WOUTERLOOD • *Department of Anatomy and Neurosciences, University Medical Center, Vrije University, Amsterdam, The Netherlands*

PEER WULFF • *Institute of Physiology, Christian Albrechts University, Kiel, Germany*

FENG ZHANG • *Department of Brain and Cognitive Sciences, McGovern Institute, Massachusetts Institute of Technology, Cambridge, MA, USA*

Chapter 1

A Survey of Current Neuroanatomical Tracing Techniques

Floris G. Wouterlood

Abstract

This chapter provides a systematic description of neuroanatomical tracing methods, with a brush of history. Tracing can be based on uptake and transport of tracer in living neurons but can also be based on physical diffusion in living neurons after intracellular injection of tracer or, in fixed tissue as is the case of Golgi silver staining, based on complex anorganic chemical reactions. Because of the special fixation status of human brain tissue, the physicochemical methods are prominent with this kind of nervous tissue. Nowadays, the transport methods enjoy popularity in animal connectivity models because they produce fast and decisive results in terms of specific connectivity of functional systems. Transport-based tracing methods are best suited to visualize long-axon projections. For the study of short projection axons and interneurons, more sophisticated methods need to be applied such as pericellular injection or intracellular filling with dye after neurophysiological recording in living slice preparations or in vivo.

The chapter discusses the current neuroanatomical tracing methods and its equipment and procedures, starting with the "mother of retrograde tracing methods": the technique of injection, uptake, and transport of the enzyme horseradish peroxidase. It continues with fluorescent dye tracing. The fluorescent compounds have two advantages: they are extremely stable and they can easily be combined with immunofluorescence to determine the neurochemical identity of the labeled neurons.

The most commonly used anterograde tracers are *Phaseolus vulgaris*-Leucoagglutinin (PHA-L), which requires immunohistochemical detection, and biotinylated dextran amine (BDA), which is detected via reaction with streptavidin conjugated to a reporter molecule. Single and multiple fluorescence methods receive much attention because they so perfectly combine with modern laser scanning microscopes and digital image processing.

Key words Neuroanatomical tracing, Anterograde transport, Retrograde transport, Fluorescence, Combined methods

1 Introduction

Neurons, the building blocks of the nervous system, express a seemingly chaotic repertoire of neurotransmitters, coactive substances, messenger systems, transporters, and receptors. In addition, they occur in a bewildering morphological and physiological variety. In the context of the complexity of the brain, the cell biologist's niche is the study of the individual building blocks, while the neuroanatomists' realm is the participation, qualitatively

Benjamin R. Arenkiel (ed.), *Neural Tracing Methods: Tracing Neurons and Their Connections*, Neuromethods, vol. 92, DOI 10.1007/978-1-4939-1963-5_1, © Springer Science+Business Media New York 2015

and quantitatively, of individual or groups of neurons in neuronal networks and in input/output connectivity. Cell biology and neuroanatomy meet each other where the neurons meet, that is, at the level of synapses and small neuronal networks. Based on the data collected by anatomists and cell biologist, a psychologist can try to explain the decisions and actions supported by the nervous system, for instance, a motor act, a mental action, or something intangible as writing a book chapter.

The emphasis of this chapter lies on methods designed to study neuronal connectivity at the gross level ("tract tracing"), but we will descend toward the domain of the cell biologist by dealing with methods aimed at tracing connections between small neuron populations and even individual neurons.

At this point, a trio of late nineteenth century scientists should be remembered who shaped today's neuroscience landscape because their work led to the concept of the nervous system as a collection of electrically interacting cells: the German cytologist Heinrich von Waldeyer-Hartz, the English physiologist Sir Charles Scott Sherrington, and Santiago Ramon y Cajal, the great Spanish histologist. Von Waldeyer is regarded as the first author to propagate (in [190]) the idea that the nervous system is made up of discrete units: the cells. Together with Sherrington's notion of the synapse (expressed for the first time in Foster's [58] textbook), the conceptual stage was set up at the turn of the nineteenth into the twentieth century for neuroanatomical tracing as we know it today. Giant, revolutionary steps forward in neuronal tracing were set in the mean time by Ramon y Cajal who pioneered the art with the revolutionary Golgi silver impregnation technique. He was the first to describe systematically neuron types in the brain based on morphological features of their cell bodies, dendrites, and axons. His two-volume *Système Nerveux de l 'Homme et des Vertébrés* [148] still impresses the reader with its incredible amount of detail and insight in brain connectivity. Ramon y Cajal's greatest scientific legacy is the neuron theory.

In this chapter, neuroanatomical tracing methods are categorized into three groups: those that trace the cell bodies back from the axon terminals (called "retrograde techniques"); techniques that trace the other way, from the cell body toward the axon terminals (called "anterograde techniques"); and a third group of techniques specifically aimed at the study of interneurons and other neurons with short axons, the microcircuit neurons whose processes generally remain inside the boundaries of a nucleus or cortical layer.

2 Retrograde Techniques

2.1 Horseradish Peroxidase (HRP)

The cytoplasm of neurons contains elaborate transport mechanisms that shuttle all sorts of macromolecules back and forth between the perikaryon and the periphery. In the case of neurons with their long neurites, the periphery can be extremely far away from the cell body, and considerable energy is spent to distribute material from the cell bodies to the terminals or motor endplates. Axonal transport was first described in [195] by Weiss and Hiscoe, yet the idea to use this phenomenon to trace neuronal connectivity arose a quarter of a century later in Göteborg, Sweden. Here, Kristensson and Olsson [107, 108] reported on a successful tract-tracing experiment: the enzyme horseradish peroxidase (HRP) was deposited into the gastrocnemius muscle of rats. After a few days postsurgery, HRP activity was detected in motoneurons in the spinal cord segment known to send spinal nerves to the muscle. In the very same year, Kristensson et al. [109] reported on a similar experiment concerning the deposition of HRP into the tongue of a rat. In this experiment, motoneurons in the hypoglossal nucleus of the brainstem became labeled. LaVail and LaVail [120] picked up the idea and demonstrated uptake and retrograde transport of HRP in the CNS itself. Thus, tract tracing in the peripheral and central nervous system based on transport of macromolecules was born.

HRP is easy to apply via pressure injection, iontophoretic injection, or crystal deposit or by means of placement of spongy material soaked in HRP solution, while its detection [75] is fast, easy, and straightforward. Macromolecules like HRP are taken up in large quantities by a rather aspecific process called fluid-phase endocytosis, and they are transported in membrane-bound vesicles to the perikaryon for further processing in lysosomes, i.e., degradation or disposal. The granular appearance of the reaction product in the light microscope is the expression of this vesicle-bound transport, and the location inside vesicular structures can be verified with electron microscopy. Thus, the tracer is mass internalized and transported as bulk inside a containment membrane (called an endosome) which is different and less sensitive than transport of molecules hooked directly into transport mechanisms of the cell. Endosomes typically show an oval or tubular shape (Figs. 1 and 2). As the brown deposit obtained via classical, diaminobenzidine-based HRP detection [75] is electron dense, the fate of taken up HRP can easily be followed in the electron microscope (Figs. 1 and 2). After the histochemical reaction, the deposit provides a visible label that can be used to completely document labeled neurons before further processing and study of the ultrastructural details of the previously light microscopically identified neurons by electron microscopy. This feature forms the basis of correlated light-electron microscopy, a term thought to be coined first by Somogyi et al. [170].

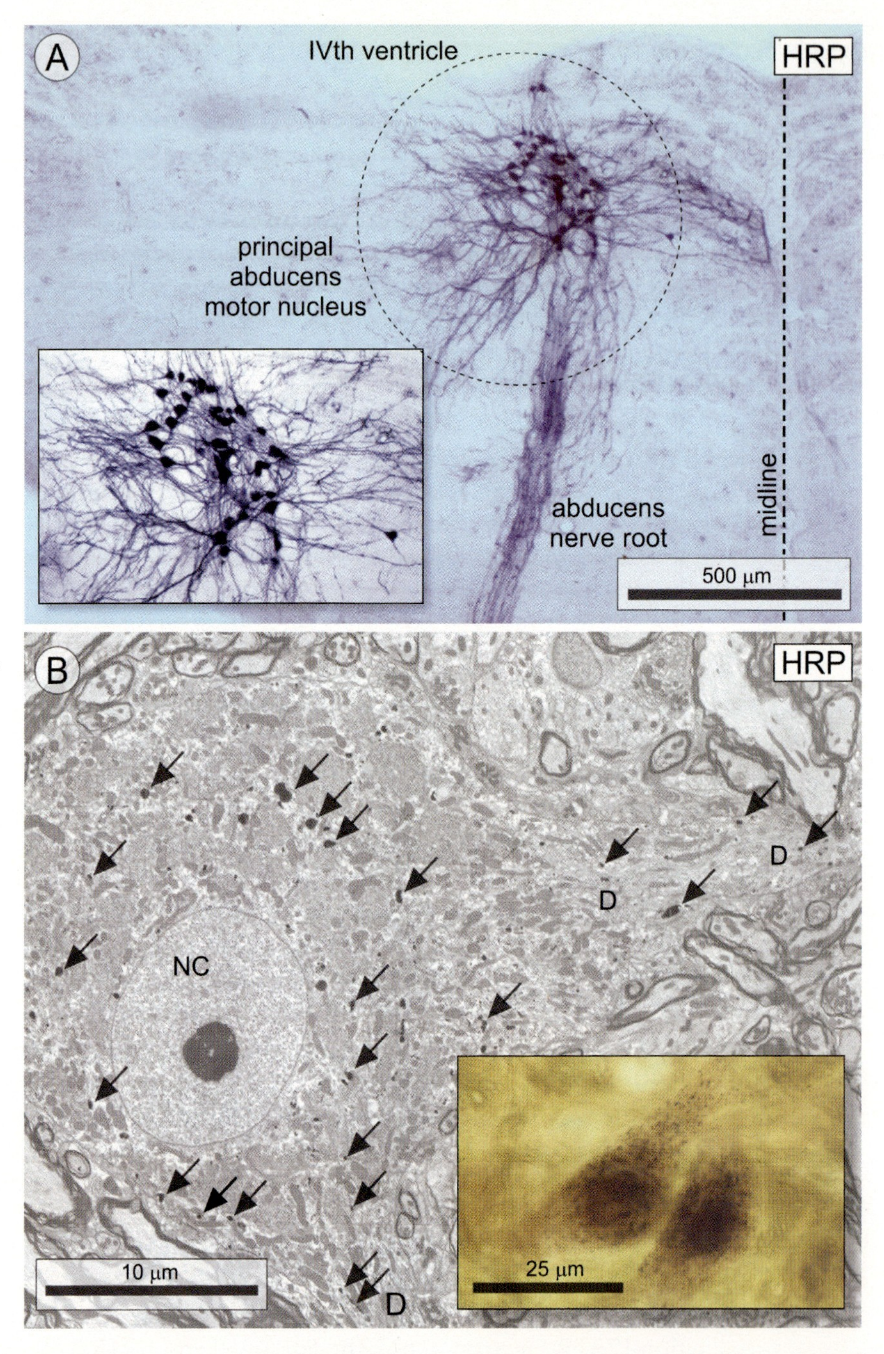

Fig. 1 Retrograde HRP tracing. (**a**) Labeling of abducens motoneurons in the reptile, *Varanus exanthematicus*, following application of a HRP soaked sponge to the cut VIth nerve in the orbita. TMB substrate. Brain stem; level of the principal VI motor nucleus. Root axons, cell bodies, and dendrites contain a dense, dark deposit. *Inset* in (**a**) shows the retrogradely labeled abducens motoneurons at higher magnification. (**b**) Electron micrograph of a retrogradely labeled abducens motoneuron, DAB substrate. The cell body is laden with endosomes: small, dark, membrane particles (*arrows*), i.e., the electron microscopic representatives of the brown granules seen in light microscopy (inset). *NC* nucleus, *D* dendrite. Full description in Barbas-Henry and Wouterlood [7]

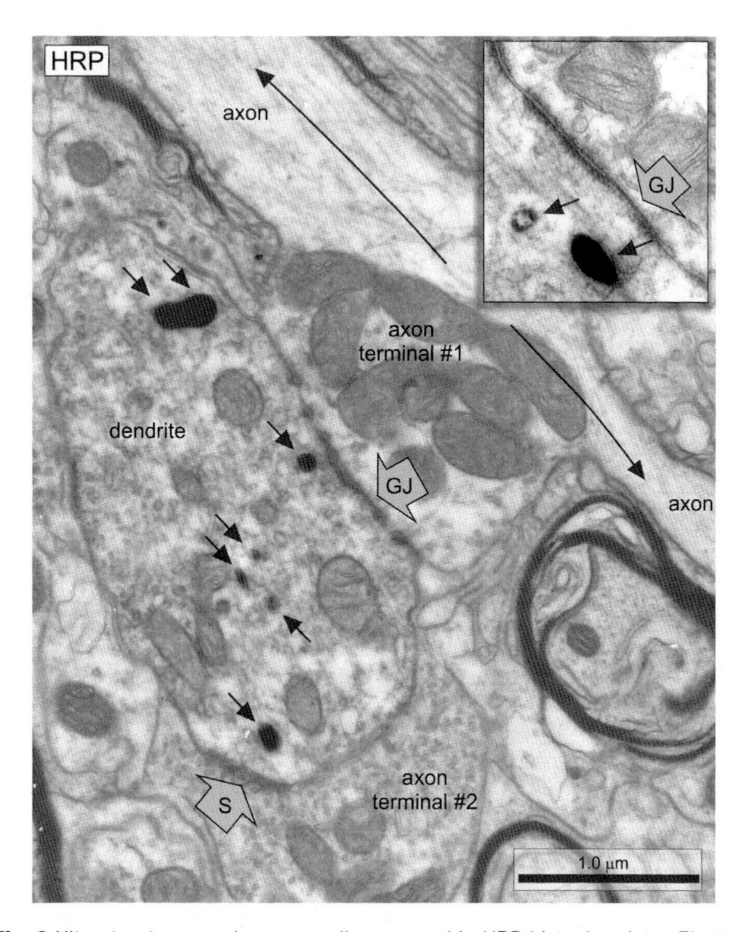

Fig. 2 Ultrastructure can be very well preserved in HRP histochemistry. Electron micrograph of an axon terminal (#1) at a node of Ranvier of a myelinated axon, forming an electrotonic synapse (*GJ* gap junction, *arrow*) with a HRP retrogradely labeled dendrite of an accessory abducens motoneuron in the reptile, *Varanus exanthematicus*. The dendrite contains dark HRP containing endosomes (*arrows*). The same dendrite receives an additional "classical" synapse from axon terminal #2 (*arrow*, S). Inset: another gap junction in this experiment, higher magnification

As internalization of HRP can be considered aspecific, it may occur anywhere along a neuron's membrane, not exclusively at axon terminals. Thus, HRP is prone to being taken up by fibers of passage in an injection site, a feature observed only infrequently with anterograde tracers. Because the uptake process depends on general endocytosis, it appears not to be limited to specific types of neuron. Another aspect of general uptake is that transport is not necessarily restricted to retrograde transport. Indeed, quite soon after the introduction of HRP as a tracer, the first reports emerged that HRP is also transported anterogradely [31, 152]. These and similar reports were later confirmed using procedures with higher sensitivity than the original HRP enzyme histochemistry [73].

2.2 Classical HRP Retrograde Tracing Procedure

The classical approach employing retrograde transport to study connectivity of the CNS is via the injection of 0.1–0.5 μl of a 20–50 % solution of HRP in saline through a Hamilton microsyringe. The most common types of HRP used are Sigma Type VI (Sigma-Aldrich, St. Louis, MO) and Boehringer (Boehringer-Ingelheim, Germany). Alternative vehicles for HRP application include foam or gel drenched in HRP solution, lyophilized HRP, pellets, and plain HRP gel. As foams, gels, and pellets are hard to inject, these "hard" vehicles are often implanted, or they are brought in close contact with the proximal stump of a transected cranial or peripheral nerve [7]. Large mechanical injections may damage the brain parenchyma at the syringe tip and cause local necrosis. For this reason, mechanical administration must be executed carefully, usually over a 10–20 min period. We leave the syringe in situ for a short period of time after the delivery of the tracer to avoid backfilling of the needle track with HRP, and then we retract very, very slowly.

As retrogradely transported HRP ends up in lysosomes, the labeling does not have a permanent character. Transport is related to distance and speed (fast transport may occur in warm-blooded animals at a rate of 100–450 mm/day) [49, 132, 141, 166, 220]. The slow component allows speeds of 0.1–20 mm/day. Transport speeds typically differ between fiber systems and often need empirical fine tuning. Anterograde and retrograde transport is maintained and guided by the intra-axonal microtubule system which also forms a critical component of the cytoskeleton. Retrograde transport is mediated by proteins of the dynein family, while anterograde transport is associated with molecules of the kinesin family [23, 83]. Thus, transport can be inhibited using a microtubule assembly arrestor such as colchicine. The optimal survival period following injection of HRP varies between 2 and 5 days, after which the histochemical staining begins to produce less dense deposit, an early sign of metabolic breakdown.

HRP catalyzes the conversion of a substrate into an insoluble precipitate, with hydrogen peroxidase as electron acceptor while the peroxide decays into water and molecular oxygen. A repertoire of electron donors (substrates) has been introduced over the years: 3,3′-diaminobenzidine (DAB) [75],[1] tetramethylbenzidine (TMB) [131], benzidine dihydrochloride (BDHC) [113], *o*-tolidine [170], *o*-dianisidine [194], paraphenylenediamine-pyrocatechol

[1] Diaminobenzidine and benzidine derivates such as BDHC are considered carcinogenic and should be treated with care, that is, in a dedicated container in a fume hood. Gloves should be worn at all times to avoid contact with the skin. All material in contact with DAB and BDHC should be decontaminated with chlorine after the experiment, and care must be taken not to contaminate the microscope used to monitor at intervals the progress of the histochemical reaction.

(PPD-PC; [25]), 1-naphthol/azur B (1-NBD; [129]), alkaline phosphatase (purple reaction product; [204]), the magenta chromogen V-VIP (Zhou and Grovofa, [215, 223]), HistoGreen [178], and nickel-enhanced DAB [79]. The electron transfer from the substrate to the peroxide oxidant can be mitigated via glucose oxidase [92].

Cells are not always filled to capacity with HRP after a tracing experiment. In case of weak staining, the enzymatic reaction can be boosted by adding additional peroxide oxidant. This forced boost should be done with caution however, since it affects the background staining which has a tendency to build up extra fast when too much substrate is available. The same holds for immunohistochemistry using the bridged antibody method with a peroxidase-antiperoxidase as the vehicle and a precipitate as the final reporter.

The typical fixative for HRP tracing includes a mixture of 0.5–1.5 % glutaraldehyde with 4 % buffered formaldehyde [133]. As this is a typical “strong” fixative used for electron microscopy, it may be no surprise that ultrastructural detail is often well preserved (Figs. 1 and 2). The best fixation of the CNS is achieved via a perfusion fixation procedure. Tissue sections are usually cut on a freezing or vibrating microtome. Endogenous peroxide activity is suppressed by “bleaching” brain sections for 1–2 min with a phosphate-buffered 1–2 % hydrogen peroxide solution. Endogenous peroxidase activity is ubiquitously present in erythrocytes and may outperform HRP activity if fixation has been poorly performed or in postmortem tissue where blood vessels still contain erythrocytes—also vascular endothelium and choroid plexus cells are notorious sources of massive endogenous peroxidase activity. The bleached sections are carefully rinsed in 50 mM Tris–HCl, pH 7.6, and preincubated for 30 min in 5 mg 3,3′-diaminobenzidine in 10 ml Tris–HCl (or with some modification with another of the available chromogens), whereafter the reaction is started by adding 3.3 μl of 30 % H_2O_2 per 10 ml of preincubation solution. The peroxidase catalyzes electron transfer from the diaminobenzidine donor molecule to oxygen ions in the peroxide oxidizer. The result is the formation of a brown diaminobenzidine precipitate, with molecular oxygen as a byproduct. When the peroxidase reaction is too fast and too strong, the formed oxygen may become gaseous and form gas bubbles that damage the histological detail. Itoh et al. [92] introduced the glucose-oxidase-peroxidase histochemical reaction to slow down the release of electrons from the peroxide in order to better control the reaction.

Contrast of the formed precipitate versus background can be enhanced by the addition of nickel-ammonium sulfate to the incubation medium (Ni-enhancement; [79]). The resulting black deposit gives the human eye better contrast against the background compared with brown, non-enhanced DAB precipitate.

Sensitivity can be increased considerably by using wheat-germ agglutinin-conjugated HRP as the tracer instead of native HRP [73, 165]. The wheat germ binds to lectin receptors on the outside of neuronal limiting membranes and in this way stimulates receptor-mediated uptake of the HRP.

2.3 Retrograde Tracing with Fluorescent Substances

While Kristensson gained instant fame through his [109] publications on retrograde tracing with HRP, he had already experimented with uptake and retrograde transport of other compounds. One year before his landmark article, he published a report on experiments in which a conjugate of albumin and the fluorescent dye, Evans Blue, had been injected into the gastrocnemius muscle of rats. After apparent retrograde transport, a fluorescence signal was detected in spinal motoneurons [106]. The most attractive of tracing with fluorescent dyes is that the label is visible in a fluorescence microscope immediately after cutting sections, without the need for histochemistry. It is a one-step event, simple, and effective. Seven years later, in Rotterdam, the Netherlands, the Kuypers group at Erasmus University were the first to demonstrate that unconjugated anorganic fluorescent dyes injected focally into the CNS "behaved" like HRP, that is, these dyes are taken up by nerve terminals and transported retrogradely to the perikaryal cytoplasm, nucleus, or specific organelles of the involved projection neurons [12, 110, 111]. Within a few years, a variety of fluorescent retrograde tracers with similar features had been identified, including Bisbenzimide, DAPI-primuline, propidium iodide [111], True Blue, Granular Blue [12], Nuclear Yellow, Fast Blue (FB; [14, 112]), Diamidino Yellow (DY; [100]), and Fluoro-Gold (FG; [161]). With the exception of FG, these compounds bind to adenine-thymine-rich nucleic acids. A special advantage of these tracers, including FG, is that they are reliable in tracing experiments where extremely long fiber tracts are the subject of investigation, e.g., corticospinal and bulbospinal projections (e.g., [155]), a feature shared with anterograde autoradiography tracers (e.g., [84]).

Several retrograde tracers, e.g., FB, FG, and DY, withstand immunofluorescence procedures [167, 168]. The implications of this feature are far-reaching. Careful selection of fluorescent tracers with different spectral characteristics makes it possible to conduct double and even triple fluorescent retrograde tracing, and the compatibility and spectral separation also enables combination of retrograde fluorescent tracing and neurotransmitter immunofluorescence. Multiple retrograde fluorescence labeling experiments were for the first time reported by Bentivoglio et al. [13]. Richmond et al. [156] systematically compared results with seven different retrograde tracers in a cat peripheral motor nerve model: FB, FG, fluorescein dextran, rhodamine dextran, fluorescent latex microspheres, HRP, and DiI. Björklund and Skagerberg [18] and Van der Kooy and Steinbusch [184] pioneered retrograde fluorescence tracing combined with detection of a neuroactive substance via histo- or immunofluorescence. Reviews dealing with early attempts to demonstrate collateralization of axonal projections via double retrograde fluorescence tracing have been published by Huisman et al. [90], Skirboll et al. [168], and Wessendorf [196].

Of the early fluorescent tracers, FB and DY remain to this day the most often used for single or double retrograde fluorescent labeling. Because the initial manufacturer of FB (Dr. Ihling, Germany) has terminated production of this compound, FB is becoming scarce and is increasingly being replaced by FG. For this reason, and also because FG is very popular among tract-tracing neuroscientists, FG deserves special attention.

2.4 Retrograde Tracing with Fluoro-Gold

Fluoro-Gold (FG, a highly purified mix of stilbene compounds) was introduced as a versatile, highly fluorescent, stable retrograde neuroanatomical tracer with low toxicity by Schmued and Fallon [161]. Wessendorf [197] identified hydroxystilbamidine as the probably active component. The dye has at pH 7.4 an excitation band between 350 and 390 nm and a broad emission band between 530 and 600 nm (company information at www.fluorochrome.com/FGProtocol.htm), providing an intense silver-white–light yellow appearance in standard fluorescence microscopy using mercury illumination (Fig. 3).

Fluoro-Gold is marketed by Fluorochrome, Inc. (Denver, CO). The powder can be dissolved in a range of buffers (*see* Sect. 2.7). We have obtained best results with 2 % FG dissolved fresh in cacodylate buffer. Injection is either mechanically via a Hamilton syringe or microiontophoretically via a glass pipette [162]. We prefer the mechanical approach. Survival periods range from 1 week up to 1 year, and tissue fixation and sectioning can be conducted along standard neuroscience laboratory procedures. With longer survival periods (more than 1 week), the retrogradely labeled neurons stand out with high contrast to the background (Fig. 3). As FG is an inherently fluorescing compound, the only procedure necessary to study transport is sectioning and fluorescence microscopy. No histology other than fixation is required. Even in fresh tissue, fluorescence is intense. The addition of an anti-fading agent to the mounting medium is not necessary since FG has a remarkable photobleaching-resistance capacity.

FG can be used either alone or in combination with other fluorescent tracers (e.g., Dolleman-van der Weel et al. [45, 52, 185]) or in combination with immunofluorescence procedures to detect additional markers (*see* below). The dye accumulates in small punctate structures, presumably lysosomes, in the cytoplasmic compartment of neuronal cell bodies (Fig. 3). At higher concentration, transported FG may fill neurons completely with intense white/yellow fluorescence. Van Bockstaele et al. [183]

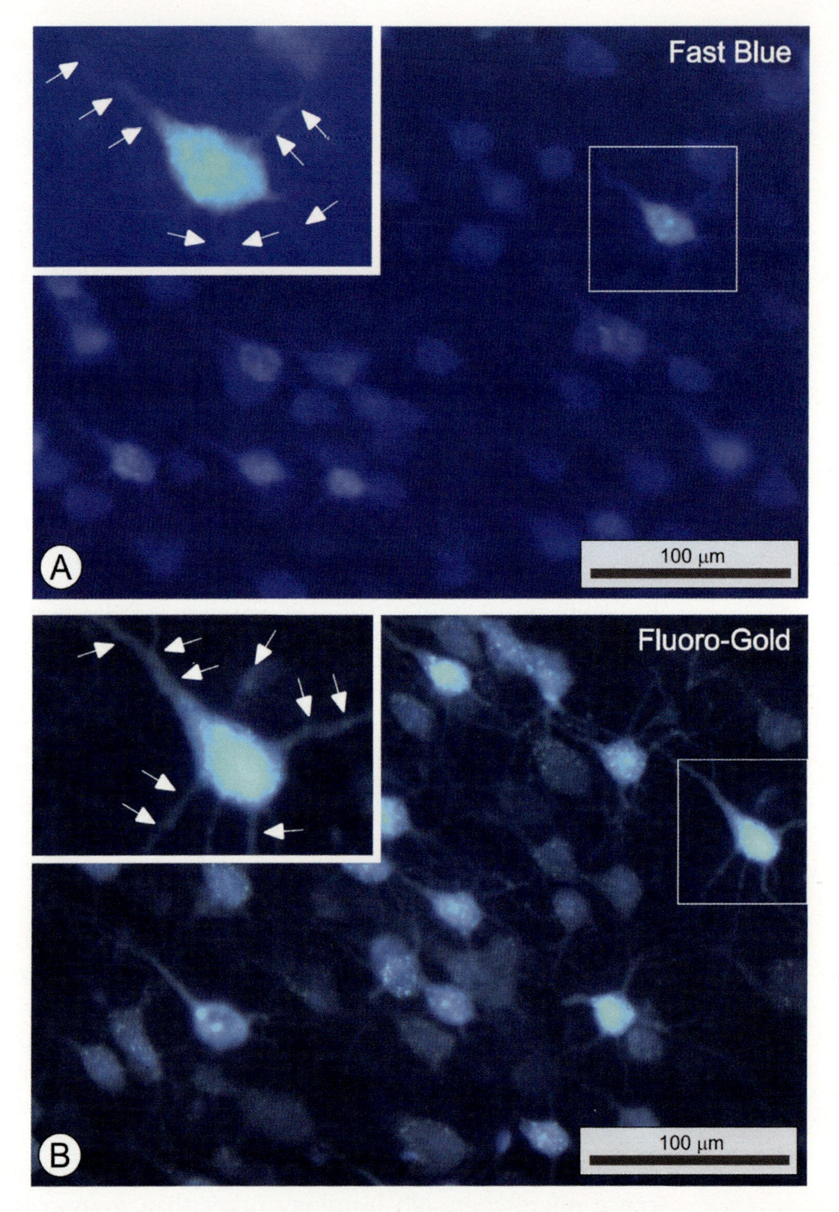

Fig. 3 Retrograde labeling of neurons in rat cerebral cortex after deposition of Fast Blue (**a**) or Fluoro-Gold (**b**) in the caudate-putamen. Standard fluorescence illumination with UV light and blue emission filter. Fast Blue labels much softer and fills the cytoplasm more evenly than Fluoro-Gold (FG) which has a granular aspect. *Insets* show the *boxed* neurons at higher magnification and, for FG, with slight contrast enhancement. Both tracers often diffuse into basal and apical dendrites (*arrows* in *insets*) which in FG-labeled neurons display more contrast with the dark background. Note that the camera's exposure time in (**b**) was half that of (**a**) because of the very bright fluorescence of FG

studied via EM immunocytochemistry the distribution of FG in neuronal perikarya after retrograde transport and found both material accumulated in lysosome-like organelles and activity dispersed in the cytoplasmic compartment.

In addition to being extremely resistant to bleaching, FG is also highly resistant to degradation in vivo. In a living animal, it may resist metabolic breakdown up to a year postinjection.

Thus, its high fluorescence intensity combined with its robustness during immunohistochemical procedures and its high resistance to fading under UV illumination has made FG the preferred tracer at the beginning of the twenty-first century for fluorescent retrograde labeling in rodents, particularly for multiple labeling in combination with other tracers. There is room for improvement, though. Its rather broad fluorescence emission spectrum renders FG fluorescence microscopy vulnerable for spectral overlap when this tracer is applied in combination with another fluorescent retrograde tracer, for instance, in the demonstration of collateralization of axonal projections. Spectral overlap brings the danger of false-positive results and must be checked by a combination of proper control experiments and by selective use of filtering or by using the characteristic that FG labels mostly cytoplasm and cytoplasmic organelles, while other fluorescent tracers label nucleic acids and therefore preferentially accumulate in the nucleus. As multilabel, multichannel confocal laser scanning requires fluorochromes with narrow spectral emission peaks, FG is less ideal for application in these instruments because of its broad emission spectrum. Furthermore, most currently installed confocal instruments lack an [expensive] UV excitation laser to excite FG. As confocal laser scanning microscopy has gained its niche in neuroscience in part because this instrument makes it possible to independently vary excitation wavelengths and illumination intensities, the availability of an antibody against FG [28] is in this respect of great help (see the combined tracing/intracellular injection/confocal scanning experiments in Sect. 5.1). Application of an anti-FG antibody makes it also possible to take material studied previously in the fluorescence microscope to the electron microscope [43, 183].

2.5 Combination of Fluoro-Gold Tract Tracing and Immunohistochemistry

When Schmued and Fallon introduced FG as a retrograde neuroanatomical tracer, they discussed the issues of its use in combination with anterograde tract tracing and its use together with immunohistochemical detection of a marker inside the retrogradely labeled neurons [161]. As the emission spectrum of FG consists of light in the blue-green domain of the visible light, a second fluorescence detection can be set up using reporter molecules with emission wavelengths in the orange-red domain, e.g., that offered by TRITC (570 nm) (emission peaks in brackets), Cy3 (570 nm), Alexa Fluor® 568 (603 nm), DyLight 594 (616 nm), Alexa Fluor 594 (617 nm), and Texas Red (620 nm), to name a few. When detection of infrared fluorescence emission is possible, e.g., via an infrared-sensitive camera or in a confocal instrument, the "infrared labels" Cy5 (670 nm), Alexa Fluor 633 (647 nm), or Alexa Fluor 647 (668 nm) can be used as well. Practical considerations such as the available filter sets in the fluorescence microscope, or the specifications of the lasers and filter sets in a confocal laser scanning microscope, guide the selection of suitable fluorochromes. The experiment described below may serve as an illustration of how to deal with a question regarding the immunohistochemical fingerprint of a neuron retrogradely labeled with Fluoro-Gold.

Neurons in one of the midline thalamic nuclei, the nucleus reuniens (NRT), are known to provide a dense fiber innervation to the hippocampus and parahippocampal cortex ([81, 206]; Dolleman-van der Weel et al. [45, 188]). NRT contains neurons rich in calcium-binding enzymes like calretinin and calbindin [2, 35]. Calcium-binding proteins are usually present in fast-spiking GABAergic interneurons. Excitatory, glutamatergic neurons in NRT have been reported on the other hand [221]. Question was whether the dense calretinin-immunopositive fiber plexus in the parahippocampal cortex can be attributed in its entirety to local circuit neurons or that at least part of the calretinin innervation is of thalamic origin, e.g., from neurons with cell bodies located in NRT. Thus, we used leftover diencephalon sections from rats from previous FG retrograde tracing experiments (Dolleman-van der Weel et al. [45]) that had been stored in a –20 °C freezer for more than 10 years. FG had been injected into NRT, via iontophoresis (2 % FG in 5 mM sodium acetate buffer, pH 5), with postsurgery survival times between 1 and 3 weeks. These sections were retrieved, washed, and incubated free-floating overnight at room temperature with monoclonal (mouse) anti-calretinin. The secondary antibody was a goat anti-rabbit IgG conjugated with Alexa Fluor™ 594. The mounted sections were studied in a Leica fluorescence microscope equipped with filter cubes A (blue; UV excitation 340–380 nm, dichroic mirror 400 nm, longpass 425 nm) and TX2 (intermediate red; excitation bandpass 560 nm; dichroic mirror 595 nm; emission bandpass 645/75 nm) (Fig. 4). Several neurons in NRT contained both retrogradely transported label and calretinin-associated immunofluorescence, providing evidence for calretinin-containing neurons in this thalamic nucleus projecting to the temporal cortex.

2.6 Fluorescent Retrograde Tracing with Compounds Other Than Fluoro-Gold

Kuypers and collaborators [13, 111] were the first to demonstrate that fluorescent dyes injected focally into the brain "behaved" similarly to HRP, that is, they are taken up by nerve terminals and transported retrogradely to the cell bodies of the involved projection neurons. Schmued and Fallon [161] also tested "their" FG against the HRP "golden retrograde tracing standard," and the electron microscopic investigation by Van Bockstaele et al. [183] confirmed Schmued and Fallon's findings. In the mean time, rhodamine B-isothiocyanate (RITC) had been proposed as a retrograde tracer in poikilothermic animals [177]. After a brief period of experimentation with numerous fluorescent compounds in nonhuman

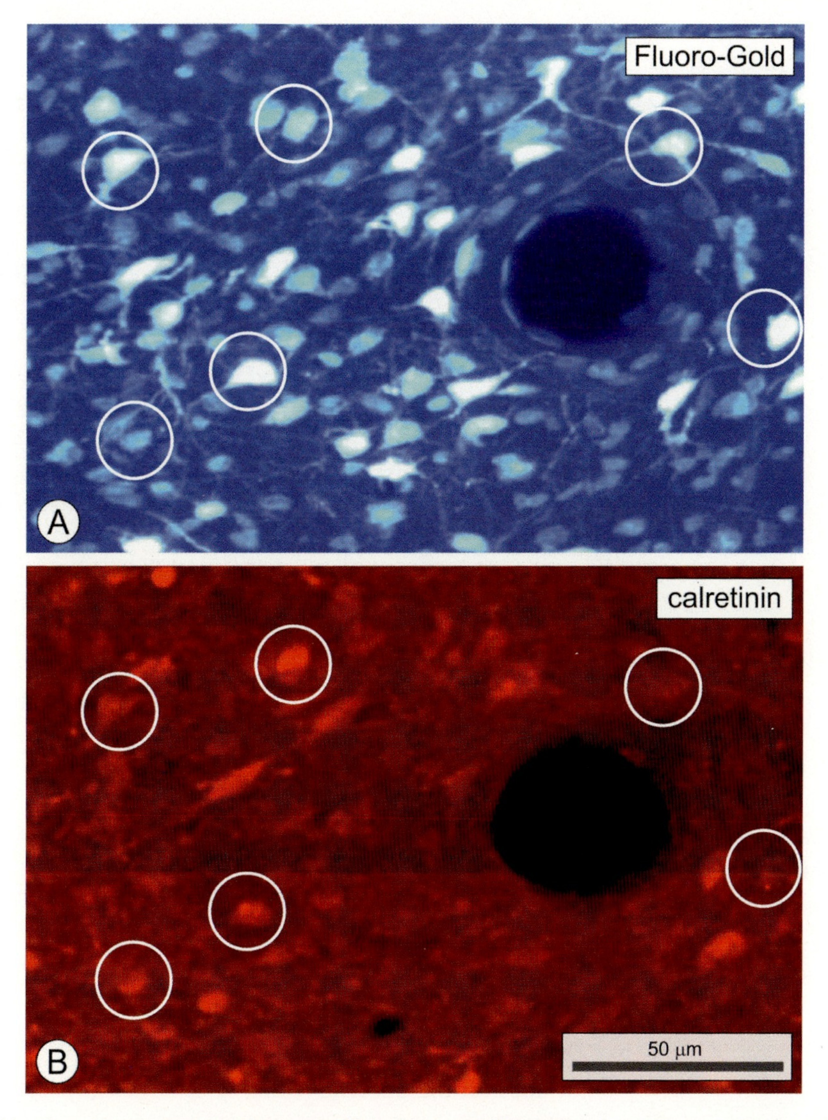

Fig. 4 Combination of retrograde tracing with FG and immunofluorescence staining. (**a**) Retrogradely labeled neurons in the nucleus reuniens thalami (NRT) of a rat following deposition of FG in the parahippocampal cortex. Sections were subjected to calretinin-immunofluorescence (**b**), using Alexa Fluor 594 as reporter fluorochrome. Sections photographed under epifluorescence illumination with the appropriate filtering (*see* text). Circled neurons show up in both filtering which is evidence for colocalization

primates, the neuroscience community settled with the combination of FB [14] and DY [100] for retrograde double labeling. In the rat, Dolleman-van der Weel et al. [44] published a study in which three retrograde fluorescent tracers were injected at various brain loci in one surgical session to determine whether all neurons of the nucleus reuniens of the thalamus (NRT; *see* previous section) project to all subfields of the hippocampal region via collateral axons or that specific groups of NRT neurons supply fibers to specific hippocampal subfields. These authors used Fast Blue (FB), FG, and Diamidino Yellow (DY) as the identifying tracers to sort out this axon collateralization issue. They found mostly single-labeled neurons, from which they concluded that individual NRT neurons project to their own individual hippocampal subfield. Using the combination of FG (deposited in prefrontal cortical areas) and Fluoro-Ruby (deposited in hippocampal loci), Hoover and Vertes [87] noted a few double-labeled neurons in NRT, suggesting that these neurons possess collateral axons projecting to both prefrontal cortex and hippocampus.

As FB and DY persist in neurons over extended periods of time, these dyes can be used for longitudinal studies, e.g., developmental remodeling of neuronal pathways (*see* [91, 130]). A comparison of several retrograde fluorescent tracers using the facial nucleus of the rat as a model was published in [89] by Horikawa and Powell.

The great advantage of FB, FG, and DY compared to the initially introduced fluorescent tracers is that tissue fixation can be handled with a standard buffered formaldehyde fixative that also allows immunohistochemistry. Since FB and DY label different features of the cell at the same excitation wavelength (the cytoplasm in blue for FB and the nucleus in yellow for DY), double-labeled neurons can be distinguished with ease.

2.7 Summary of the Most Widely Used Retrograde Fluorescent Tracers and Their Properties

Fast Blue (FB) [14]

- Injection: mechanical, iontophoretic. Track labeling: yes.
- Uptake by damaged fibers of passage: yes.
- Vehicle: 3 % in distilled water; 2 % in 100 mM phosphate buffer, pH 7.4.
- Toxic: no.
- Post-application survival time: 1–2 weeks; primates up to 5 weeks.
- Tissue fixation: standard neutral 4 % phosphate-buffered formalin solution.
- Fluorescence (neutral mounting medium): UV illumination, blue filter, soft blue cytoplasmic label. Label diffuses into primary dendrites (*see* Fig. 3a).
- Compatible with EM: no.
- Compatible with other fluorescent tracers and immunofluorescence: yes.
- Resistance to fading: poor.

Fluoro-Gold (FG) [161]

- Injection: mechanical, iontophoretic. Track labeling; yes. Necrosis at (large) injection site: yes.
- Uptake by damaged fibers of passage: yes.

- Vehicles: 2 % in distilled water; 2 % in saline, 2 % in 5 mM Na-acetate buffer, pH 5.0; 2 % in 100 mM phosphate buffer, pH 7.4; 2 % in 100 mM cacodylate buffer (2.14 g sodium cacodylate in 100 ml distilled water), pH 7.3.
- Toxic: no.
- Post-application survival time: 1 week to 1 year.
- Tissue fixation: standard neutral 4 % phosphate-buffered formalin solution; compatible with range of (heavy metal-free) fixatives [161].
- Fluorescence (neutral mounting medium): UV illumination, blue filter, high contrast silver-white-yellow granular label in cytoplasm. Nucleus sometimes labeled. Label diffuses into primary dendrites (*see* Fig. 3b).
- Compatible with EM: via antibody and peroxidase reaction [183].
- Compatible with other fluorescent tracers and immunofluorescence: yes.
- Resistance to fading: very good.

Diamidino Yellow Dihydrochloride (DY) [100]

- Injection: mechanical. Track labeling: no. Necrosis at (large) injection site: yes.
- Uptake by damaged fibers of passage: yes.
- Vehicle: suspension in distilled water (sonicate; dissolves poorly); 200 mM phosphate buffer, pH 7.4.
- Toxic: no.
- Post-application survival time: rodents, 1 week; primates, up to 5 weeks.
- Tissue fixation: standard neutral 4 % phosphate-buffered formalin solution. Sectioning should be done fast after fixation to avoid diffusion.
- Fluorescence (neutral mounting medium): UV illumination, blue filter, yellow nuclear fluorescence mostly confined to the nucleus.
- Compatible with EM: no.
- Compatible with other fluorescent tracers and immunofluorescence: yes.
- Resistance to fading: relatively poor (half time 69 s; [143]).

2.8 Retrograde Tracing with Micro- and Nanoparticles

As the endocytosis of HRP can be conceived as a bulk process, why then not offer to axon terminals complete particles for uptake and retrograde transport? Such an approach would eliminate histochemistry with all its pitfalls, and the particles would diffuse minimally at the site of injection and after transport hopefully persist for long in the cell bodies. Electron dense particles would be very welcome since these could serve to provide an exclusive label in the electron microscope. Katz et al. [99] implemented this idea through the mechanical application of minuscule (diameter 20–200 nm) fluorescent, rhodamine carrying latex beads. These microbeads were taken up and transported retrogradely. Combined light-electron microscopic examination of retrogradely transported fluorescent microbeads was for the first time achieved by Egensperger and Holländer [50]. Katz and coworkers observed that small particles are not internalized indiscriminately but that they have to fulfill certain physical and chemical conditions [99]. Physical size and chemical "attractiveness" for membranes apparently does matter. Thus, gold nanoparticles (diameter 10–12 nm) coated with wheatgerm conjugated apoHRP (WGA-apoHRP-gold; [8]) are taken up by axon terminals and retrogradely transported to the parent cell bodies where they were rendered visible for light microscopic viewing through a silver enhancement step. The nanoparticles used by Basbaum and Menétrey were big enough to be viewed directly in the electron microscope. By contrast, latex beads dissolve during embedding in plastic and appear in electron microscopic preparations as electron lucent cavities. Compatibility of gold particles coated with WGA-apoHRP exists with other neuroanatomical tracing methods [8] and also with non-tracing histochemical procedures, e.g., in situ hybridization [95]. Gold particles are particularly suitable markers in electron microscopy applications because of their distinct geometry and electron absorptive characteristics. A quite new development is the combination of retrograde tracing with fluorescent latex microbeads followed with labeling with carbocyanine dye of target neurons whose cell bodies have accumulated the transported microbeads [139]. The future of nanoparticles seems assured since hollow nanoparticles have been introduced successfully as vehicles to deliver viral vectors into neurons (30 nm diameter silica particles; [16]). While delivery succeeded, it remained doubtful whether the nanoparticles had actually been transported over significant distances inside neuronal processes of the transfected neurons. Further development of this branch of nanoparticle delivery is necessary to obtain nanoparticles that travel inside axons via coupling to internal transport systems before delivering their payload. The payload itself could be anything interesting to neuroscientists, cell biologists, and clinicians alike, e.g., drugs, permanent fluorescent dyes, or photosensitive dyes for photodynamic cancer treatment [159].

2.9 Retrograde Tracing with Conjugated or Unconjugated Cholera Toxin B Fragment (CTB)

Trojanowski et al. [181, 182] introduced HRP conjugated to the nontoxic B fragment of cholera toxin (choleragenoid) (CTB-HRP) as a tracer. Sensitivity was claimed to be much higher than with native HRP. While native HRP is apparently taken up indiscriminately by any fiber of passage, undamaged axons of passage have only a limited capacity to internalize CTB-HRP. The retrograde transport of CTB-HRP labels a greater number of neurons and reveals the dendritic arbors of labeled neurons better than does native HRP. The superior sensitivity of CTB-HRP is due to three circumstances. First, uptake is receptor mediated, i.e., it reflects binding of CTB to specific receptors on neuronal membranes, which is not affected by conjugation of the toxin fragment with HRP (reviewed by [180]). Such binding results in the uptake of the CTB-HRP conjugate by adsorptive endocytosis, a more efficient process than the fluid-phase "bulk" endocytosis through which native HRP is internalized. Second, the detection of transported conjugate is based on a very sensitive immunohistochemical procedure. Third, CTB-HRP is less rapidly eliminated from retrogradely labeled neurons than native HRP [193]. CTB-HRP is transported bidirectionally, like WGA-HRP, and this can be an advantage or a disadvantage, depending on the system under investigation. The speed of retrograde axonal transport of CTB-HRP compared with native HRP is fast, e.g., about 108 mm/day for WGA-HRP [180].

The next step in the evolution of CTB tracing consisted of applying unconjugated CTB followed by the immunohistochemical detection of the transported marker [125]. Compared with HRP tracing, a big advantage of the unconjugated CTB tracing procedure is that peroxidase, next to being a tracer, is also present intrinsically in neurons in peroxisomes and lysosomes. As a consequence, background staining in HRP histochemical procedures must be continuously checked and endogenous peroxidase activity suppressed by additional measures. As CTB is exogenous to the brain, its immunohistochemical detection is independent from background staining. Today, most CTB applications include either immunoperoxidase or immunofluorescence detection, via an intermediary step of incubation with an antibody against CTB [42]. Apart from the immunohistochemical detection procedures, transported CTB can be observed directly in a fluorescence microscope if the applied CTB tracer molecule is conjugated to a fluorescent reporter molecule, e.g., FITC or TRITC [126] or one of the extremely photostable and intense Alexa Fluor™ dyes [32, 33].

2.10 Use of Retrograde and Transneuronal Transport of Viruses

The success story of tracing using live or attenuated viruses is described in the chapter by José Lanciego in this volume.

3 Anterograde Tracing Techniques

3.1 Note on Tritiated Amino Acid Uptake and Transport

The metabolism of a cell requires the uptake of amino acids, sugars, lipids, and trace substances in order to incorporate these in macromolecules. Some of these metabolites are transported from the neuronal cell body through the axon toward the axon terminals by means of what is called "orthograde" or "anterograde" transport. The uptake, incorporation, and transport of amino acids like ^{3}H leucine and ^{3}H proline into proteins had been extensively studied in the 1950s and early 1960s (e.g., [46, 47]). In vivo anterograde transport of radioactive isotope-labeled proteins was for the first time fully and exclusively exploited as a neuroanatomical tracing tool by Cowan et al. [37]. In brief, tritiated amino acids are injected, taken up by neuronal cell bodies, used by ribosomes to manufacture polypeptides, assembled in the endoplasmic reticulum into proteins, attached to the cellular transport system, and subsequently transported to the axonal endings. As axons and axon terminals lack ribosomes and endoplasmic reticulum, they are dependent on the perikaryon for a regular supply of polypeptides and proteins. This mechanism implies that neuroanatomical tracing based on this principle is exclusively anterograde. After a certain transport time, the tissue is fixed, and a decay of the isotopes is detected via autoradiography. Slides with brain sections are coated with a thin layer of silver bromide emulsion and kept for a long time in the dark to allow alpha particles spinning off decaying isotopes to exchange energy with the ionic silver atoms that become metallic silver. A photographic development procedure then builds more silver up and around individual silver atoms, and an image is obtained in which places where decay is concentrated are covered with silver grains at high density. The product of autoradiographical tracing thus consists of an *indirect* label located in the photographic emulsion *outside* the labeled neurons, while a histochemically or immunohistochemically applied label, albeit indirect as well, is much more appropriately located, notably *inside* the labeled neurons. Safety and containment concerns with respect to tritiated compounds and additional concerns about being responsible for the production and management of long-living radioactive waste have continuously dampened the enthusiasm among neuroscientists for this tracing method. Nevertheless, the impact on neuroscience of this method has been so immense that it deserves mentioning in the present review. Two technical "tricks" govern orthograde axoplasmic transport – autoradiographical tracing (abbreviated OOAT; [82]). The main trick is the application of amino acids that contain radioisotopes, while the second is the application of radioisotopes with short (^{32}P, ^{35}S) or long (^{3}H, ^{14}C) decay times. The radioisotope that offered the best trade-off between sensitivity and resolution of the autoradiographical detection on the one hand and safety concerns for the investigator, technicians, students, and the general public on the other (not to mention costs) was the low-energy and medium-energy long-living beta particle-emitter ^{3}H (decay half time 12.3 years). However, a beta particle travels several tens of microns from the polypeptide molecule that spawned it through the tissue section and into the emulsion layer before hitting by chance a silver bromide molecule. As a consequence, the resolution of the detection depends on a statistical function consisting of two components: one spatial distribution function describing the traveled distance from its origin to the upper surface of the section where the photographic emulsion begins, and one that describes the chance that a beta particle hits a silver bromide molecule somewhere along its track in the photographic emulsion. The result is a so-called point spread function (PSF) of the decaying isotope, analogous to the point spread function of photons emitted from a fluorochrome molecule in a modern confocal microscope (*see* [11]). Thus, image formation through autoradiograms is based on a statistical process and is always a silver grain "cloud' distribution hovering above the spot in the section where the radioactive protein occurs. Pinpointing, that is, calculating from a location of a silver grain in an image the exact location of the "invisible" radioactive protein molecule (in the tissue section) where the radiation has come from, equals retracing the statistical distribution from a point source (the location of the radioisotope producing the radioactivity) onto a target plane (the photographic emulsion) [5]. The procedures involved in OOAT are explained at length by Cowan and Cuénod [38]. A thorough discussion of the advantages and caveats of the OOAT method has been published by Swanson [174].

3.2 Dextran Amines

Glover and coworkers published in [70] a paper in which they described neuroanatomical tract tracing with a novel agent: dextran amine conjugated with the fluorescent dye, rhodamine. This application of dextrans rapidly spawned an entire family of dextran-based tracers, notably dextran amine-Lucifer yellow (LYD) [27, 137], dextran amine-tetramethylrhodamine (FR) (also called "Fluoro-Ruby"; [61, 163]), dextran amine-FITC [61, 137], and, most importantly, biotinylated dextran amine (BDA; [22, 187], review by [149]). Contrary to the bidirectional transport of HRP, WGA, and CTB, or the predominant retrograde transport of many other tracers (*see* sections above), dextrans are taken up and transported predominantly in the anterograde direction. A word of caution should be expressed here. In our experience with injections of BDA in the caudate-putamen of rats, we experienced in several cases quite complete retrograde labeling of cerebral cortical pyramidal neurons. Injections of BDA in temporal cortical areas of rats in some cases produced retrogradely labeled pyramidal cells in the contralateral temporal cortex. Reiner and Honig [149] reported similar cases of retrograde transport in rat visual cortex. In extreme cases, retrograde transport of BDA can be so complete that false anterograde labeling can be expected, that is, complete labeling of axons belonging to the retrogradely labeled neurons [30]. The mechanism of uptake of BDA is unknown [151, 149]. BDA is rapidly taken up by neurons within the injection site, fills the cells evenly, and is axonally transported at a speed of 15–20 mm/week [150]. The anterograde transport may be mediated by kinesin-type molecular motor proteins [23]. Label becomes distributed evenly along their entire trajectories, revealing all the exquisite details that axons possess: ramifications, collaterals, terminal branches, terminal rosettes, and boutons. In rats, a survival time of 1 week is usually in the safe range to study short and long projections. BDA remains stable in the rodent brain up to 4 weeks postinjection, while in primates it may remain detectable up to 7 weeks after application. The tracer works equally good with unmyelinated and myelinated fibers. The histological detection procedure is based on the strong, irreversible biotin-avidin interaction which opens the opportunity to combine neuroanatomical tracing with a host of (multiple) immunohistochemical detection systems. The biotin-avidin reaction is also sufficiently robust to allow extension to single-label electron microscopy [210] and, in combination with a second technique, even double-label electron microscopy [153]. This wide versatility, together with the ease of application and the reliability of the tracer, brought BDA instant popularity among neuroscientists [149].

3.3 Procedures for Biotinylated Dextran Amine Labeling and Staining

Biotinylated dextran amine (BDA) conjugated to lysine (MW 5 or 10 kDa, Invitrogen-Molecular Probes, Eugene, OR) is dissolved at 5 % (rodents) or 10 % (primates) in 10 mM phosphate buffer, pH 7.25, and under stereotaxic guidance applied to the brain. Both iontophoretic and mechanical injections have been reported [151]. We routinely use (in rat) iontophoretic application of 5 % BDA in 10 mM phosphate buffer, pH 7.25, via glass micropipettes and 5 μA 7 sec on-off positive direct current to inject the BDA (inner tip diameter of the pipette 20–40 μm) [74, 210]. Injection spots under these conditions have diameters of approximately 250 μm and seldomly show sign of necrosis (Fig. 5). Survival time is in proportion to the length of the projection under study. For instance, corticobulbar projections in the rat require a 7-day survival time, but the survival time can be extended if necessary for several weeks without significant loss of staining (maximum of 4 weeks). BDA is compatible with a wide range of fixatives which implies that in multilabel experiments, the choice for a particular fixative can be made depending on the most demanding marker. In rodents, we prefer a fixative that allows additional immunocytochemistry: buffered 4 % formaldehyde, 0.1 % glutaraldehyde, and 0.25 % of a saturated aqueous picric acid solution in the final volume. After fixation, the brain can be cut with any of the available

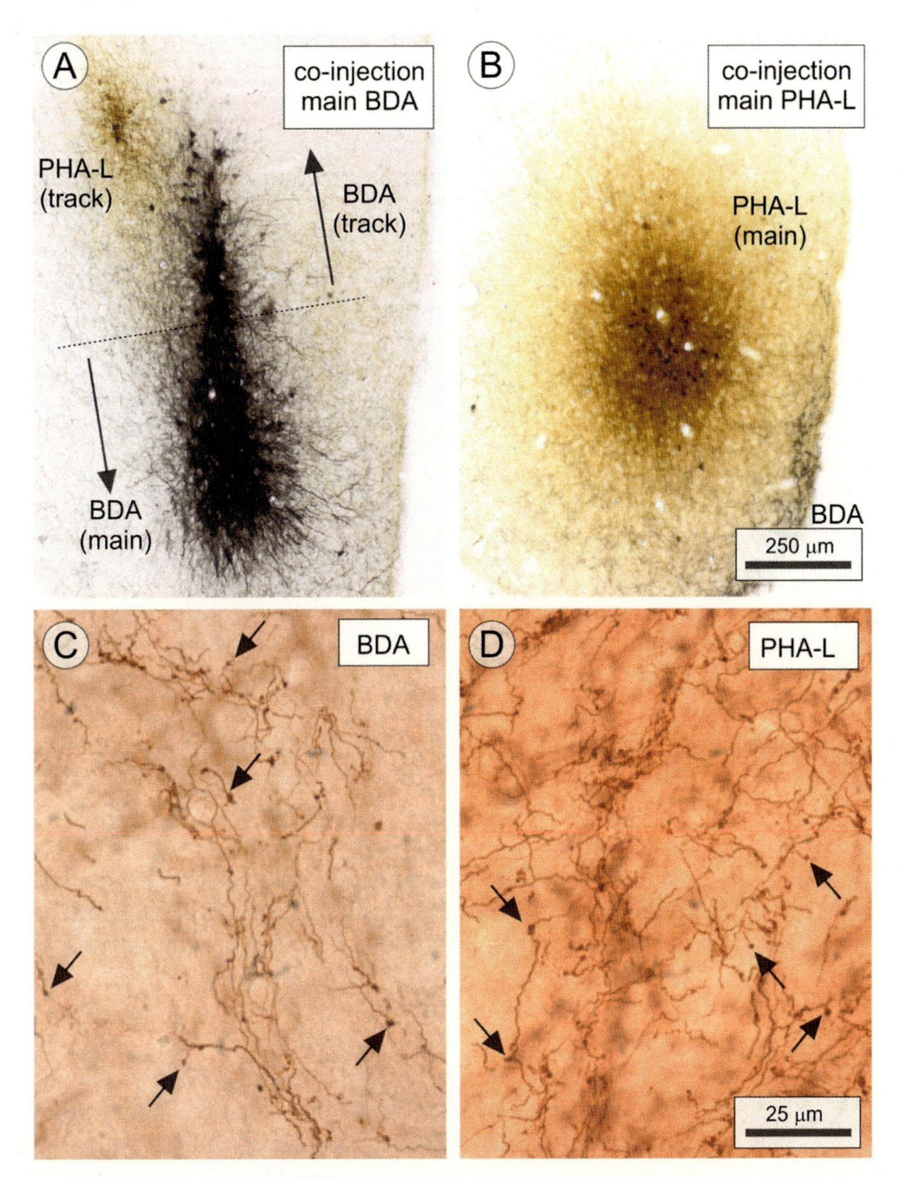

Fig. 5 (**a**, **b**) Co-injection for parallel anterograde BDA-PHA-L tracing. Single surgery session; each tracer injected via a separate penetration and injection. Double staining for BDA (DAB-Ni, *black color*) and PHA-L (DAB, *brown color*). Injection in prefrontal cortex of a rat (case 2004046; two adjacent sections showing the injection sites; slide courtesy Dr. Y. van Dongen). The streptavidin-DAB-Ni detection of the BDA provides a high contrast and low background; the immunohistochemical detection of PHA-L produces some background. Note the presence of many completely labeled neurons in the BDA and PHA-L injection spots. Both PHA-L and BDA show track labeling. Magnification in (**a**) is the same as in (**b–d**), Terminal labeling of fibers with BDA (**c**) and PHA-L (**d**). DAB visualization. A few of the terminal boutons are indicated with *arrows*. Magnification in (**c**) is the same as in (**d**)

sectioning methods. Brain sections with a thickness between 40 and 400 μm can be processed. Thinner sections may be required when a second or third marker is visualized in addition to BDA. We have successfully worked with free-floating sections as thin as 10 μm [215]. Visualization of transported BDA is simple: reaction with a standard [strept-]avidin-biotin complex (e.g., Vectastain kit, Vector, Burlingame, CA) followed by incubation in a standard 3,3′-diaminobenzidine-H_2O_2 mixture. The procedures are as follows: rinse 3× for 10 min in 50 mM Tris–HCl, pH 7.6, and react with the DAB solution for 10–25 min ("plain" DAB: 5 mg 3,3′-diaminobenzidine in 10 ml 50 mM Tris–HCl, pH 7.6, filtered and 3.3 μl of 35 % H_2O_2 added; Ni-enhanced DAB: add 15 mg of Ni-ammonium sulfate to the "plain DAB" reaction medium).The reaction is terminated via several rinses in 50 mM Tris–HCl, pH 7.6. Sections are then mounted on gelatinized slides, dried, counterstained if desired, and coverslipped with mounting medium.

An alternative way of visualizing transported BDA is via reaction with streptavidin conjugated to a fluorochrome. This is an easy one-step visualization: as a rule of thumb, incubation of sections for 1.5 h in a 1:400 diluted fluorochrome-streptavidin conjugate is sufficient. An entire repertoire of streptavidins conjugated to fluorescent reporter molecules is available from several suppliers. Preparations stained with a streptavidin-fluorescent marker conjugate can be studied in the fluorescence microscope or in a confocal instrument.

As BDA "fills" the neuronal cytoplasm within the cell's limiting membranes, the staining product forms a homogeneous label inside neurons against a perfectly clear background and, in this way, resembles classical Golgi silver impregnation. All the details of the labeled neurons in the injection sites, e.g., dendritic spines, are available for inspection (Fig. 5). A major advantage of the even distribution of the label along the entire trajectories of the fibers allows extremely precise mapping of fiber tracts, analysis of the compartmentation of large fascicles and association bundles, and the study of terminal projection patterns. 3D computer reconstruction of BDA-labeled neurons is supported as well by this feature as several software programs use the density distribution of label as parameter for the determination of cell boundaries [213, 214, 216]. In the electron microscope, the label generated by BDA processing has always a cytoplasmic location, leaving the nucleus in most cases free of label. Detail in the electron microscope can be preserved so well that the pre- and postsynaptic membrane densities of labeled axon terminals can be appreciated (Fig. 6) (cf. [210]). Retrograde transport of BDA may occur, resulting sometimes in a granular deposit of the tracer in a limited number of neuronal perikarya. Infrequently, "dense" labeling is present of cell bodies and dendrites in areas known to project to the site of injection [150]. When this apparent retrograde transport occurs, then "false" anterograde labeling may occur of collateral

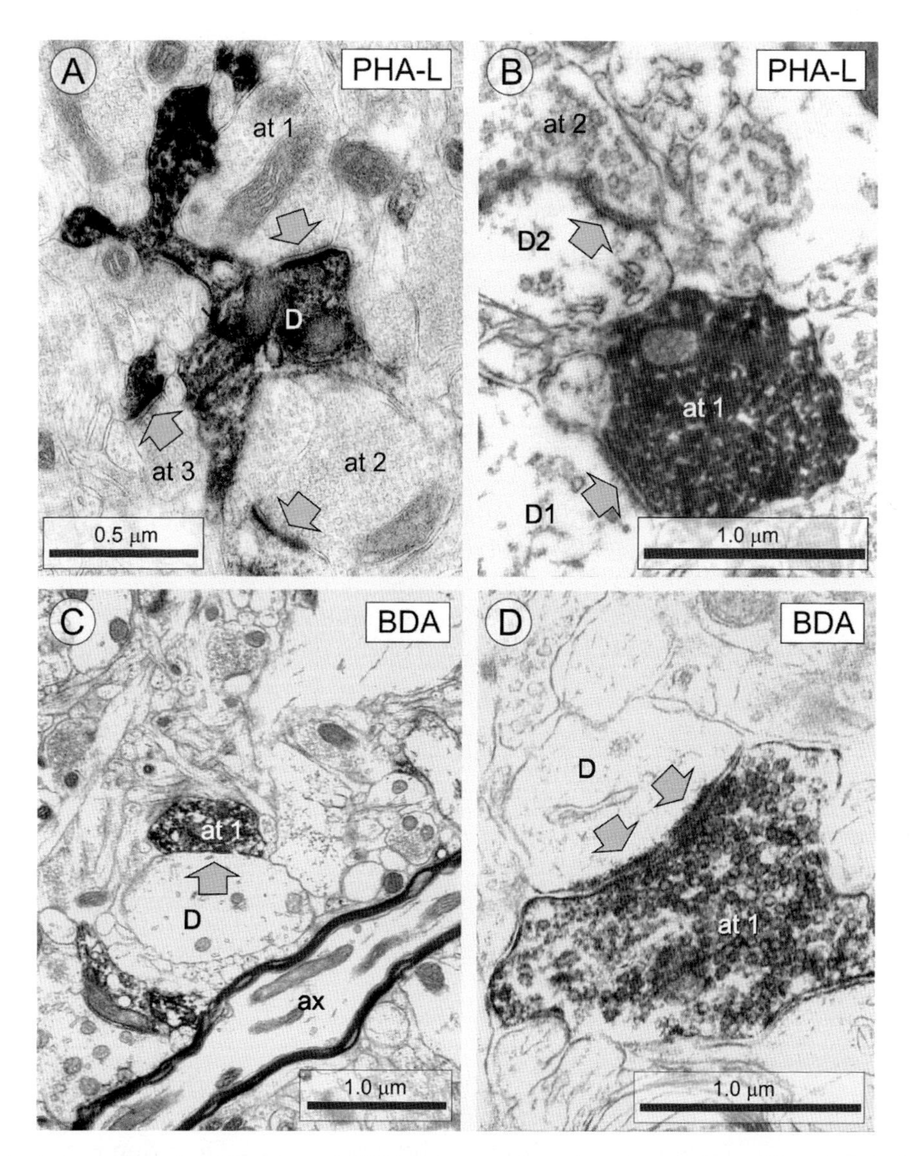

Fig. 6 Electron micrographs of PHA-L- and BDA-labeled structures in various regions of the rat brain. (**a**) PHA-L-filled dendrite (**d**) of a medium-sized spiny striatal neuron close to the injection site. Mild detergent treatment. Several unlabeled axon terminals (at 1 through at 3) form synapses (*arrows*) with spines of the labeled dendrite. (**b**) PHA-L-labeled axon terminal (at 1) in hippocampal region CA1 after injection of the label in the midline thalamus. Freeze-thaw treatment of the tissue. (**c**) BDA detection without addition of detergent to the incubation medium. Ultrastructural detail is much better preserved than in PHA-L preparations. A BDA-labeled axon terminal (at 1) forms a synapse (*arrow*) with an unlabeled dendrite (**d**). *ax* myelinated axon. *D* At high magnification, the ultrastructural details of BDA-labeled terminals (at 1) can be fully appreciated as well as that of the big dendritic spine (**d**) that receives a synapse from the labeled terminal (*arrows*)

fibers and even axon terminals belonging to these initially retrogradely labeled neurons [30]. The retrograde transport component of BDA was exploited fully by Bácskai et al. [6] through application of fluorescent derivatives of dextran amine (tetramethylrhodamine dextran amine (RDA)) and fluorescein dextran amine (FDA), contralateral to each other to the cut end of the hypoglossal nerve in the frog, *Rana esculenta*. The aim of this particular experiment was to study cellular aspects of the relationships of hypoglossal motoneurons across the midline of the brain stem.

As the label is located in the neuron's interior, the detection reporter molecule (streptavidin HRP) must, in the conventional procedure of free-floating incubation, penetrate into the sections and cross cellular membranes to bind to its target. This penetration process takes some time to happen. However, the detection of transported BDA does not require antibodies. Because a molecule of streptavidin conjugated to a fluorochrome has a modest molecular weight compared with an antibody, it does not suffer from poor penetration to the degree encountered with, for example, lectin tracing immunocytochemistry. The addition of a detergent to enhance the penetration of reagents into tissue sections (e.g., 0.3–0.5 % Triton X-100) is not always necessary and can be replaced, for instance, in a procedure optimized for electron microscopy, by a freeze-thaw procedure or can even be omitted completely.

3.4 Phaseolus vulgaris-Leukoagglutinin (PHA-L)

Phaseolus vulgaris (the red kidney bean) is a plant whose beans are a rich source of a phytohemagglutinating glycoprotein whose subunits exhibit either leucocyte agglutinating (L-subunit) or erythrocyte agglutinating (E-subunit) activity [78]. A combination of four L-subunits produces a stable compound that, after deposition into the CNS, appears to be taken up nearly exclusively by neuronal cell bodies [67]. In rodents, injection is via iontophoretic application of a 2.5 % solution in 50–100 mM phosphate buffer, pH 7.4, via a glass micropipette (tip diameter is critical in this respect, between 10 and 25 μm). We have in our experience never noted "false" anterograde axonal transport to a degree such as experienced with BDA. Uptake is thought to be receptor mediated, and after uptake, the lectin spreads after a while evenly through all the processes of the neuron, especially the axon (Fig. 5). Detection is via immunohistochemistry with a primary antibody raised in rabbit or goat against the lectin and a secondary antibody conjugated to biotin or a fluorescent compound [67]. For the first time after the delicate silver impregnation preparations made by Ramon y Cajal, nearly 100 years before, a different way of visualization was achieved of the most exquisite details of entire neurons, most importantly the fibers and the complete axon including the terminal branches of the axon all the way downstream to the terminal boutons (Fig. 5d). In contrast to Ramon y Cajal's stainings, the PHA-L tracing method works equally well in young and adult brains and is not restricted to neurons with unmyelinated axons. Good light microscopic visualization, that is, with all the visible cellular detail, requires the addition of 0.3–0.5 % of an aggressive detergent to the incubation media, e.g., Triton X-100. As detergents remorselessly extract all lipids from the tissue, the quality of preservation of ultrastructural detail is lost in a detergent regime. Measures to find a balanced compromise between light microscopic visibility and preservation of ultrastructure include very short detergent treatment, use of a mild detergent, and freeze-thaw procedures in liquid nitrogen-cooled isopentane [205, 208].

The researchers who introduced the tracing method immediately understood that detection based on an immunohistochemical procedure would open the door widely for combinations with parallel procedures, e.g., multiple immunostaining [67, 68], and with other tracing methods, for instance, retrograde transport of a fluorescent tracer or even anterograde tracing with a noncompetitive tracer substance.

3.5 Parallel Anterograde Tracing

Gerfen and Sawchenko described in their original [67] publication an experiment combining an injection of PHA-L and tritiated amino acids in which they successfully localized both tracers in the appropriate pathways. Unfortunately, they did not illustrate this important experiment. In those early days, the only alternative and technically compatible anterograde tracing technique was making a lesion and tracking down degenerating axon terminals in the electron microscope [198]. The first report on successful application of two comparable anterograde tracing compounds in a light microscopic study to identify convergence of fibers of different origins on target neurons was published by Antal et al. [1]. These authors employed PHA-L as the first tracer and biotinylated or thioninolated PHA-L as the parallel tracer. Bevan et al. [15] combined PHA-L tracing with WGA-HRP tracing in a parallel anterograde tracing experiment. Soon after the introduction of BDA as reliable and detailed anterograde tracer, Lanciego and Wouterlood [114] followed up by introducing parallel anterograde tracing using PHA-L as the first tracer and BDA as the parallel tracer. Such a combination has the big advantage of exploiting two completely different and very sensitive detection systems. Some years later, Lanciego combined expanded parallel tracing with retrograde tracing [116, 118] which he also introduced in primates [117]. Morecraft et al. [135] beautifully applied multiple parallel anterograde tracing in primates (five different tracers) to study compartmentation of corticofugal projections in the corona radiata and internal capsule. Wright et al. [217] expanded parallel PHA-L-BDA tracing with electron microscopy. The use of modern confocal laser scanning microscopes makes it much easier, faster, and more decisive to identify multiple markers than with double-colorimetric staining methods. It can be safely argued today that double- and triple-fluorescence methods have superseded multiple-colorimetric anterograde tracing.

A good example in our lab of a full-fledged multifluorescence technique was a study by Kajiwara et al. [98]. Presumed, single target pyramidal neurons in hippocampal area CA1 were literally "lighted up" via intracellular injection of the fluorescent marker Alexa Fluor™ 555 in a lightly fixed slice preparation. One week prior to the intracellular injection, the rats had received an injection with PHA-L in the ipsilateral temporal cortex to label cortical afferent fibers and a small injection of BDA in area CA3 on the ipsilateral side to anterogradely label the Schaffer collaterals (*see* scheme, Fig. 7a). PHA-L was visualized via a primary anti-PHA-L antibody and a secondary antibody conjugated with Alexa Fluor™ 488. The transported BDA was visualized via streptavidin conjugated with Alexa Fluor™ 633. In a confocal laser scanning microscope, the AF555-labeled dendrites of the CA1 pyramidal neurons were identified. Z-scanning at high magnification (63× immersion; 8× electronic zoom) of regions of interest followed by computer processing and 3D reconstruction made it possible to visualize and quantify appositions between boutons on either PHA-L-labeled fibers or BDA-labeled fibers and the dendritic shafts and dendritic of the AF555-labeled CA1 pyramidal neurons (Fig. 7), thus establishing evidence for convergence of projections onto individual, identified neurons.

3.6 Colocalization of an Anterograde Tracer and a Second Marker

Gerfen and Sawchenko envisaged in their original publication combinations of their PHA-L tracing method with other tracing and immunohistochemical methods, and they reported on an experiment using double-fluorescence microscopy [67]. However, the epifluorescence "hardware" technology of that day was such that reliable estimation of colocalization of fluorescence was only possible in structures the size of neuronal cell bodies. The introduction of the confocal laser scanning microscope (CLSM) in the early 1990s brought the increase in optical resolution necessary to reliably study colocalization of markers inside very small structures such as fibers and axon terminals. Resolution of the standard CLSM depends on the quality of the objective lens and is approximately 200 nm radially, while axial resolution is 3–4 times less [11]. Technological developments such as stimulated emission depletion microscopy (STED) and stochastic optical reconstruction microscopy (STORM/PALM) offer the resolution in the 20–60 nm range that is really necessary to study markers attached to cellular organelles and vesicles involved in uptake, transport, and secretion as well as components of synapses and proteins involved in receptor anchoring and messaging systems [17, 53]. These novel optical methods open an entire field for future study, that of identified fibers (via neuroanatomical tracing) and their contents including that of axon terminals. We call this superresolution-based localization "pinpoint tracing." In anticipation of such "pinpoint-tracing" equipment, we attempt to determine

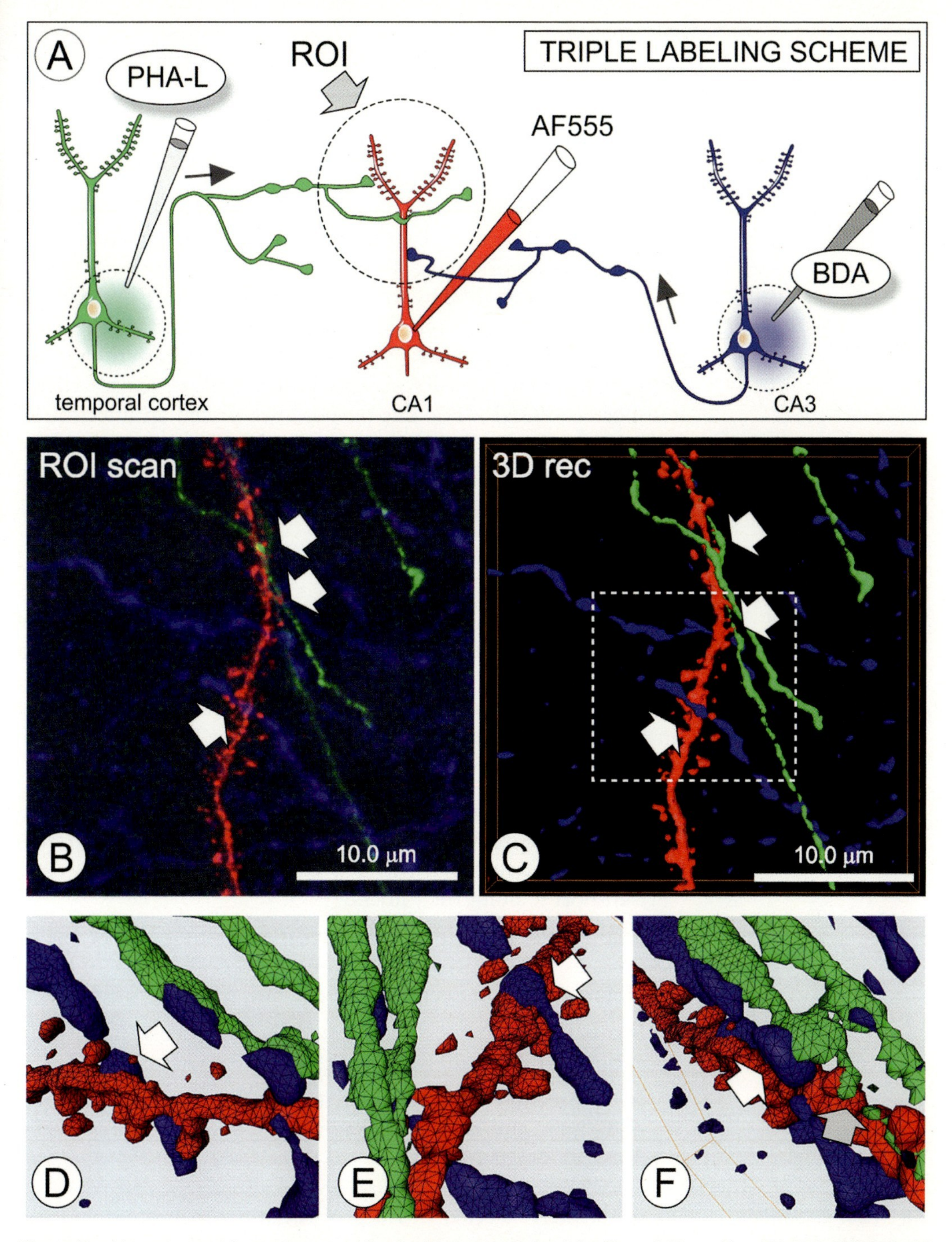

Fig. 7 Double anterograde tracing combined with intracellular injection of Alexa Fluor™ 555 (AF555). (**a**) Experimental setup of this study: convergence of fibers belonging to two projection systems onto single target neurons. Two sources of innervation are labeled: temporal cortex via focal injection of PHA-L and CA3 via focal injection of BDA. The target neurons are penetrated with a micropipette filled with the fluorescent dye Alexa Fluor™ 555 (code colored *red*). PHA-L is detected via immunofluorescence (488 nm fluorochrome; code colored *green*) and BDA via streptavidin-633 nm fluorochrome (code colored *blue*). The AF 555 is inherent fluorescent. (**b**) In a confocal laser scanning instrument, the region of interest (ROI) is Z-scanned with three laser/detection channels: 488, 543, and 633 nm. (**c**) 3D computer reconstructions reveal appositions (*arrows*) of PHA-L- and BDA-labeled fibers with the dendrites of the AF555 injected neuron. (**d**–**f**) are frames selected from a 360° 3D reconstruction rotation to better identify the spatial relationships of the appositions (contacts indicated with *arrows*)

colocalization inside thin fibers and in compartments of dendrites and axon terminals via statistical mapping of overlaying distribution patterns of different markers. Images of each marker are acquired in unique, marker-associated channels in a standard CLSM. We provide here a state-of-the-art example of multifluorescence work done in collaboration with various American and European laboratories (Fig. 8). The scientific question was to find evidence supporting the hypothesis that in the dentate gyrus of the rat, a set of presynaptic terminals exist of fibers that take their origin in the supramammillary area and that contain both the vesicular glutamate transporter 2 (VGluT2) and vesicular GABA transporter (VGAT). For this purpose, we placed an injection of BDA in the supramammillary region. After 1-week survival, the rats were sacrificed, brain sections cut, and a triple staining performed as reported in Sect. 3.7.

3.7 Procedure for Anterograde Tracing Combined with Detection of Two Additional Markers

The purpose is to investigate whether or not an anterogradely labeled fiber projection expresses one or two additional markers. In this example, we combine neuroanatomical tracing with immunofluorescence determination of VGluT2 and VGAT (*see* Sect. 3.6), both vesicular transporter molecules that are integrated in the walls of synaptic vesicles and for that matter are expressed in axon terminals. Any other specific neuroactive substance can be looked for as well, e.g., postsynaptic density-associated proteins (e.g., [211, 215]).

- Surgery: stereotaxically inject biotinylated dextran amine into the appropriate brain locus. A safe survival time allowing transport (in rodents) is usually 1 week. Neurons take up label and transport it along their fibers.
- Sacrifice the animal, and section on a vibrating microtome or on a freezing microtome. If one of the additional markers requires very thin sections, then section with a cryostat.
- Incubation nr. 1: cocktail of anti-VGluT2 (raised in rabbit) and anti-VGAT (mouse) (overnight, room temperature).
- Incubation nr. 2: cocktail of goat anti-rabbit Alexa Fluor™ 488, goat anti-mouse Alexa Fluor™ 633, and streptavidin Alexa Fluor™ 546.

3.8 Post-incubation Processing of Triple-Stained Sections

We deliberately selected this trio of Alexa fluorochromes and this combination of secondary antibodies and fluorochromated streptavidin (a) to pertinently exclude any spectral overlap between the vesicular transporter markers and (b) because these fluorochromes perfectly match the laser/channel combinations available in our confocal instrument. After finishing the incubations, sections were mounted on gelatin-coated glass slides, dried, and coverslipped with a standard mounting medium (e.g., Entellan (Merck), DPX (Fluka), or Aquamount (Gurr)) (if tissue shrinking is a matter of concern).

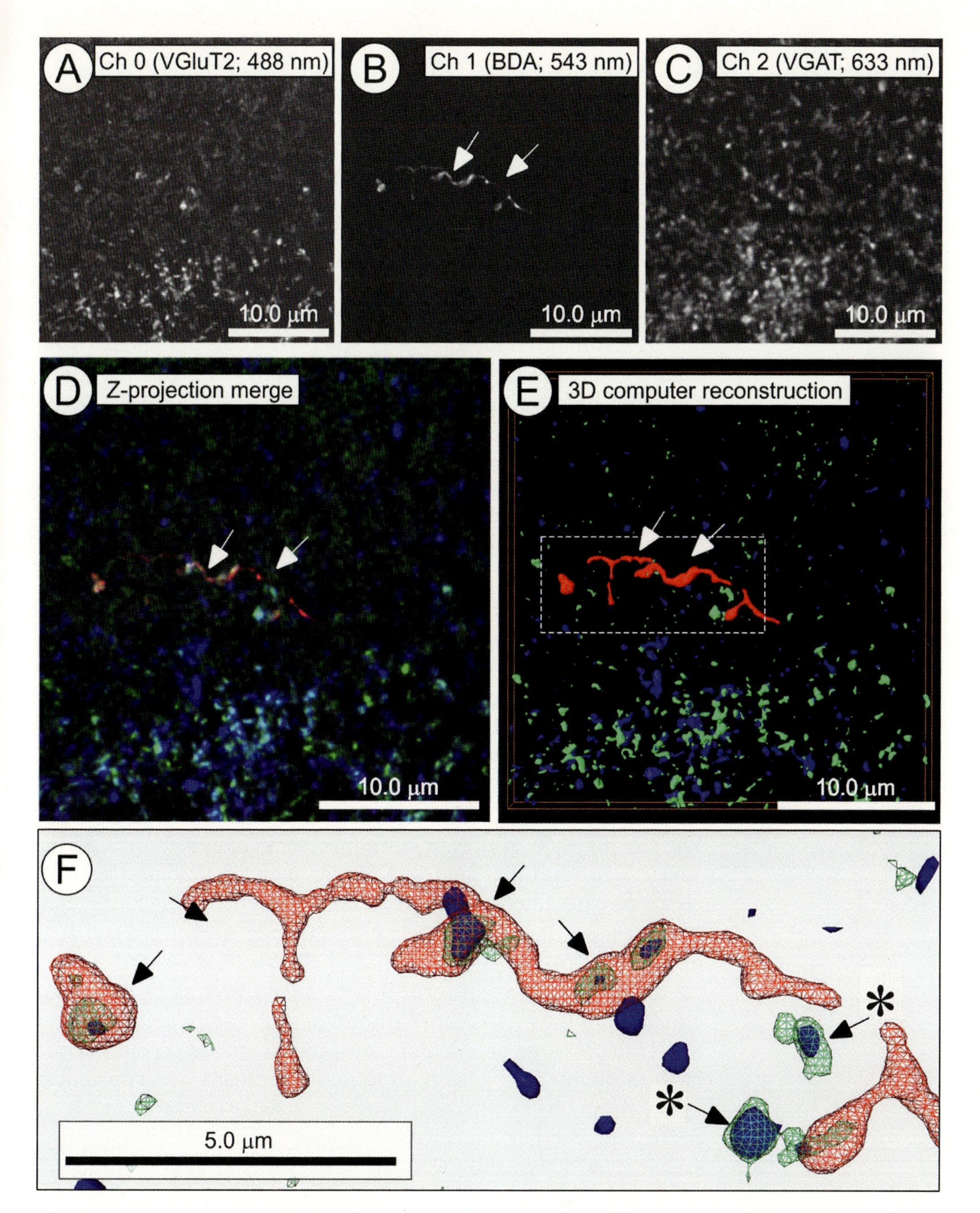

Fig. 8 Anterograde tracing with BDA combined with immunofluorescence detection of vesicular glutamate transporter 2 (VGluT2) and vesicular GABA transporter (VGAT). CLSM imaging. BDA injected in the supramammillary region of a rat; image acquisition in the dentate gyrus. (**a**–**c**) Z-projections of the images obtained in each of the three channels configured in the CLSM: Ch0 = VGluT2; Ch1 = BDA; Ch2 = VGAT. *Arrows* in (**b**) indicate BDA-labeled fiber. (**d**) Merge image with information from each channel color-coded: Ch0 green, Ch1 red, and Ch2 blue. Light blue occurs in case of overlap of pixels of channels 0 and 3; orange in case of overlap of channels 0 and 1; magenta if pixels in all three channels overlap. *Arrows* indicate the BDA-labeled fiber. (**e**, **f**) 3D reconstruction of the merged channels. The BDA-labeled fiber (*red*) and VGluT2 particles (*green*) are rendered transparent, while VGAT particles (*blue*) are rendered solid. "Overlap" (colocalization) of VGluT2 and VGAT immunofluorescence occurs both in separate particles (*asterisk*) as well as in the BDA-labeled fiber, suggesting mixed glutamate-GABA innervation from the supramammillary region

The slides are stored in the dark at −20 °C to prolong fluorescence life. In our experience, the fluorescence in the sections remains under these conditions crisp and excellent up to 10 years after mounting.

Positive, accurate detection of colocalization of small aggregates of axon-terminal-associated proteins (e.g., small populations of synaptic vesicles) inside axon terminals that are labeled with the anterograde tracer cannot be accomplished with a routine (epi) fluorescence microscope but needs a confocal scanning instrument and Z-scanning conditions [11, 200]. We took our slides to the CLSM and performed triple-label Z-scanning, taking all the conditions and precautions in mind that are associated with the detection of multiple labels in one and the same small structure (i.e., channel configuration, controls, and checks with single-stained sections; *see* [11, 200]). The scanning procedure produced for each sample three Z-stacks of 8-bit grayscale bitmapped images, that is, one stack for each channel. Image acquisition was done via scanning first with the laser/filter/detector configuration belonging to channel 1, then with the channel 2 laser/filter/detector configuration, and finally with channel 3 laser/filter/detector configuration after which the microscope stage was moved one increment in the Z-direction and the image acquisition repeated (so-called sequential, frame-by-frame Z-scanning). As biological structures are represented in 2D bitmapped images by pixels, in 3D the same biological structures are represented as aggregates of high-intensity voxels with specific locations in a stacked, 3D matrix. We ran for each sample a computer script controlling a SCIL_Image algorithm that analyzed the three corresponding stack pixels for pixel to determine aggregates of pixels occurring in the same XYZ position in the 3D matrix in all three stacks. We call such occurrence "overlap" [11, 213, 214]. Overlap is governed by optical resolution which, in turn, is determined by Abbe's diffraction equation $r = 0.61\,\lambda/NA_{obj}$. Without optical equipment such as a STED microscope or stochastic imaging, statistical spread of overlap is inevitable, and as resolution in our standard CLSM prohibits us to distinguish individual synaptic vesicles, we are unable to determine whether one small axon terminal contains two types of synaptic vesicle (separate "VGluT2-vesicles" and "VGAT-vesicles") or whether all synaptic vesicles inside the terminal contain both types of vesicular transporter. The statistical nature of overlapping voxels explains why we often observe *partial overlap* between channels of aggregates of voxels. The procedure with SCIL_Image analysis struggles with this phenomenon. An alternative approach such as the center-distance algorithm [3] may offer some help, but a step forward in resolution is obviously a better way. Anyway, we could statistically underpin our visual findings, notably that of overlap of a fraction of the fibers, VGluT2-immunofluorescent structures, and VGAT-immunofluorescent structures (Fig. 8d–f). Colocalization of VGluT2 and VGAT was independently corroborated via dual label electron microscopy [21].

3.9 Note on Lesion Methods

The first experimental neuroanatomical tract-tracing experiments [192, 127] preceded the work by Von Waldeyer, Sherrington, and Ramon y Cajal and therefore predate the neuron theory. However, these early lesion experiments interfered with major myelinated tracts through surgical lesions, and the results were interpreted in gross macroscopic terms only.

Building upon Waller's legacy, a whole subdiscipline blossomed in experimental neuroanatomical tract tracing, supported in its heyday by a most powerful neuroanatomical histological technique, the selective silver staining, originally developed by Nauta [138] and refined by Fink and Heimer [57]. After the introduction of transport-based methods, the experimental lesion-degeneration methods have been replaced in neuroanatomical tracing by those based on anterograde and retrograde transport while lesions are still being used by neuroscientists in depletion and behavioral studies as well as for supplying an independent, electron dense "label" for axon terminals in the electron microscope [212]. A complete historical and technical review of lesion-silver-degeneration methods can be found in Giolli and Karamanlidis [69].

3.10 Methods Employing Neurotoxicity

As they leave a trace of destruction in the brain, neuron-killing substances ("neurotoxins") can be used as neuroanatomical "tracers" (e.g., [124, 191]). Neurotoxins have been found useful to eliminate specific types of neuron whose limiting membranes contain the particular receptor or molecule with which the toxin reacts. The popularity of tracing based on neurotoxins is limited, not because of lack of potency but because next to specialized personnel also specialized laboratory and experimental animal housing facilities are required to responsibly handle this kind of biohazardous substance and the animals treated or infected with them.

Predating the use of viral and bacterial neurotoxins is the use of excitotoxic lesioning for functional and structural brain mapping. Excitotoxins, e.g., glutamate, kainic acid, ibotenic acid, 6-hydroxy dopamine, MPTP, or *N*-methyl-D-aspartate (NMDA), do not require highly specialized laboratory and animal housing facilities. Applications range from making electrolytic or thermal lesions or radiofrequency lesions to penetrations with a microknife. The advantage of the focal injection of an excitotoxic substance is the sparing of fibers of passage. This subject is not further discussed in this chapter. Reviews on the subject can be found in Kobayashi [105], Moore [134], Jonsson [96], Jarrard [94], and Kirby et al. [103].

3.11 Bidirectional Tracing

As in living neurons molecular motors carry substances in all directions along the cytoskeleton microtubule network, particular tracers might have affinity to several types of molecular motors and therefore may be transported bidirectionally. Some lectins and neurotoxins indeed do. Schwab et al. [164] and Gonatas et al. [73] introduced HRP conjugated to a lectin, wheat-germ agglutinin (WGA-HRP), as a tracer with anterograde and retrograde capabilities.

Soon afterwards, Trojanowski et al. [181, 182] published their report on uptake and bidirectional transport of HRP conjugated to the nontoxic B fragment of cholera toxin (choleragenoid) (CTB-HRP). Sensitivity of tracing based on this uptake/transport is high, both anterogradely and retrogradely. Advantages of this method are that uptake is receptor mediated and that undamaged axons of passage in the injection spot have only a limited capacity to internalize these probes. Trojanowski reviewed the matter in [180].

Anterograde axonal transport of WGA-HRP occurs fast (e.g., about 108 mm/day). By comparison, native HRP is anterogradely transported at a rate of 288–432 mm/day [180].

Recently, fluorochromated CTB has become commercially available, with a spectral repertoire of Alexa Fluor™ molecules as reporter. These compounds are directly visible in a fluorescence microscope, and because the Alexas are particularly photostable, they are well suited for inspection in a confocal laser scanning microscope. Thus, lectins are here to stay for a while as neuroanatomical tracers. The fascinating issue of lectin tracing is covered in detail elsewhere in this book.

4 Local Circuit Tracing Techniques

The orthograde and retrograde neuroanatomical tracing techniques listed in the previous sections have in common that they label long projections. In the context of brain organization, "long" means that the neurons involved give rise to axons leaving the cortical field or subcortical nucleus to synapse with neurons located elsewhere. Microcircuit neurons and interneurons, in general those neurons whose axon stays within a space with a diameter equal to or smaller than the injection site of a tracer, escape being detected because their axons blend with the injection spot. This characteristic of microcircuit neurons and interneurons makes application of retrograde tracers unattractive (injection sites tend to be large; retrograde tracers accumulate mostly in cell bodies and do not label axons), but also widely used tracers such as PHA-L and BDA still provide too large injection spots. Obviously, making progressively smaller tracer injections renders microcircuit neurons and interneurons better exposed, while the ultimate possibility is the injection of tracer into single neurons. As we have some experience with these methods, we will discuss a few of these below, from top (single-neuron labeling) to bottom (small populations of neurons).

4.1 Intracellular Filling of Single Neurons

Progress in micromanipulation, electronics, and chemistry made it possible in the 1960s to approach individual neurons with a hollow electrode, penetrate its limiting membrane, record electrical events, and eject a dye into the cell to reveal the details of its morphology [173]. Initially, this work was the domain of neurophysiologists working with living cells. Two decades later, Rho and Sidman [154] and Buhl and Lübke [24] extended the technique to neurons in slices of lightly fixed brain, an innovation that made the method instantly attractive to neuroanatomists. Until then, the only way to study single cells morphologically was through one of the notoriously unpredictable Golgi silver impregnation procedures. Histologically, the lightly fixed brain procedure is compatible with most anterograde and retrograde tracing techniques described in this chapter. We illustrate this with an example of striato-mesostriatal connectivity. The substantia nigra pars compacta (SNc) contains cell bodies of dopaminergic neurons that innervate the neostriatum, while the fibers from striatal neurons that innervate the substantia nigra end in the pars reticulata (SNr) [66] which lies adjacent to the SNc. Spatially, there is separation between the terminal field of the striatal projection in SNr and the cell bodies in SNc. However, the physical distance between the striatal endings and the dopaminergic cell bodies may be bridged by the dendrites of the SNc neurons penetrating SNr. The anatomical approach to study the synaptic relationships of striatal endings in SNr and dopaminergic nigrostriatal neurons involves the combination of four techniques: (a) retrograde labeling of the nigrostriatal cells, (b) anterograde labeling of striatonigral projection fibers, (c) intracellular injection of retrogradely labeled cells in the SNc to visualize the dendrites of these cells penetrating SNr, and (d) immunofluorescence with an antibody against tyrosine hydroxylase to verify that the intracellularly injected cells are dopaminergic. The procedure (in the rat) was as follows (scheme of the concept in Fig. 9):

- Surgery: focal injection under stereotaxic guidance of an anterograde tracer (BDA) in the caudate-putamen (i.e., striatum).
- In the same surgical session: deposition of a retrograde tracer (Fast Blue or Fluoro-Gold) in the striatum in a location different from that of the BDA injection.
- Postsurgery survival to allow transport to proceed: 7 days.
- Sacrifice of the animal via transcardial perfusion with a light fixative: 4 % buffered formaldehyde, pH 7.4.
- Dissection of the brain and sectioning of the midbrain into 500 μm-thick slices with a vibrating microtome.
- Inspection of the slices in a fluorescence microscope to identify retrogradely labeled neurons in the ventral mesencephalon.
- Approach and penetration of retrogradely fluorescent cells with a micropipette and filling the impaled cells with Lucifer yellow (technical details in [207]).
- Post-fixation overnight of the slices with 4 % buffered formaldehyde, pH 7.4, with 0.1 % glutaraldehyde, in a refrigerator.

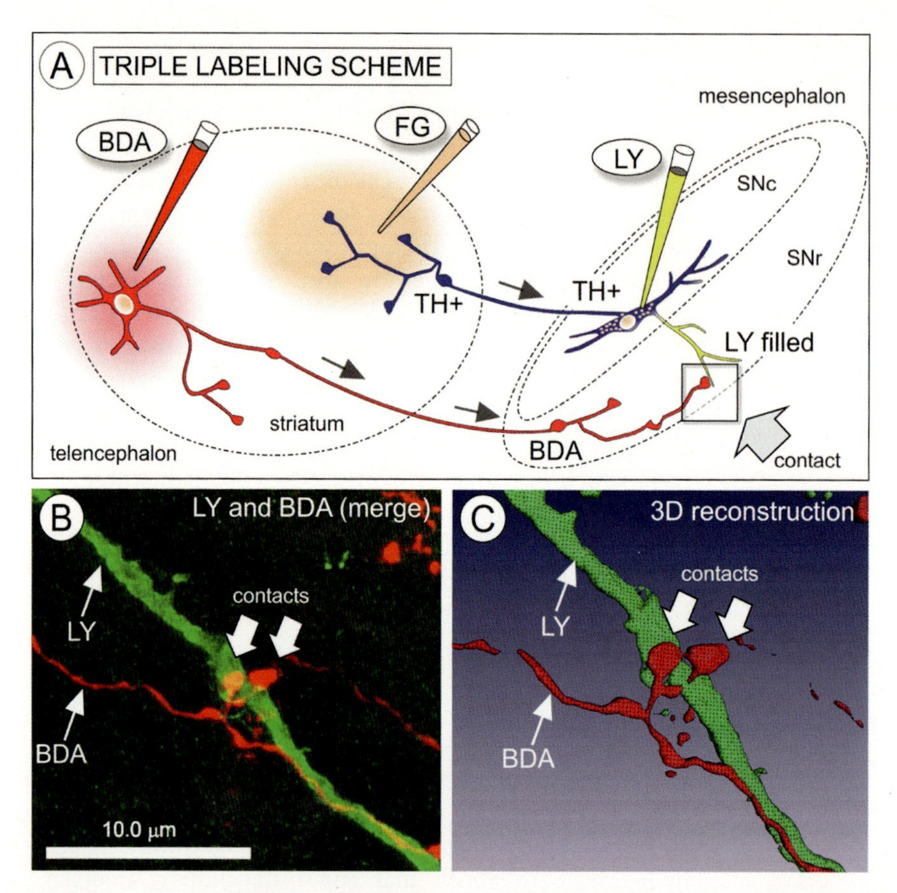

Fig. 9 Anterograde tracing with BDA combined with retrograde tracing with Fluoro-Gold (FG), intracellular injection of Lucifer yellow (LY), and immunofluorescence detection of tyrosine hydroxylase (TH; marker for dopaminergic structures). (**a**) Scheme showing all manipulations. BDA labels striatonigral fibers in the mesencephalon. FG, injected simultaneously, retrogradely labels neurons in the substantia nigra pars compacta (SNc). After survival, LY is in slices of the mesencephalon injected intracellularly into cell bodies of retrogradely labeled SNc cells and fills their dendritic trees. (**b**, **c**) After resectioning the slices followed by BDA-, LY-, TH-immunofluorescence and confocal laser scanning, the LY-filled dendrites are color-coded green (in **b**) and 3D reconstructed (in **c**). The BDA-labeled fibers are color-coded *red*. In **b** and **c**, two contacts between a BDA-labeled striatonigral fiber and a dendrite of a nigrostriatal, dopaminergic SNc neuron are visible (*big arrows*)

- Cryoprotection with 20 % glycerin and 2 % dimethyl sulfoxide (DMSO) in phosphate buffer [158], freezing, and resectioning into 40 μm thick sections.
- Triple staining as outlined in Sect. 3.7: first in a cocktail of an antibody against Lucifer yellow (raised in rabbit) and an antibody raised in mouse against tyrosine hydroxylase and then in a cocktail of goat anti-mouse Alexa Fluor™633, goat anti-rabbit Alexa Fluor™488, and streptavidin Alexa Fluor™ 546.
- Mounting, drying, and coverslipping.

- Z-scanning in a confocal laser scanning microscope with three laser/filter channels, 488, 543, and 633 (green, red, and infra-red), and associated 3D reconstruction.
- Determination of apposition of axon terminals ("boutons") of BDA-labeled striatal fibers with dendrites containing Lucifer yellow (identified, by virtue of the labeling of their perikaryon with retrogradely transported fluorescent dye as belonging to a nigrostriatal neuron).

In frames b and c of Fig. 9, the results of the BDA tracing and LY injection are shown. BDA label is color-coded red, while LY appears color-coded green. On several occasions, varicosities of BDA-labeled fibers can be seen in contact in SNr with the LY-labeled dendrites. Correlative light and electron microscopy has shown good correlation between swellings (boutons) on anterogradely labeled fibers as the light microscopic representations of axon terminals [203]. Because these dendrites belong to retrogradely labeled SNc neurons projecting to the striatum, we are witnessing here contacts between striatonigral fibers and nigrostriatal neurons. Through imaging in the infrared channel of the confocal instrument, we verified whether this dendrite contained infrared label, i.e., immunopositivity for tyrosine hydroxylase indicating that this dendrite belongs to a dopaminergic neuron. In fact this method is a confocal laser scanning microscope refinement of an epifluorescence forerunner protocol of this method that we had published 20 years ago [209]. This "epifluorescence combination" consisted of retrograde tracing with Fast Blue, intracellular injection with Lucifer yellow of Fast Blue tagged cells, anterograde PHA-L tracing, and classical TRITC-immunofluorescence of the cell bodies of the intracellularly Lucifer yellow injected labeled SN neurons [209]. The next step will be to take the preparation to the electron microscope, for which the technology is operational [97].

One drawback of intracellular filling of neurons in slices of fixed brain is that the injected dye diffuses freely into the dendrites while it does not travel into the axon. Intracellular injection in living neurons does not suffer from this restriction. The intracellular deposition of biocytin [65] or BDA [147] into single living neurons, following intracellular recording, makes it possible to trace complete dendritic patterns and to follow as well the complete distribution of axons, axon collaterals, and axon terminals of individual neurons. Collaboration of neuroanatomists and neurophysiologists is required here.

4.2 Pericellular Injection

One confusing aspect of classical tract-tracing procedures is that they involve large injections with tracer and produce labeling of massive populations of neurons. Most "gross" tracing studies interpret a number of individual tracing experiments into a generalized connectivity pattern at the neuron population level. Study of microcircuits and interneurons requires a more refined approach with preparations that contain the smallest number of labeled neurons possible. Peri- or juxtacellular deposition of a small amount of Neurobiotin immediately next to a neuron's cell body via a micropipette used to record physiological activity from that cell ([144]; reviewed by Duque and Záborszky, [48]) results under proper conditions of mild stimulation in uptake of the tracer by that cell [218] or by small clusters of cells [144, 176]. The labeled cells then display all their morphological details in a Golgi silver impregnation like fashion, including their entire axon trajectories and terminal ramifications down to the synaptic boutons. A very sophisticated example of microcircuit tracing through small, juxtacellular injections was provided by [142]. Finally, the characteristics of the pericellular/juxtacellular tracers Neurobiotin and biocytin favor their application in combined experiments dealing with short-distance connectivity, e.g., cortical layer-to-layer tracing and to single-neuron visualization (e.g., [186, 212]).

4.3 Neurobiotin, Biocytin, and Derivatives: Light and Electron Microscopy

The *N*-(2-aminoethyl)biotinamide derivative of biotin, "Neurobiotin" (trademark of Vector, Burlingame, CA), was introduced by Kita and Armstrong [104] basically as an alternative for HRP, Procion yellow, and Lucifer yellow to mark neurons after intracellular neurophysiological recording. After the neurophysiology session, the morphological features of the recorded cells can thus be documented. In this particular application, Neurobiotin has enjoyed throughout the years a steady popularity. It is being applied both in mammals (rat, rabbit; as exemplified recently by [160]), amphibians [222], and invertebrates [56]. Neurobiotin can also be used as an anterograde neuroanatomical tracer.

Biotin conjugated to lysine ("biocytin") can be applied both in intracellular physiological recording-labeling purposes in single living neurons [88] and for tracing [102]. Transport of biocytin is mostly anterograde, especially when the injection spots are small. King et al. [102] reported some injection track labeling and also retrograde transport following large, mechanical injections (up to 1.3 μl with a Hamilton syringe). An extensive comparison between the characteristics of Neurobiotin and biocytin was published by Lapper and Bolam [119]. These authors confirmed King et al.'s [102] observation that both Neurobiotin and biocytin show a tendency of being transported retrogradely.

Postinjection survival times after biocytin or Neurobiotin injections need to be relatively short (less than 4 days) because both compounds are metabolized quickly. Both substances therefore are being used exclusively for short-distance tracing and for the visualization via intracellular injection of single neurons. Li et al. [122] reported successful application of biocytin as an intracellular-injection marker in neurons in slices of fixed brain. Delivering the colorless biocytin into cells in fixed brain slices offers quite a challenge (Li et al. [122] used micropipettes loaded with biocytin mixed with Lucifer yellow). This challenge can be addressed by using a colored derivative of biocytin, named biotin-dextran miniruby [123] or by doping the biocytin with Lucifer yellow.

Neurobiotin and biocytin can be easily applied in electron microscopy procedures [36, 119, 175]. In this respect, the characteristics of these tracers resemble those of BDA. In order to generate a second electron dense label for double labeling at the electron microscope level, biocytin tracing has been combined with the anterograde degeneration method that also requires short postsurgery survival times (2–3 days in rat) [212].

Finally, a special application of Neurobiotin is light microscopic tracing in slices of postmortem human brain [39, 40].

4.4 Golgi Silver Impregnation

The superb detail of neuronal morphology reflected in drawings and prints of silver-impregnated neurons by Ramón y Cajal [148] and by the original discoverer Camillo Golgi [72] still raises respect and admiration for this technique and its early practicers. The essence of Golgi silver impregnation is that, given the proper sequence of immersion in the inorganic solutions and a little bit of luck a stable, black amorphous deposit (i.e., silver chromate; [59]) forms inside individual, isolated neurons. With some additional chemistry, it is possible to embed the material in plastic, ultrasection it, and study the stained neurons in the electron microscope [20, 55] (reviewed by [54, 199]). Thus, the light microscopic description of a neuron can be followed up in a correlative light-electron microscopy approach with the study of its subcellular details such as synaptic vesicle and pre- and postsynaptic membranes. This is of importance given the superb single nanometer resolution offered by an electron microscope. Images acquired with confocal laser scanning microscopy, even the best ones, are so diffraction limited that the existence of a synapse can be made plausible only.

Golgi silver impregnation can be used as a tracing tool. In well-impregnated neurons, the main axon and its ramifications contain the precipitate that identifies these structures as processes belonging to that neuron, even when the axon has traveled far away from the parent cell body. However, the electron microscope reveals that as soon as a myelin sheath develops around an axon, the silver chromate precipitate in that axon decreases sharply over a short trajectory [55, 202]. Golgi silver impregnation therefore only reveals unmyelinated axons. Axon collaterals can be traced down to their distal axon terminals as long as they have not developed a myelin sheath and, of course, as long as they are present in the tissue blocks subjected to the impregnation procedure. Retrograde HRP tracing offers considerable help at this point. Somogyi et al. [170] successfully combined retrograde HRP tracing and Golgi silver impregnation and studied the preparations via a correlative light-electron microscopy procedure.

4.5 Golgi-Rapid Procedure

Many Golgi silver impregnation recipes are available in the literature. We have used for correlative light-electron microscopy with success the following Golgi-rapid procedure on blocks of perfusion-fixed brain [202]:

- Place brain blocks, slabs, or entire fixed brains in a mordant consisting of 4 % formaldehyde, 3 % potassium dichromate, and 2 % chloral hydrate, 3–4 days, at room temperature.
- Transfer to 0.75 % silver nitrate (keep under dark conditions) 2–3 days, at room temperature.

Numerous improvements of Golgi silver impregnation for electron microscopy purposes have been published: anterograde degeneration following a lesion combined with Golgi silver impregnation [19, 62, 170, 201], impregnation of brain slabs or thick sections [41, 60, 64, 93], HRP retrograde tracing combined with Golgi silver impregnation [60, 169, 170], postembedding neurotransmitter immunocytochemistry in combination with correlative light-electron microscopy of Golgi silver-impregnated neurons [171], and combination of Golgi silver impregnation with immunofluorescence processing to reveal multiple neuronal markers [172]. A decisive technological breakthrough in this field was gold toning [55] to stabilize the precipitate for easier preparation of ultrathin sections.

5 Other Methods

5.1 Diffusion of Cobalt and Nickel Compounds

Intracellular filling of complete neurons in whole-mount insect ganglion preparations with cobalt chloride was reported for the first time by Pitman et al. [145]. Fuller and Prior [63] extended the technique to the vertebrate brain. A variation with a cobalt-lysine complex was used by Lázár [121] in demonstrating retinotectal connectivity in a chronically anesthetized frog preparation. Although cobalt-lysine tracing has been superseded by the techniques that utilize transport of macromolecules or lectins, it is still being used with cold-blooded vertebrate animal species (e.g., [9, 128, 157, 179]) and, more frequently, in invertebrate brains [4, 29, 76]. Baba [4] reported on tracing with NiCl next to cobalt ions.

5.2 Tract-Specific Intrinsic Fluorescence in Transgenic Mice

What if neurons could be genetically manipulated such that they metabolize fluorescent compounds with their own intrinsic protein manufacturing machinery? Mice have been genetically tinkered with to reach this goal. Neocortical pyramidal neurons, hippocampal neurons, and neurons in the amygdala of Thy1-eYFP-H strain mice produce the protein eYFP which is a red-shifted spectral variant of GFP, the green fluorescent protein [146]. The eYFP manufacturing neurons subsequently distribute the eYFP through their efferent fibers, making it possible in these animals to study a number of tracts directly in the fluorescence microscope, without additional histological staining or immunohistochemistry. GFP can be used as a tool to visualize the details of the morphology of isolated mouse neurons [26, 80].

5.3 Tract Tracing in Human Brain: In Vivo and Postmortem

The rapid development in the last decade of diffusion tensor magnetic resonance imaging (DTI) techniques has led to a completely new type of analysis of gross white matter connectivity in the human brain (reviewed in [77]). The analysis of DTI datasets, through a set of computer algorithms collectively called "tractography," reveals the trajectories of myelinated white matter fiber bundles [34, 140]. The same has also been done for the rat brain [219]. White matter tract tracing via DTI is quite some achievement since tract tracing in humans has traditionally been dominated by manual dissection [71], by silver staining in material obtained from patients with brain lesions, or with the slow diffusion of strongly lipophilic carbocyanine dyes along axon trajectories (see the chapter in this book on carbocyanine dye tracing). DTI is limited to myelinated tracts and furthermore offers no indication with respect to the polarity of the fiber tracts (the direction of information flow). Resolution of DTI tract tracing is very poor compared with histological images. The cause of this low resolution is the large brain volume sampled in MRI in one voxel (pixels approximately 250 × 250 μm and slice thickness 1,000 μm). However, advances in MRI technology together with ongoing refinement of computer software will undoubtedly cause an increase of resolution leading to much more detail in the knowledge of fiber connectivity in the human brain.

In postmortem human brain, BDA and Neurobiotin can be applied provided that the tissue is unfixed and obtained after a relatively short postmortem delay (3–8 h). Brain slices obtained via autopsy are first incubated with artificial cerebrospinal fluid following which spot injections can be made with BDA or Neurobiotin [39, 40]. As the uptake and transport of BDA and Neurobiotin rely on active metabolism, these tracers fail to do their work in fixed human brain tissue. Carbocyanine dyes on the other hand diffuse in the myelin sheaths of both living and fixed brain [189] and, on the basis of this characteristic, have been applied with success in postmortem human brain tissue ([136], reviewed by [85]). Diffusion of DiI in myelinated axons is bidirectional. Finally, several investigators have successfully injected Lucifer yellow intracellularly into neurons in slices of fixed human brain (e.g., [10, 51]), but these experiments served primarily for cell identification and not for tracing per se.

5.4 Lipophilic Carbocyanine Dye Tracing

Labeling of individual neurons can be produced via the application of a strongly lipophilic, dialkylcarbocyanine, or dialkyl aminostyryl dye (DiI, DiA; [86]). These carbocyanine and styryl dyes diffuse exclusively along the lipid portions of neuronal membranes and label the involved cells completely including their entire dendritic arborization and especially including the delicate dendritic appendages. In this sense, carbocyanine dye labeling resembles Golgi silver impregnation. DI labeling in all its variations are described in a separate chapter.

Acknowledgments

It is a pleasure to acknowledge the continuous support provided by technicians who skillfully assisted us: Barbara Jorritsma-Byham, Annaatje Pattiselanno, Peter Goede, Amber Boekel, John Bol, Eveline Timmermans, Luciënne Baks-te Bulte, Yvon Galis, and Angela Engel. Their professional attitude, skill, and enthusiasm made the day. I am also indebted to many colleagues who at some point gave advice and contributed ideas and improvements, to name a few, Jochen Staiger, José Lanciego, Riichi Kajiwara, and Jean-Luc Boulland. It should be mentioned that a score of graduate students assisted in various projects. Finally, computer and other digital assistance by Nico Blijleven and Jeroen Beliën were of vital support.

References

1. Antal M, Freund TF, Somogyi P, McIlhinney RA (1990) Simultaneous anterograde labelling of two afferent pathways to the same target area with *Phaseolus vulgaris* leucoagglutinin and *Phaseolus vulgaris* leucoagglutinin conjugated to biotin or dinitrophenol. J Chem Neuroanat 3:1–9
2. Arai R, Jacobowitz DM, Deura S (1994) Distribution of calretinin, calbindin-D28k, and parvalbumin in the rat thalamus. Brain Res Bull 33:595–614
3. Ausdenmoore BD, Markwell ZA, Ladle DR (2011) Localization of presynaptic inputs on dendrites of individually labeled neurons in three dimensional space using a center distance algorithm. J Neurosci Meth 200:129–143
4. Baba Y (2000) New methods of dye application for staining motor neurons in an insect. J Neurosci Meth 98:165–169
5. Bachmann L, Salpeter MM (1969) Resolution in electron microscope radioautography. J Cell Biol 41:1–32
6. Bácskai T, Veress G, Halasi G, Matesz C (2010) Crossing dendrites of the hypoglossal motoneurons: possible morphological substrate of coordinated and synchronized tongue movements of the frog, Rana esculenta. Brain Res 1313:89–96
7. Barbas-Henry HA, Wouterlood FG (1988) Synaptic connections between primary trigeminal afferents and accessory abducens motoneurons in the monitor lizard, *Varanus exanthematicus*. J Comp Neurol 267:387–397
8. Basbaum AI, Menétrey D (1987) Wheat germ agglutinin–apoHRP gold: a new retrograde tracer for light- and electron-microscopic single and double-label studies. J Comp Neurol 261:306–318
9. Bazer GT, Ebbesson SO (1984) A simplified cobalt-lysine method for tracing axon trajectories in the central nervous system of vertebrates. Neurosci Lett 51:315–318
10. Belichenko PV, Dahlström A (1994) Dual channel confocal laser scanning microscopy of lucifer yellow-microinjected human brain cells combined with Texas red immunofluorescence. J Neurosci Meth 52:111–118
11. Beliën JAM, Wouterlood FG (2012) Confocal laser scanning: of instrument, computer processing and men. In: Wouterlood FG (ed) Cellular imaging techniques for neuroscience and beyond. Academic, New York, pp 2–34
12. Bentivoglio M, Kuypers HG, Catsman-Berrevoets CE, Dann O (1979) Fluorescent retrograde neuronal labeling in rat by means of substances binding specifically to

adenine-thymine rich DNA. Neurosci Lett 12:235–240

13. Bentivoglio M, Van der Kooy D, Kuypers HGJM (1979) The organization of the efferent projections of the substantia nigra in the rat. A retrograde fluorescent double labeling study. Brain Res 174:1–17
14. Bentivoglio M, Kuypers HG, Catsman-Berrevoets CE, Loewe H, Dann O (1980) Two new fluorescent retrograde neuronal tracers which are transported over long distances. Neurosci Lett 18:25–30
15. Bevan MD, Crossman AR, Bolam JP (1994) Neurons projecting from the entopeduncular nucleus to the thalamus receive convergent synaptic inputs from the subthalamic nucleus and the neostriatum in the rat. Brain Res 659:99–109
16. Bharali DJ, Klejbor I, Stachowiak EK et al (2005) Organically modified silica nanoparticles: a nonviral vector for in vivo gene delivery and expression in the brain. Proc Natl Acad Sci U S A 102:11539–11544
17. Bianchini P, Mondal PP, Dilipkumar S, Cella Zanacchi F et al (2012) Multiphoton microscopy advances towards super resolution. In: Wouterlood FG (ed) Cellular imaging techniques for neuroscience and beyond. Academic Press, New York, pp 121–140
18. Björklund A, Skagerberg G (1979) Simultaneous use of retrograde fluorescent tracers and fluorescence histochemistry for convenient and precise mapping of monoaminergic projections and collateral arrangements in the CNS. J Neurosci Meth 1:261–277
19. Blackstad TW (1965) Mapping of experimental axon degeneration by electron microscopy of Golgi preparations. Z Zellforsch Mikrosk Anat 67:819–834
20. Blackstad TW (1975) Electron microscopy of experimental axonal degeneration in photochemically modified Golgi preparations: a procedure for precise mapping of nervous connections. Brain Res 95:191–210
21. Boulland J-L, Jenstad M, Boekel A, Wouterlood FG et al (2009) Vesicular glutamate and GABA transporters sort to distinct sets of vesicles at a symmetric synapse. Cerebral Cortex 19:241–248
22. Brandt HM, Apkarian AV (1992) Biotin-dextran: a sensitive anterograde tracer for neuroanatomic studies in rat and monkey. J Neurosci Meth 45:35–40
23. Brown A (2003) Axon transport of membranous and nonmembranous cargoes: a unified perspective. J Cell Biol 160:817–821
24. Buhl EH, Lübke J (1989) Intracellular lucifer yellow injection in fixed brain slices combined with retrograde tracing, light and electron microscopy. Neuroscience 28:3–16
25. Carson KA, Mesulam MM (1982) Electron microscopic demonstration of neural connections using horseradish peroxidase: a comparison of the tetramethylbenzidine procedure with seven other histochemical methods. J Histochem Cytochem 30:425–435
26. Chakravarthy S, Keck T, Roelandse M et al (2008) Cre-dependent expression of multiple transgenes in isolated neurons of the adult forebrain. PLoS One 3:e3059
27. Chang HT (1991) Anterograde transport of Lucifer yellow-dextran conjugate. Brain Res Bull 26:813–816
28. Chang HT, Kuo H, Whittaker JA, Cooper NG (1990) Light and electron microscopic analysis of projection neurons retrogradely labeled with Fluoro-Gold: notes on the application of antibodies to Fluoro-Gold. J Neurosci Meth 35:31–37
29. Chase R, Tolloczko B (1993) Tracing neural pathways in snail olfaction: from the tip of the tentacles to the brain and beyond. Microsc Res Tech 24:214–230
30. Chen S, Aston-Jones G (1998) Axonal collateral-collateral transport of tract tracers in brain neurons: false anterograde labelling and useful tool. Neuroscience 82: 1151–1163
31. Colman DR, Scalia F, Cabrales E (1976) Light and electron microscopic observations on the anterograde transport of horseradish peroxidase in the optic pathway in the mouse and rat. Brain Res 102:156–163
32. Conte WL, Kamishina H, Reep RL (2009) The efficacy of the fluorescent conjugates of cholera toxin subunit B for multiple retrograde tracing in the central nervous system. Brain Struct Funct 213:3667–3673
33. Conte WL, Kamishina H, Reep RL (2009) Multiple neuroanatomical tract-tracing using fluorescent Alexa Fluor conjugates of cholera toxin subunit B in rats. Nat Protoc 4:1158–1166
34. Conturo TE, Lori NF, Cull TS et al (1999) Tracking neuronal fiber pathways in the living human brain. Proc Natl Acad Sci U S A 96:10422–10427
35. Coveñas R, De León M, Narváez JA, Aguirre JA, González-Barón S (1995) Calbindin D-28K-immunoreactivity in the cat diencephalon: an immunocytochemical study. Arch Ital Biol 133:263–272
36. Cowan RL, Sesack SR, Van Bockstaele EJ et al (1994) Analysis of synaptic inputs and targets of physiologically characterized neurons in rat frontal cortex: combined in vivo

intracellular recording and immunolabeling. Synapse 17:101–114

37. Cowan WM, Gottlieb DI, Hendrickson AE et al (1972) The autoradiographic demonstration of axonal connections in the central nervous system. Brain Res 37:21–51
38. Cowan WM, Cuénod M (1975) The use of axonal transport for the study of neural connections: a retrospective survey. In: Cowan MW, Cuénod M (eds) The use of axonal transport for studies of neuronal connectivity. Elsevier, Amsterdam, pp 1–24
39. Dai J, Swaab DF, Buijs RM (1998) Recovery of axonal transport in "dead neurons". Lancet 351:499–500
40. Dai J, Van Der Vliet J, Swaab DF, Buijs RM (1998) Postmortem anterograde tracing of intrahypothalamic projections of the human dorsomedial nucleus of the hypothalamus. J Comp Neurol 401:16–33
41. Dall'Oglio A, Ferme D, Brusco J, Moreira JE, Rasia-Filho AA (2010) The "single-section" Golgi method adapted for formalin-fixed human brain and light microscopy. J Neurosci Meth 189:51–55
42. Dederen PJ, Gribnau AA, Curfs MH (1994) Retrograde neuronal tracing with cholera toxin B subunit: comparison of three different visualization methods. Histochem J 26:856–862
43. Deller T, Naumann T, Frotscher M (2000) Retrograde and anterograde tracing combined with transmitter identification and electron microscopy. J Neurosci Meth 103:117–126
44. Dolleman-van der Weel MJ, Wouterlood FG, Witter MP (1994) Multiple anterograde tracing, combining Phaseolus vulgaris leucoagglutinin with rhodamine- and biotin-conjugated dextran amine. J Neurosci Meth 51:9–21
45. Dolleman-van Der Weel MJ, Witter MP (1996) Projections from the nucleus reuniens thalami to the entorhinal cortex, hippocampal field CA1, and the subiculum in the rat arise from different populations of neurons. J Comp Neurol 364:637–650
46. Droz B, Leblond CP (1962) Migration of proteins along the axons of the sciatic nerve. Science 137:1047–1048
47. Droz B, Leblond CP (1963) Axonal migration of proteins in the central nervous system and peripheral nerves as shown by radioautography. J Comp Neurol 121:325–346
48. Duque A, Zaborszky L (2006) Juxtacellular labeling of individual neurons in vivo: from electrophysiology to synaptology. In: Zaborszky L, Wouterlood FG, Lanciego JL (eds) Neuroanatomical tract-tracing 3: molecules – neurons – systems. Springer, New York, pp 197–236
49. Edwards SB, Hendrickson A (1981) The autoradiographic tracing of axonal connections in the central nervous system. In: Heimer L, RoBards M (eds) Neuroanatomical tract-tracing methods. Plenum, New York, pp 171–205
50. Egensperger R, Holländer H (1988) Electron microscopic visualization of fluorescent microspheres used as a neuronal tracer. J Neurosci Meth 23:181–186
51. Einstein G (1988) Intracellular injection of lucifer yellow into cortical neurons in lightly fixed sections and its application to human autopsy material. J Neurosci Meth 26:95–103
52. Erro E, Lanciego JL, Giménez-Amaya JM (1999) Relationships between thalamostriatal neurons and pedunculopontine projections to the thalamus: a neuroanatomical tract-tracing study in the rat. Exp Brain Res 127:162–170
53. Ewers H (2012) Nano resolution optical imaging through localization microscopy. In: Wouterlood FG (ed) Cellular imaging techniques for neuroscience and beyond. Academic Press, New York, pp 81–100
54. Fairén A (2005) Pioneering a golden age of cerebral microcircuits: the births of the combined Golgi–electron microscope methods. Neuroscience 136:607–614
55. Fairén A, Peters A, Saldanha J (1977) A new procedure for examining Golgi impregnated neurons by light and electron microscopy. J Neurocytol 6:311–337
56. Fan RJ, Marin-Burgin A, French KA, Otto FW (2005) A dye mixture (Neurobiotin and Alexa 488) reveals extensive dye-coupling among neurons in leeches; physiology confirms the connections. J Comp Physiol A 191: 1157–1171
57. Fink RP, Heimer L (1967) Two methods for selective silver impregnation of degenerating axons and their synaptic endings in the central nervous system. Brain Res 4:369–374
58. Foster M, Sherrington CS (1897) A textbook of physiology. Macmillan, New York, p 1252, www.archive.org/stream/textbookofphysio1897fost#page/n5/mode/2up
59. Fregerslev S, Blackstad TW, Fredens K, Holm MJ (1971) Golgi potassium-dichromate silver-nitrate impregnation. Nature of the precipitate studied by x-ray powder diffraction methods. Histochemie 26:289–304
60. Freund TF, Somogyi P (1983) The section-Golgi impregnation procedure. 1. Description of the method and its combination with histochemistry after intracellular iontophoresis or retrograde transport of horseradish peroxidase. Neuroscience 9:463–474

61. Fritzsch B, Sonntag R (1991) Sequential double labelling with different fluorescent dyes coupled to dextran amines as a tool to estimate the accuracy of tracer application and of regeneration. J Neurosci Meth 39:9–17
62. Frotscher M, Rinne U, Hassler R, Wagner A (1981) Termination of cortical afferents on identified neurons in the caudate nucleus of the cat. a combined Golgi-EM degeneration study. Exp Brain Res 41:329–337
63. Fuller PM, Prior DJ (1975) Cobalt iontophoresis techniques for tracing afferent and efferent connections in the vertebrate CNS. Brain Res 88:211–220
64. Gabbott PL, Somogyi J (1984) The 'single' section Golgi-impregnation procedure: methodological description. J Neurosci Meth 11: 221–230
65. Gauthier J, Parent M, Lévesque M, Parent A (1999) The axonal arborisation of single nigrostriatal axons in rats. Brain Res 834: 228–232
66. Gerfen CR (2004) Basal ganglia. In: Paxinos G (ed) The rat nervous system, 3rd edn. Elsevier, Oxford, pp 455–508
67. Gerfen CR, Sawchenko PE (1984) An anterograde neuroanatomical tracing method that shows the detailed morphology of neurons, their axons and terminals: immunohistochemical localization of an axonally transported plant lectin, *Phaseolus vulgaris* leucoagglutinin (PHA-L). Brain Res 290:219–238
68. Gerfen CR, Sawchenko PE (1985) A method for anterograde axonal tracing of chemically specified circuits in the central nervous system: combined *Phaseolus vulgaris-leucoagglutinin* (PHA-L) tract tracing and immunohistochemistry. Brain Res 343:144–150
69. Giolli RA, Karamanlidis AN (1978) The study of degenerating nerve fibers using silver-impregnation methods. In: Robertson RT (ed) Neuroanatomical research techniques (Methods in psychology, Thompson RF (ed)). Academic Press, New York, pp 212–240
70. Glover JC, Petursdottir G, Jansen JKS (1986) Fluorescent dextran amines used as axonal tracers in the nervous system of chicken embryo. J Neurosci Meth 18:243–254
71. Gluhbecovic N, Williams TH (1980) The human brain. A photographic guide. Harper & Row Hagerstown, MD, p 176
72. Golgi C (1873) Sulla struttura della sostanza grigia del cervello (Comunicazione preventiva). Gaz Med Ital (Lombardia) 33:244–246
73. Gonatas NK, Harper C, Mizutani T, Gonatas JO (1979) Superior sensitivity of conjugates of horseradish peroxidase with wheat germ agglutinin for studies of retrograde axonal transport. J Histochem Cytochem 27: 728–734
74. Gonzalo N, Moreno A, Erdozain MA et al (2001) A sequential protocol combining dual neuroanatomical tract-tracing with the visualization of local circuit neurons within the striatum. J Neurosci Meth 111:59–66
75. Graham RC Jr, Karnovsky MJ (1965) The histochemical demonstration of monoamine oxidase activity by coupled peroxidatic oxidation. J Histochem Cytochem 13:604–605
76. Hackney CM, Altman JS (1982) Cobalt mapping of the nervous system: how to avoid artifacts. J Neurobiol 13:403–411
77. Hagmann P, Cammoun L, Gigandet X et al (2010) MR connectomics: principles and challenges. J Neurosci Meth 194:34–45
78. Hamelryck TW, Dao-Thi M-H, Poortmans F et al (1996) The crystallographic structure of phytohemagglutinin-L. J Biol Chem 271: 20479–20485
79. Hancock MB (1986) Two-color immunoperoxidase staining: visualization of anatomic relationships between immunoreactive neural elements. Am J Anat 175:343–352
80. Harvey AR, Ehlert E, de Wit J et al (2009) Use of GFP to analyze morphology, connectivity, and function of cells in the central nervous system. In: Hicks BW (ed) Methods in molecular biology, viral applications of green fluorescent protein, vol 515. Humana, New York, pp 63–95
81. Herkenham M (1978) The connections of the nucleus reuniens thalami: evidence for a direct thalamo-hippocampal pathway in the rat. J Comp Neurol 177:589–610
82. Hendrickson AE (1982) The orthograde axoplamic transport autoradiographic tracing technique and its implications for additional neuroanatomical analysis of the striate cortex. In: Chan-Palay V, Palay SL (eds) Cytochemical methods in neuroanatomy. Alan R. Liss, New York, pp 1–16
83. Hirokawa N, Takemura R (2005) Molecular motors and mechanisms of directional transport in neurons. Nat Rev Neurosci 6:201–214
84. Holstege JC, Kuypers HG (1987) Brainstem projections to lumbar motoneurons in rat. I. An ultrastructural study using autoradiography and the combination of autoradiography and horseradish peroxidase histochemistry. Neuroscience 21:345–367
85. Honig M (1993) DiI labelling. Neurosci Protoc 93-050-16-01-20
86. Honig MG, Hume RI (1986) Fluorescent carbocyanine dyes allow living neurons of identified origin to be studied in long-term cultures. J Cell Biol 103:171–187

87. Hoover WB, Vertes RP (2012) Collateral projections from nucleus reuniens of thalamus to hippocampus and medial prefrontal cortex in the rat: a single and double retrograde fluorescent labeling study. Brain Struct Funct 217:191–209
88. Horikawa K, Armstrong WE (1988) A versatile means of intracellular labeling: injection of biocytin and its detection with avidin conjugates. J Neurosci Meth 25:1–11
89. Horikawa K, Powell EW (1986) Comparison of techniques for retrograde labeling using the rat's facial nucleus. J Neurosci Meth 17: 287–296
90. Huisman AM, Ververs B, Cavada C, Kuypers HG (1984) Collateralization of brainstem pathways in the spinal ventral horn in rat as demonstrated with the retrograde fluorescent double-labeling technique. Brain Res 300: 362–367
91. Innocenti GM, Clarke S, Kraftsik R (1986) Interchange of callosal and association projections in the developing visual cortex. J Neurosci 6:1384–1409
92. Itoh K, Konishi A, Nomura S et al (1979) Application of coupled oxidation reaction to electron microscopic demonstration of horseradish peroxidase: cobalt-glucose oxidase method. Brain Res 175:341–346
93. Izzo PN, Graybiel AM, Bolam JP (1987) Characterization of substance P- and [Met] enkephalin-immunoreactive neurons in the caudate nucleus of cat and ferret by a single section Golgi procedure. Neuroscience 20:577–587
94. Jarrard LE (1989) On the use of ibotenic acid to lesion selectively different components of the hippocampal formation. J Neurosci Meth 29:251–259
95. Jongen-Rêlo AL, Amaral DG (2000) A double labeling technique using WGA-apoHRP-gold as a retrograde tracer and non-isotopic in situ hybridization histochemistry for the detection of mRNA. J Neurosci Meth 101:9–17
96. Jonsson G (1983) Chemical lesioning techniques: Monoamine neurotoxins. In: Björklund A, Hökfelt T (eds) Handbook of chemical neuroanatomy, vol I, Methods in chemical neuroanatomy. Elsevier, New York, pp 463–507
97. Jorritsma-Byham B, Witter MP, Wouterlood FG (1994) Combined anterograde tracing with biotinylated dextran-amine, retrograde tracing with Fast blue and intracellular filling of neurons with Lucifer yellow. An electron microscopic method. J Neurosci Meth 52:153–160
98. Kajiwara R, Wouterlood FG, Sah A et al (2008) Convergence of entorhinal and CA3 inputs onto pyramidal neurons and interneurons in hippocampal area CA1. An anatomical study in the rat. Hippocampus 18:266–280
99. Katz LC, Burkhalter A, Dreyer WJ (1984) Fluorescent latex microspheres as a retrograde neuronal marker for in vivo and in vitro studies of visual cortex. Nature 310:498–500
100. Keizer K, Kuypers HG, Huisman AM, Dann O (1983) Diamidino yellow dihydrochloride (DY 2HCl); a new fluorescent retrograde neuronal tracer, which migrates only very slowly out of the cell. Exp Brain Res 51:179–191
101. King MA, Louis PM, Hunter BE, Walker DW (1989) Biocytin: a versatile anterograde neuroanatomical tract-tracing alternative. Brain Res 497:361–367
102. Kirby ED, Jensen K, Goosens KA, Kaufer D (2012) Stereotaxic surgery for excitotoxic lesion of specific brain areas in the adult rat. J Vis Exp 65:4079
103. Kita H, Armstrong W (1991) A biotin-containing compound N-(2-aminoethyl)biotinamide for intracellular labeling and neuronal tracing studies: comparison with biocytin. J Neurosci Meth 37:141–150
104. Kobayashi RM (1978) Neurochemical effects of lesions. In: Robertson RT (ed) Neuroanatomical research techniques. Methods in physiological psychology, vol 2. Academic Press, New York, pp 317–336
105. Kristensson K (1970) Transport of fluorescent protein in peripheral nerves. Acta Neuropathol 16:293–300
106. Kristensson K, Olsson Y (1971) Uptake and retrograde transport of peroxidase in hypoglossal neurons. Electron microscopical localization in the neuronal perikaryon. Acta Neuropathol 19:1–9
107. Kristensson K, Olsson Y (1971) Retrograde axonal transport of protein. Brain Res 29: 363–365
108. Kristensson K, Olsson Y, Sjöstrand J (1971) Axonal uptake and retrograde transport of exogenous proteins in the hypoglossal nerve. Brain Res 32:399–406
109. Kuypers HGJM, Catsman-Berrevoets CE, Padt RE (1977) Retrograde axonal transport of fluorescent substances in the rat's forebrain. Neurosci Lett 6:127–135
110. Kuypers HG, Bentivoglio M, van der Kooy D, Catsman-Berrevoets CE (1979) Retrograde transport of bisbenzimide and propidium iodide through axons to their parent cell bodies. Neurosci Lett 12:1–7
111. Kuypers HG, Bentivoglio M, Catsman-Berreveoets CE, Bharos AT (1980) Double retrograde neuronal labeling through divergent

axon collaterals, using two fluorescent tracers with the same excitation wavelength which label different features of the cell. Exp Brain Res 40:383–392

112. Lakos S, Basbaum AI (1986) Benzidine dihydrochloride as a chromogen for single- and double-label light and electron microscopic immunocytochemical studies. J Histochem Cytochem 34:1047–1056
113. Lanciego JL, Wouterlood FG (1994) Dual anterograde axonal tracing with *Phaseolus vulgaris* leucoagglutinin (PHA-L) and biotinylated dextran amine (BDA). Neurosci Protoc 94-050-06-01-13
114. Lanciego JL, Goede PH, Witter MP, Wouterlood FG (1997) Use of peroxidase substrate Vector VIP for multiple staining in light microscopy. J Neurosci Meth 74:1–7
115. Lanciego JL, Wouterlood FG, Erro E, Giménez-Amaya JM (1998) Multiple axonal tracing: simultaneous detection of three tracers in the same section. Histochem Cell Biol 110:509–515
116. Lanciego JL, Luquin MR, Guillén J, Giménez-Amaya JM (1998) Multiple neuroanatomical tracing in primates. Brain Res Protoc 2:323–332
117. Lanciego JL, Wouterlood FG, Erro E et al (2000) Complex brain circuits studied via simultaneous and permanent detection of three transported neuroanatomical tracers in the same histological section. J Neurosci Meth 103:127–135
118. Lapper SR, Bolam JP (1991) The anterograde and retrograde transport of neurobiotin in the central nervous system of the rat: comparison with biocytin. J Neurosci Meth 39:163–174
119. LaVail JH, LaVail MM (1972) Retrograde axonal transport in the central nervous system. Science 176:1416–1417
120. Lázár G (1978) Application of cobalt-filling technique to show retinal projections in the frog. Neuroscience 3:725–736
121. Li D, Seeley PJ, Bliss TVP, Raisman G (1990) Intracellular injection of biocytin into fixed tissue and its detection with avidin-HRP. Neurosci Lett Suppl 38:581
122. Liu WL, Behbehani MM, Shipley MT (1993) Intracellular filling in fixed brain slices using Miniruby, a fluorescent biocytin compound. Brain Res 608:78–86
123. Llewellyn-Smith IJ, Martin CL, Arnolda LF, Minson JB (2000) Tracer-toxins: cholera toxin B-saporin as a model. J Neurosci Meth 103:83–90
124. Luppi PH, Fort P, Jouvet M (1990) Iontophoretic application of unconjugated choleratoxin B subunit (CTB) combined with immunohistochemistry of neurochemical substances: a method for transmitter identification of retrogradely labeled neurons. Brain Res 534:209–224
125. Lyckman AW, Fan G, Rios M, Jaenisch R, Sur M (2005) Normal eye-specific patterning of retinal inputs to murine subcortical visual nuclei in the absence of brain-derived neurotrophic factor. Vis Neurosci 22:27–36
126. Marchi V, Algeri EG (1886) Sulle degenerazioni discendenti consecutive a lesioni in diverse zone della corteccia cerebrale. Riv Sper Freniatr Med Leg Alien Ment 14:1–49
127. Matesz C (1994) Synaptic relations of the trigeminal motoneurons in a frog (Rana esculenta). Eur J Morphol 32:117–121
128. Mauro A, Germano I, Giaccone G et al (1985) 1-Naphthol basic dye (1-NBD), an alternative to diaminobenzidine (DAB) in immunoperoxidase techniques. Histochemistry 83:97–102
129. Meissirel C, Dehay C, Berland M, Kennedy H (1991) Segregation of callosal and association pathways during development in the visual cortex of the primate. J Neurosci 11:3297–3316
130. Mesulam MM (1978) Tetramethyl benzidine for horseradish peroxidase neurohistochemistry: a non-carcinogenic blue reaction product with superior sensitivity for visualizing neural afferents and efferents. J Histochem Cytochem 26:106–117
131. Mesulam MM, Mufson EJ (1980) The rapid anterograde transport of horseradish peroxidase. Neuroscience 5:1277–1286
132. Mesulam MM (1982) Principles of horseradish peroxidase neurochemistry and their applications for tracing neural pathways-axonal transport, enzyme histochemistry and light microscopic analysis. In: Mesulam MM (ed) Tracing neural connections with horseradish peroxidase; IBRO handbook series: methods in the neurosciences. Wiley, NY, USA, pp 1–551
133. Moore RY (1978) Surgical and chemical lesion techniques. In: Iversen LL, Iversen SD, Snyder SH (eds) Handbook of psychopharmacology, vol 9. Plenum, New York, pp 1–39
134. Morecraft RJ, Herrick JL, Stilwell-Morecraft KS et al (2002) Localization of arm representation in the corona radiata and internal capsule in the non-human primate. Brain 125:176–198
135. Mufson EJ, Brady DR, Kordower JH (1990) Tracing neuronal connections in postmortem human hippocampal complex with the carbocyanine dye DiI. Neurobiol Aging 11: 649–653

136. Nance DM, Burns J (1990) Fluorescent dextrans as sensitive anterograde neuroanatomical tracers: applications and pitfalls. Brain Res Bull 25:139–145
137. Nauta WJ (1952) Selective silver impregnation of degenerating axons in the central nervous system. Stain Technol 27:175–179
138. Neely MD, Stanwood GD, Deutch AY (2009) Combination of diOlistic labeling with retrograde tract tracing and immunohistochemistry. J Neurosci Meth 184:332–336
139. Oishi K, Zilles K, Amunts K et al (2008) Human brain white matter atlas: identification and assignment of common anatomical structures in superficial white matter. Neuroimage 43:447–457
140. Oztas E (2003) Neuronal tracing. Neuroanatomy 2:2–5
141. Parent M, Parent A (2007) The microcircuitry of primate subthalamic nucleus. Parkinsonism Relat Disord 13(Suppl 3):S292–S295
142. Payne JN (1987) Comparisons between the use of true blue and diamidino yellow as retrograde fluorescent tracers. Exp Brain Res 68: 631–642
143. Pinault D (1996) A novel single-cell staining procedure performed in vivo under electrophysiological control: morpho-functional features of juxtacellularly labeled thalamic cells and other central neurons with biocytin or Neurobiotin. J Neurosci Meth 65:113–136
144. Pitman RM, Tweedle CD, Cohen MJ (1972) Branching of central neurons; intracellular cobalt injection for light and electron-microscopy. Science 176:412–414
145. Porrero C, Rubio-Garrido P, Avendaño C, Clascá F (2010) Mapping of fluorescent protein-expressing neurons and axon pathways in adult and developing Thy1-eYFP-H transgenic mice. Brain Res 1345:59–72
146. Prensa L, Parent A (2001) The nigrostriatal pathway in the rat: a single-axon study of the relationship between dorsal and ventral tier nigral neurons and the striosome/matrix striatal compartments. J Neurosci 21:7247–7260
147. Ramón y Cajal S (1909–1911) Histologie du Système Nerveux de l'Homme et des Vertébrés 2 volumes. Maloine, Paris, facsimile reprint, 1972
148. Reiner A, Honig MG (2006) Dextran amines: versatile tools for anterograde and retrograde studies of nervous system connectivity. In: Zaborszky L, Wouterlood FG, Lanciego JL (eds) Neuroanatomical tract-tracing 3: molecules, neurons, and systems. Springer, New York, pp 304–335
149. Reiner A, Veenman CL, Honig MG (1993) Anterograde tracing using biotinylated dextran amine. Neurosci Protoc 93-050-14
150. Reiner A, Veenman CL, Medina L et al (2000) Pathway tracing using biotinylated dextran amines. J Neurosci Meth 103:23–37
151. Repérant J (1975) The orthograde transport of horseradish peroxidase in the visual system. Brain Res 85:307–312
152. Reyes BA, Carvalho AF, Vakharia K, Van Bockstaele EJ (2011) Amygdalar peptidergic circuits regulating noradrenergic locus coeruleus neurons: linking limbic and arousal centers. Exp Neurol 230:96–105
153. Rho JH, Sidman RL (1986) Intracellular injection of lucifer yellow into lightly fixed cerebellar neurons. Neurosci Lett 72:21–24
154. Rice CD, Weber SA, Waggoner AL et al (2010) Mapping of neural pathways that influence diaphragm activity and project to the lumbar spinal cord in cats. Exp Brain Res 203:205–211
155. Richmond FJ, Gladdy R, Creasy JL et al (1994) Efficacy of seven retrograde tracers, compared in multiple-labelling studies of feline motoneurones. J Neurosci Meth 53:35–46
156. Rooney D, Døving KB, Ravaille-Veron M, Szabo T (1992) The central connections of the olfactory bulbs in cod, *Gadus morhua L.* J Hirnforsch 33:63–75
157. Rosene DL, Roy NJ, Davis BJ (1986) A cryoprotection method that facilitates cutting frozen sections of whole monkey brains for histological and histochemical processing without freezing artifact. J Histochem Cytochem 34:1301–1315
158. Roy I, Ohulchanskyy TY, Pudavar HE et al (2003) Ceramic-based nanoparticles entrapping water-insoluble photosensitizing anticancer drugs: a novel drug-carrier system for photodynamic therapy. J Am Chem Soc 125: 7860–7865
159. Ruigrok TJ, Hensbroek RA, Simpson JI (2011) Spontaneous activity signatures of morphologically identified interneurons in the vestibulocerebellum. J Neurosci 31:712–724
160. Schmued LC, Fallon JH (1986) Fluoro-Gold: a new fluorescent retrograde axonal tracer with numerous unique properties. Brain Res 377:147–154
161. Schmued LC, Heimer L (1990) Iontophoretic injection of Fluoro-Gold and other fluorescent tracers. J Histochem Cytochem 38: 721–723
162. Schmued L, Kyriakidis K, Heimer L (1990) In vivo anterograde and retrograde axonal

transport of the fluorescent rhodamine-dextran-amine, Fluoro-Ruby, within the CNS. Brain Res 526:127–134

163. Schwab ME, Javoy-Agid F, Agid Y (1978) Labeled wheat germ agglutinin (WGA) as a new, highly sensitive retrograde tracer in the rat hippocampal system. Brain Res 152:145–150
164. Schwab ME, Agid I (1979) Labeled wheat germ agglutinin and tetanus toxin as highly sensitive retrograde tracers in the CNS: the afferent fiber connections of the rat nucleus accumbens. Int J Neurol 13:117–126
165. Schwartz JH (1979) Axonal transport: components, mechanisms and specificity. Annu Rev Neurosci 2:467–504
166. Skirboll L, Hökfelt T, Norell G et al (1984) A method for specific transmitter identification of retrogradely labeled neurons: immunofluorescence combined with fluorescence tracing. Brain Res Rev 8:99–127
167. Skirboll LR, Thor K, Helke C et al (1989) Use of retrograde fluorescent tracers in combination with immunohistochemical methods. In: Záborszky L, Heimer L (eds) Neuroanatomical tract-tracing methods 2. Plenum Press, London, pp 5–8
168. Somogyi P, Smith AD (1979) Projection of neostriatal spiny neurons to the substantia nigra. Application of a combined Golgi-staining and horseradish peroxidase transport procedure at both light and electron microscopic levels. Brain Res 178:3–15
169. Somogyi P, Hodgson AJ, Smith AD (1979) An approach to tracing neuron networks in the cerebral cortex and basal ganglia. Combination of Golgi staining, retrograde transport of horseradish peroxidase and anterograde degeneration of synaptic boutons in the same material. Neuroscience 4:1805–1852
170. Somogyi P, Freund TF, Hodgson AJ et al (1985) Identified axo-axonic cells are immunoreactive for GABA in the hippocampus and visual cortex of the cat. Brain Res 332: 143–149
171. Spiga S, Puddu MC, Pisano M, Diana M (2005) Morphine withdrawal-induced morphological changes in the nucleus accumbens. Eur J Neurosci 22:2332–2340
172. Stretton AO, Kravitz EA (1968) Neuronal geometry: determination with a technique of intracellular dye injection. Science 162: 132–134
173. Swanson LW (1981) Tracing central pathways with the autoradiographic method. J Histochem Cytochem 29:117–124
174. Szabadics J, Varga C, Brunner J et al (2010) Granule cells in the CA3 area. J Neurosci 30:8296–8307
175. Taverna S, van Dongen YC, Groenewegen HJ, Pennartz CM (2004) Direct physiological evidence for synaptic connectivity between medium-sized spiny neurons in rat nucleus accumbens in situ. J Neurophysiol 91:1111–1121
176. Thanos S, Bonhoeffer F (1983) Investigations on the development and topographic order of retinotectal axons: anterograde and retrograde staining of axons and perikarya with rhodamine in vivo. J Comp Neurol 219:420–430
177. Thomas MA, Lemmer B (2005) HistoGreen: a new alternative to 3.3'-diaminobenzidine-tetrahydrochloride-dihydrate (DAB) as a peroxidase substrate in immunohistochemistry? Brain Res Protcol 14:107–118
178. Tóth P, Lázár G, Wang SR et al (1994) The contralaterally projecting neurons of the isthmic nucleus in five anuran species: a retrograde tracing study with HRP and cobalt. J Comp Neurol 346:306–320
179. Trojanowski JQ (1983) Native and derivatized lectins for in vivo studies of neuronal connectivity and neuronal cell biology. J Neurosci Meth 9:185–204
180. Trojanowski JQ, Gonatas JO, Gonatas NK (1981) Conjugates of horseradish peroxidase (HRP) with cholera toxin and wheat germ agglutinin are superior to free HRP as orthograde transported markers. Brain Res 223:381–385
181. Trojanowski JQ, Gonatas JO, Steiber A, Gonatas NK (1982) Horseradish peroxidase (HRP) conjugates of cholera toxin and lectins are more sensitive retrograde transported markers than free HRP. Brain Res 231:33–50
182. Van Bockstaele EJ, Wright AM, Cestari DM, Pickel VM (1994) Immunolabeling of retrogradely transported Fluoro-Gold: sensitivity and application to ultrastructural analysis of transmitter-specific mesolimbic circuitry. J Neurosci Meth 55:65–78
183. Van der Kooy D, Steinbusch HWM (1980) Simultaneous fluorescent retrograde axonal tracing and immunofluorescent characterization of neurons. J Neurosci Res 5:479–484
184. VanderWerf F, Aramideh M, Ongerboer de Visser BW et al (1997) A retrograde double fluorescent tracing study of the levator palpebrae superioris in the cynomolgus monkey. Exp Brain Res 113:174–179
185. van Haeften T, Baks-te-Bulte L, Goede P et al (2003) Morphological and numerical analysis of synaptic interactions between neurons in the deep and superficial layers of the entorhinal cortex of the rat. Hippocampus 13:943–952
186. Veenman CL, Reiner A, Honig MG (1992) Biotinylated dextran amine as an anterograde

tracer for single- and double-label studies. J Neurosci Meth 41:239–254

187. Vertes RP, Hoover WB, Do Valle AC et al (2006) Efferent projections of reuniens and rhomboid nuclei of the thalamus in the rat. J Comp Neurol 499:768–796
188. von Bartheld CS, Cunningham DE, Rubel EW (1990) Neuronal tracing with DiI: decalcification, cryosectioning, and photoconversion for light and electron microscopic analysis. J Histochem Cytochem 38:725–733
189. von Waldeyer-Hartz HWG (1891) Ueber einige neuere Forschungen im Gebiete der Anatomie des Centralnervensystems. Deutsch Med Wochenschr Berlin 17:1213–1218, 1244–1246, 1287–1289, 1331–1332, 1350–1356
190. Walaas I, Fonnum F (1979) The effects of surgical and chemical lesions on neurotransmitter candidates in the nucleus accumbens of the rat. Neuroscience 4:209–316
191. Waller A (1850) Experiments on the sections of glossopharyngeal and hypoglossal nerves of the frog and observations of the alterations produced thereby in the structure of their primitive fibers. Phil Trans R Soc Lond 140: 423–429
192. Wan XS, Trojanowski JQ, Gonatas JO (1982) Cholera toxin and wheat germ agglutinin conjugates as neuroanatomical probes: their uptake and clearance, transganglionic and retrograde transport and sensitivity. Brain Res 243:215–224
193. Warr WB, de Olmos JS, Heimer L (1981) Horseradish peroxidase: the basic procedure. In: Heimer L, RoBards M (eds) Neuroanatomical tract-tracing methods. Plenum, New York, pp 207–262
194. Weiss P, Hiscoe HN (1948) Experiments on the mechanism of nerve growth. J Exp Zool 107:315–339
195. Wessendorf MW (1990) Characterization and use of multi-color fluorescence microscopic techniques. In: Björklund A, Hökfelt T, Wouterlood FG, van der Pol AN (eds) Handbook of chemical neuroanatomy, vol 8. Elsevier, Amsterdam, pp 1–45
196. Wessendorf MW (1990) Fluoro-Gold: composition, and mechanism of uptake. Brain Res 553:135–148
197. Wouterlood FG (1986) Study of CNS microcircuits by a combination of *Phaseolus vulgaris-leucoagglutinin* (PHA L) tracing, anterograde degeneration and electron microscopy: target neurons of fornix terminals in the septum of the rat. Neurosci Lett Suppl 26:S603
198. Wouterlood FG (1992) Techniques for converting Golgi precipitate in CNS neurons into stable electron microscopic markers. Microsc Res Tech 23:275–288
199. Wouterlood FG (2006) Combined fluorescence methods to determine synapses in the light microscope: multilabel confocal laser scanning microscopy. In: Zaborszky L, Wouterlood FG, Lanciego JL (eds) Neuroanatomical tract-tracing 3: molecules – neurons systems. Springer, New York, pp 394–435
200. Wouterlood FG, Nederlof J (1983) Termination of olfactory afferents on layer II and III pyramidal cells in the entorhinal area: degeneration Golgi EM study in the rat. Neurosci Lett 36(105):110
201. Wouterlood FG, Mugnaini E (1984) Cartwheel neurons of the dorsal cochlear nucleus. A Golgi electron microscopic study in the rat. J Comp Neurol 227:136–157
202. Wouterlood FG, Groenewegen HJ (1985) Neuroanatomical tracing by use of *Phaseolus vulgaris*-leucoagglutinin (PHA -L): electron microscopy of PHA L filled neuronal somata, dendrites, axons and axon terminals. Brain Res 326:188–191
203. Wouterlood FG, Bol JGJM, Steinbusch HWM (1987) Double- label immunocytochemistry: combination of anterograde neuroanatomical tracing with *Phaseolus vulgaris*-leucoagglutinin and enzyme immunocytochemistry of target neurons. J Histochem Cytochem 35:817–823
204. Wouterlood FG, Sauren YMHF, Pattiselanno A (1988) Compromises between penetration of antisera and preservation of ultrastructure in pre embedding electron microscopic immunocytochemistry. J Chem Neuroanat 1:65–80
205. Wouterlood FG, Saldana E, Witter MP (1990) Projection from the nucleus reuniens thalami to the hippocampal region: light and electron microscopic tracing study in the rat with the anterograde tracer *Phaseolus vulgaris-leucoagglutinin*. J Comp Neurol 296:589–610
206. Wouterlood FG, Jorritsma-Byham B, Goede PH (1990) Combination of anterograde tracing with *Phaseolus vulgaris*-leucoagglutinin, retrograde fluorescent tracing and fixed-slice intracellular injection of lucifer yellow. J Neurosci Meth 33:207–217
207. Wouterlood FG, Groenewegen HJ (1991) The *Phaseolus vulgaris*-leucoagglutinin tracing technique for the study of neuronal connections. Progr Histochem Cytochem 22:1–78
208. Wouterlood FG, Goede PH, Arts MPM, Groenewegen HJ (1992) Simultaneous characterization of efferent and afferent connectivity, neuroactive substances and morphology of neurons. J Histochem Cytochem 40:457–465
209. Wouterlood FG, Jorritsma-Byham B (1993) The anterograde neuroanatomical tracer biotinylated dextran amine: comparison with the tracer PHA-L in preparations for electron microscopy. J Neurosci Meth 48:75–87

210. Wouterlood FG, Böckers T, Witter MP (2003) Synaptic contacts between identified neurons visualized in the confocal laser scanning microscope. Neuroanatomical tracing combined with immunofluorescence detection of post-synaptic density proteins and target neuron-markers. J Neurosci Meth 128:129–142
211. Wouterlood FG, van Haeften T, Eijkhoudt M et al (2004) Input from the presubiculum to dendrites of layer-V neurons of the medial entorhinal cortex of the rat. Brain Res 1013:1–12
212. Wouterlood FG, Boekel AJ, Meijer GA, Beliën JAM (2007) Computer assisted estimation in the CNS of 3D multimarker overlap or touch at the level of individual nerve endings. A confocal laser scanning microscope application. J Neurosci Res 85:1215–1228
213. Wouterlood FG, Boekel AJ, Kajiwara R, Beliën JAM (2008) Counting contacts between neurons in 3D in confocal laser scanning images. J Neurosci Meth 171:296–308
214. Wouterlood FG, Aliane V, Boekel AJ et al (2008) Origin of calretinin containing, vesicular glutamate transporter 2- coexpressing fiber terminals in the entorhinal cortex of the rat. J Comp Neurol 506:359–370
215. Wouterlood FG, Beliën JAM (2014) Translation, touch and overlap in multi-fluorescence confocal laser scanning microscopy to quantitate synaptic connectivity. In: Bakota L, Brandt R (eds) Laser scanning microscopy and quantitative image analysis of neuronal tissue, Neuromethods 87. Humana, New York, NY, pp 1–36
216. Wright AK, Norrie L, Ingham CA, Hutton EA, Arbuthnott GW (1999) Double anterograde tracing of outputs from adjacent "barrel columns" of rat somatosensory cortex. Neostriatal projection patterns and terminal ultrastructure. Neuroscience 88:119–133
217. Wu Y, Richard S, Parent A (2000) The organization of the striatal output system: a single-cell juxtacellular labeling study in the rat. Neurosci Res 38:49–62
218. Xue R, van Zijl PC, Crain BJ et al (1999) In vivo three-dimensional reconstruction of rat brain axonal projections by diffusion tensor imaging. Magn Reson Med 42:1123–1127
219. Záborszky L, Heimer L (1989) Combination of tracer techniques, especially HRP and PHA-L, with transmitter identification for correlated light and electron microscopic studies. In: Heimer L, Záborszky L (eds) Neuroanatomical tract tracing methods 2. Plenum Press, New York, Recent Progress, pp 49–96
220. Zhang DX, Bertram EH (2002) Midline thalamic region: widespread excitatory input to the entorhinal cortex and amygdala. J Neurosci 22:3277–3284
221. Zhang J, Zhang AJ, Wu SM (2006) Immunocytochemical analysis of GABA-positive and calretinin-positive horizontal cells in the tiger salamander retina. J Comp Neurol 499:432–441
222. Zhou M, Grofova I (1995) The use of peroxidase substrate Vector VIP in electron microscopic single and double antigen localization. J Neurosci Meth 62:149–158

Chapter 2

Wheat Germ Agglutinin (WGA) Tracing: A Classic Approach for Unraveling Neural Circuitry

Sabrina L. Levy, Joshua J. White, and Roy V. Sillitoe

Abstract

Neuroanatomical tracing is a fundamental technique that has long been considered the primary method for visualizing brain networks in all areas of neuroscience. Although there are many new approaches for tracing neuronal connections, the lectin-based wheat germ agglutinin (WGA) tracing approach is still widely used, and it is firmly regarded as a classic method in the field. WGA has been used extensively to unravel both simple and complex neural networks in the central and peripheral nervous systems. It is reliable and versatile, as projections are labeled in the anterograde and retrograde directions. It is robust enough for tracking fine pathways in small animals, and it is stable enough for long-term tracing of neurons in large species. In some systems, WGA can even travel transynaptically to label the connected neurons. In this chapter, we outline the technical and conceptual details that have made WGA a powerful tool, and we discuss practical considerations for effectively using WGA. We also discuss the recent use of an Alexa conjugated WGA approach for multicolor labeling of different tracts in the same animal.

Key words WGA tracing, Projections, Connectivity, Circuits, Topography, Activity, Cerebellum

1 The Development of Lectin Tracers Revolutionized Brain Anatomy Studies

The idea of visualizing neural connections to understand the brain is not new; in fact, it stems back to the seminal work of Ramon y Cajal [7] (Cajal 1909–1911). Methods such as Camillo Golgi's Golgi stain, originally referred to as the "black reaction" (la reazione nera), were invaluable for defining the basic components of the brain—that circuits were constructed out of interconnected but individual cells. However, the desire to better understand circuits, and not just cells, quickly arose, sparking the development of new connectivity approaches such as lesion-induced mapping. In this technique, investigators made small lesions into localized regions of the brain, and then after a certain period of survival, they sacrificed the animal and mapped the "degenerative" terminals in the brain. This approach produced many remarkable and valuable insights into brain anatomy, both at the level of inter-region

Benjamin R. Arenkiel (ed.), *Neural Tracing Methods: Tracing Neurons and Their Connections*, Neuromethods, vol. 92, DOI 10.1007/978-1-4939-1963-5_2, © Springer Science+Business Media New York 2015

pathways and also at the level of distinguishing patterns of intrinsic circuit connectivity in particular brain locations [72, 76]. The discipline of neuroanatomy was also accelerated by the development of selective silver impregnation of degenerating axons [17, 81]. Meanwhile, the field still continued to grow with the application of radioactive amino acids, which were very sensitive and detectable via autoradiography. But then, the sophistication of the techniques advanced exponentially with the introduction of neural tracers [11, 15, 25, 81], and thus, a new era was born. In particular, it was by the 1980s that the use of plant lectins had really caught on and revolutionized the field because they offered, for the first time, specificity and selectivity in how neurons were labeled and also how they were visualized [44]. This chapter discusses one particular lectin, WGA, which quickly rose in popularity because of its superb ability to reveal neuronal and circuit architecture in great detail and, on the practical side, its simplicity of use that allowed just about any neuroscience lab to adapt the protocols.

2 Wheat Germ Agglutinin (WGA) Is a Versatile Tracer for Revealing Neural Connectivity

The most widely used lectin is called wheat germ agglutinin (WGA). WGA is isolated from the wheat, *Triticum vulgaris*. In solution, WGA has a molecular weight of 38 kDa [14, 21, 38, 44, 47]. It binds specifically to *N*-acetyl-D-glucosamine and *N*-acetylneuraminic acid (sialic acid) residues, which are both ubiquitous in all neural membranes [44]. The specific affinity of WGA for neural membranes makes it an excellent tracer for tracking neural connections in any region of the brain, spinal cord, and peripheral nervous system [4, 71]. Moreover, this membrane affinity enables it to travel in the anterograde and retrograde directions within the cell. Once it is taken up, it is not only transported bidirectionally, but in some cases, it jumps transganglionically (in peripheral circuits) or transynaptically (in central circuits) [10, 14, 44, 61]. Moreover, it can be used by itself or conjugated to HRP and fluorophores (Fig. 1; *see* below Sects. 6 and 7). In its conjugated forms, WGA travels rapidly, often labeling projections within a day or two after injection [56, 57, 63]. WGA can be injected by pressure (as simple as using a Hamilton syringe), or because it is a charged molecule, a small current applied by iontophoresis can also be used to enhance delivery, especially when nanoliter quantities need to be delivered into very discrete brain or spinal cord nuclei.

But, the process of WGA labeling requires more than just the receptors. Following its injection into tissue, which often requires a small amount of tissue damage to be induced for the most effective uptake into cells to occur (*see* [63, 57] for conjugates), WGA is taken up by diffusion near the injection site and also by endocytosis

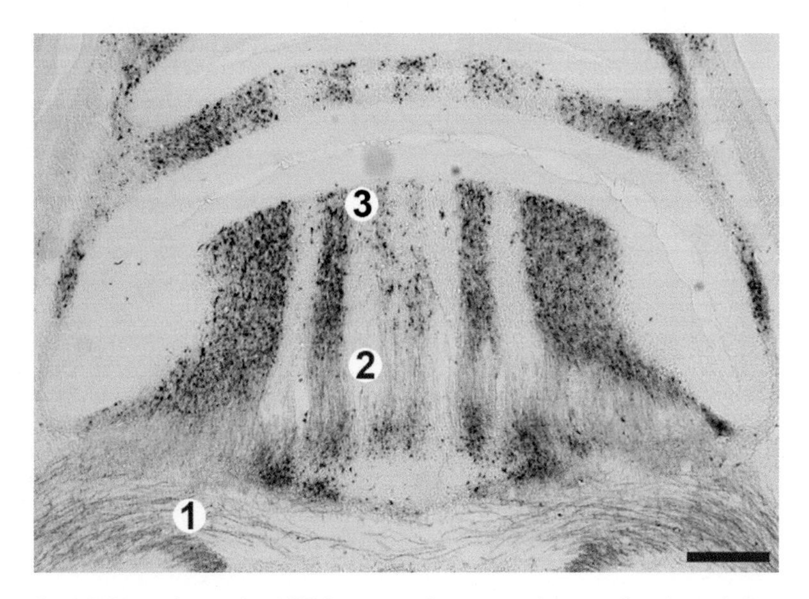

Fig. 1 WGA conjugated to HRP is an excellent tracer for revealing the trajectory of large tracts (labeled **1**; cerebellar peduncle), topographic axon projections (labeled **2**, cerebellar white matter), and axonal terminals (labeled **3**, cerebellar mossy fibers). The tissue section was acquired from an adult mouse cerebellum and cut coronally at 40 μm. The WGA-HRP was revealed using the TMB method. Scale bar = 500 μm

in the surrounding cells. WGA and other lectins are incorporated into cells via adsorptive-mediated endocytosis upon binding to the surface receptors, which is typically a more effective process than bulk endocytosis [14, 44]. WGA is then actively transported in vesicles, allowing it to incorporate into all regions of a cell, including the axons and dendrites. It thus can travel within the cell in both the anterograde and retrograde directions, and eventually, it also can cross the cell's synaptic compartments and enter into connected cells from either end of the neuron (terminals on axons or synapses on dendrites) [44]. By reacting the WGA to produce a visible chemical product, the resulting transport and location of the WGA in the cell provides exquisite visualization of neurons and the circuits that they function within.

3 WGA Has Been Used Widely to Study Neural Projections in a Variety of Species and Ages

In addition to its commercial availability, flexibility in how it can be delivered, and lack of toxicity to neurons after they are labeled in different brain systems, inter alia, WGA is commonly used in all branches of neuroscience because it can be effectively used essentially in any animal model, including vertebrates and invertebrates, with relative ease and limited equipment. In particular, the effectiveness of WGA tracing has been demonstrated in a range of vertebrate species including mouse [57, 63, 79], rat [51, 69, 70], cat [29, 33, 34], nonhuman primate [9, 73, 74], and even frog [36, 78, 82]. Interestingly, some studies even used WGA to label projections in fixed tissue from the human brain [35] and eye [1].

WGA has also proven to be very useful for studying the nervous system in an invertebrate genetic model organism, the fruit fly *Drosophila melanogaster*. Here, Yoshihara et al. [79] used the power of modern genetic approaches for tracing circuits by genetically encoding WGA and then expressing it in vivo in *Drosophila* in order to simultaneously label specific genetically defined subsets of neurons and their targets by exploiting the ability of WGA to cross synapses [79, 80]. Thus, the idea of tracing circuits evolved from purely examining anatomical connections to also examining the lineages of the component cells of the circuits (please *see* below in Sect. 6 for an additional discussion about genetically encoded WGA). The same investigators again used the *Drosophila* model to demonstrate the possibility that the transsynaptic transport of WGA may occur by an exocytosis-endocytosis process [68]. This work supported the idea that the ability of WGA and its conjugates to travel so efficiently in vivo rested in its capacity to harness the endogenous cell biological machinery to move in, around, and out of/between cells.

Our recent studies show the power of WGA-based tracing not only for studies of adult circuitry [19, 58] but also for the investigation of developing circuits in the perinatal [63] and postnatal developing mammalian brain [57]. WGA-based tracing has also been used extensively as a neuroanatomical tracer for revealing early developing brain networks and specifically as means of studying axonal transport in chick [39], which is a classic model organism in neuroembryology [24] and also a well-established model in the field of neuroanatomy [18, 40, 48, 53]. As in other vertebrate model organisms including rodents, cats, and monkeys [16, 27, 31, 50, 55], WGA conjugate tracers (Fig. 1; *see* below Sect. 4) were also demonstrated to travel transneuronally in the chicken nervous system [20].

4 WGA Conjugates: Increased Sensitivity for Improved Visualization of Neurons

Some original problems with WGA were the difficulties in attaining reliable staining, and one major reason for this was that by itself, sensitivity was in fact relatively low. To overcome this problem, WGA was conjugated to HRP, horseradish peroxidase [22, 44]. HRP itself has been exploited as an easy to use and effective tracer [43], but its conjugation to form WGA-HRP produced spectacular results ([44]; Figs. 1 and 2). Since then, many studies have used WGA-HRP (some examples from cerebellum include

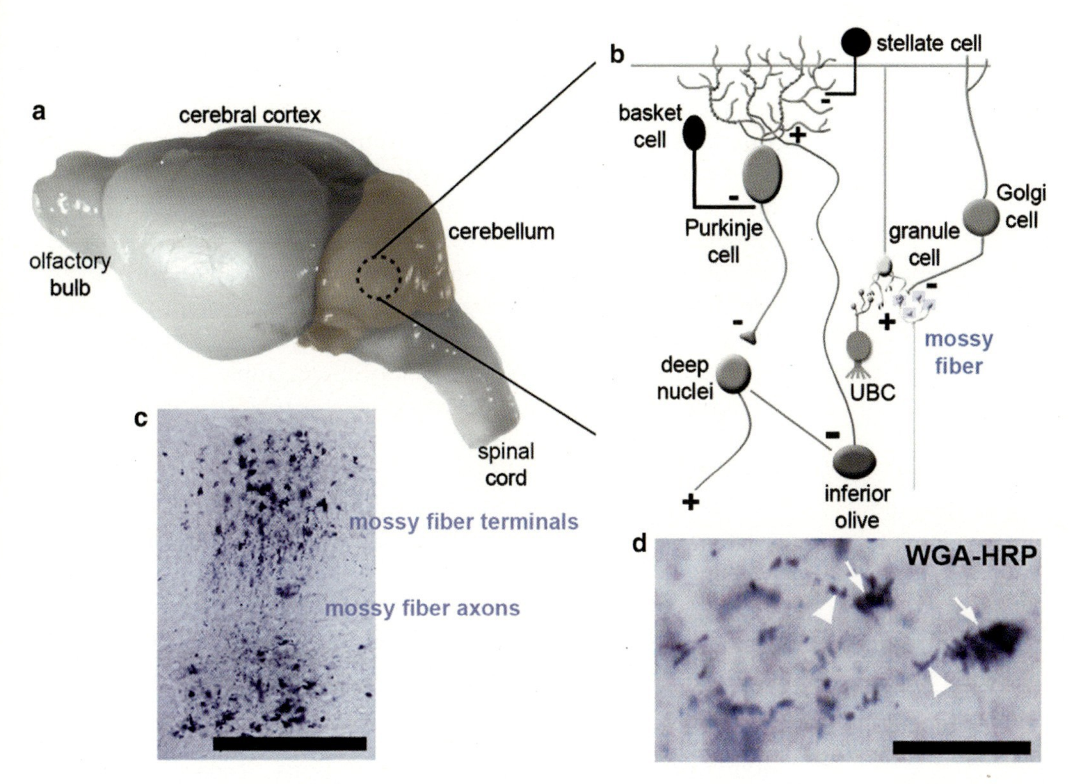

Fig. 2 (**a**) Schematic of an adult mouse brain, with the cerebellum highlighted in peach. (**b**) The basic circuitry of the cerebellum, with the mossy fiber glomerulus colored *blue/purple*. The excitatory synapses are labeled with a "+" and the inhibitory terminals "−." *UBC* unipolar brush cell. (**c**) WGA-HRP tracing showing mossy fiber axons and terminals stained using the TMB method. (**d**) The mossy fiber axons (*arrowheads*) and terminals (within the glomerulus; *arrows*) shown at higher power. Scale bar = 125 μm in (**c**) and 25 μm in (**d**). Schematic adapted from Reeber et al. (2013) Front Syst Neurosci 7:83

[63, 75]), especially when revealed using the intense blue/purple reaction product of the tetramethylbenzidine (TMB; Fig. 1) method ([12, 26, 44–46]). The reagents for the TMB reaction are safer than other more carcinogenic chromogen reaction procedures, and the deep blue product photographs beautifully, clearly revealing somata, axons, and terminals ([57, 63]; Figs. 1 and. 2).

The increased sensitivity of the WGA conjugates and their ability to better mark neurons may reside in their differential capacity to be taken up by cells. Neurons readily endocytose and transport macromolecules, which are packaged into vesicles, and it appears that WGA and its variants essentially utilize the same machinery to move in and around cells. HRP uses an endocytic pathway and is uptaken by bulk endocytosis. WGA is transported more efficiently. This potentially is the case because it uses adsorptive endocytosis [8, 28, 30, 44], in which a macromolecule binds to surface receptors that are then internalized into vesicular structures and transported through the cell [14]. In this respect, WGA-HRP is most superior, but WGA is more effective than HRP alone, the common factor for efficiency of course being the WGA. The conjugation of WGA to HRP thus enables more efficient transport that can occur even when lower concentrations and volumes of tracer are injected into the brain, and thus, its effective uptake can occur at lower extracellular concentrations [14]. Additionally, adsorptive endocytosis is apparently less influenced by the levels of neural activity compared to bulk endocytosis [44, 66]. This is an important consideration for attaining clean tract tracing, particularly in brain regions that exhibit spontaneous high frequency firing in many of its neuronal populations—the cerebellum is a prime example (*see* Fig. 6 for an example of spontaneous activity of a cerebellar nuclear neuron recorded in vivo, in mouse).

HRP, WGA, and WGA-HRP are all transported in the retrograde, anterograde, and transganglionic directions. On the practical side of choosing a tracer, one must note that there are two notable differences between HRP and WGA transport, however. First, as stated above, retrograde and transganglionic tracing by WGA-based tracers is far more extensive and effective than HRP alone [44]. This is an important practical consideration because it enables the use of smaller and more localized injections of WGA-HRP, after which effective labeling is still achieved and indeed with higher precision of marking the projections. Second, WGA tracers can travel transsynaptically, while HRP does not [44]. This is a major advantage over other injectable tracers because it offers the possibility of marking and mapping neural circuits as they are connected in vivo and also the opportunity to identify the locations and cellular architecture of higher order neurons in the circuit. But, note that the mode of WGA uptake and its transport in the cells may differ between the peripheral and central nervous systems, as evidenced by a radiolabeled version of the tracer [65]. Therefore, when using WGA it is essential to design experiments based specifically on the region of interest, because specific features of the tracer should be considered when interpreting the data. Additional conjugates of WGA with previously used and also newer fluorescent molecules are discussed separately below (Sect. 7).

5 WGA Conjugates: Increased Diversity for Broader Neuroscience Applications

Over the last three decades there have been many excellent and creative modifications of WGA, and these are based on the conjugate that has been linked to it. One such modification is the very useful pairing of the WGA approach with ultrastructural techniques, in which it has been conjugated to gold particles. This conjugate is called WGA-apo-HRP-gold, and it can be linked to particles of 10, 15, 20 nm, etc., and therefore, two tracers with different particle sizes can be injected to define circuit trajectory, convergence, and divergence [32, 37, 53]. Importantly, this tracer can be used not only for standard light microscopy but also for electron microcopy [2, 3]. Interestingly, WGA may also be an effective mediator/vehicle for efficient drug delivery. Several studies have been interested in understanding the transcellular transport mechanism of wheat germ agglutinin-functionalized nanoparticles (WGA-NP) [64]. Transcellular transport of WGA-NP occurs via a cytoskeleton-dependent and clathrin-mediated mechanism, and its intracellular transport is in part dependent on the endolysosome pathway [64]. Using this endogenous machinery, WGA-NP apparently can enhance the delivery of peptides into the brain following intranasal administration [62]. In contrast to these therapeutic strategies, WGA has also been used to kill neurons in experimental investigations. For instance, a WGA-ricin A-chain conjugate was used as a "suicide transport" agent that retrogradely transported toxin to produce anatomically selective lesions in rat vagal motor neurons [49]. From these studies, it is clear that WGA is not only a superb tool for "looking" at the brain, but it has massive potential for trying to get other molecules of interest "into" the brain and specifically into specific cells that can be targeted with designer drug molecules for therapy.

6 Genetically Encoded WGA

In this chapter, we will not discuss in detail the advantages and problems with genetically encoded WGA; however, we will present just the basics of the approach, in order to highlight the ever-growing utility of this powerful tracer. WGA has been widely used in transgenic mice by directing its expression with neuron-specific promoters [79]. This approach has been widely adapted for use throughout neuroscience, in both vertebrates and invertebrates. Yoshihara et al. [79] were among the first groups to use a genetic approach to introduce WGA into neurons by controlling its expression with specific promoter elements in order to target unique cell populations. Then, pairing WGA tracing with the Cre-recombinase genetic system enabled additional control over expression allowing an even more powerful analysis of the underlying neuroanatomy in various systems by achieving spatial and temporal resolution. For example, Braz et al. developed a transgenic mouse with Cre-mediated induction of WGA in selective neuronal populations [6]. In this study they targeted WGA to the cerebellar sensorimotor system [6]. They demonstrated the ability of genetically encoded WGA to travel anterogradely and transneuronally, as confirmed by WGA expression in connected cerebellar nuclear and red nuclear neurons that are downstream from Purkinje cells, the neurons in which recombination took place (therefore, second-order cerebellar nuclear and third-order red nuclear neurons were labeled, respectively). In an elegant follow-up study, they found that peripheral nerve axotomy could increase Cre-mediated recombination and expression of WGA [5]. In these models, Cre-recombinase excises a marker gene (*LacZ*, which encodes ß-galactosidase) and a stop codon to induce the expression of WGA. Braz et al. demonstrated two means of introducing Cre: by crossing the WGA "reporter" mice with a Cre-expressing mouse line or by injecting a Cre-expressing virus directly into the brain of the WGA mice [6]. A similar approach was also used to genetically track taste receptor sensory maps [67], and an additional transgenic WGA mouse was generated to examine hippocampal connectivity in vivo [77]. The advantages of this genetic approach are severalfold, including the ability to generate reproducible manipulations across experiments, being able to genetically trace neurons and circuit connectivity without requiring invasive surgical injections, and the opportunity to use multicolor molecular labeling to define the exact cellular and genetic identities of the traced neurons and also cells in the surrounding circuits.

7 Fluorescent-Tagged WGA: An Invaluable Multicolor Tool for Revealing Circuit Patterns

Despite the incredible advances in the last two decades that have catapulted our understanding of brain structure and function, still, there is a pressing need to resolve how different circuits are organized and how the cells in each circuit are altered after specific manipulations. To address such systems level questions, one method would be to simultaneously label multiple pathways in the same animal and then view the precise trajectories of the different pathways at high resolution in three dimensions. To start to address this problem, we developed a WGA-based approach to trace circuit connectivity in the cerebellum. For our approach to be beneficial and ideally suited for systems analysis, we considered three major criteria: (1) rapid labeling of short- and long-range tracts, (2) clear visibility of individual axons and their terminals (e.g., the "rosette" in the mossy fiber terminals; Fig. 2), and (3) bright fluorescent tag for visualizing multiple fiber types even at lower magnification and for combinatorial analysis of multiple tracts in the same animal. Using our established surgical protocol for accessing the spinal cord and labeling spinocerebellar fibers [63], we tested the potential of a relatively new line of WGA conjugated Alexa fluorophore tracers (WGA-Alexa 555 *red*, WGA-Alexa 488 *green*, and WGA-Alexa 350 *blue*; Invitrogen/Life Technologies, NY) for labeling the spinocerebellar tract [57]. We found that in adults, the long lumbar spinocerebellar axons and terminals could be traced in their entirety in less than 6 h. Moreover, individual axons were visible and the glomerular structure of rosettes was revealed [58]. Multiple pathways could be labeled and viewed in the same animal (Fig. 3; [57]), and surface imaging of adult spinocerebellar terminals, which reside more than 150 μm below the pial surface, was possible in whole fixed cerebella (Fig. 4). Importantly, the "fibers of passage" problem [44], which plaques the use of many tracers for combinatorial analysis, is not an issue with WGA-Alexa (Fig. 5). We found that within the cerebellar peduncles, axons labeled with each of the colors, and passing in very close proximity to one another, did not cause inappropriate transfer of tracer between axons ([57]; Fig. 5). Like WGA-HRP, WGA-Alexa also travels transynaptically and clearly labels connected neurons [23]. However, it is important to note that WGA-Alexa is not the only fluorescent conjugate available, since this spectacularly flexible molecule has also been conjugated to tetramethylrhodamine isothiocyanate-dextran (WGA-TRITC), which was also shown to be a useful tracer in brain [60].

Because of these features, we applied this tracing strategy in several different studies to address distinct questions. For example, the fluorescent-based nature of this tracing model makes it an ideal approach for neural systems analysis because we can effectively pair the bright WGA-Alexa labeling with molecular markers by immunolabeling [19, 58]. We have also successfully applied the technique to developing circuits [56]. And, we have recently combined the WGA-Alexa approach with genetic fate mapping techniques in order to lineage trace-specific genetically defined neurons while simultaneously tracking the circuits that those cells integrate into during early postnatal development [59]. Moreover, we demonstrate here an example application whereby single cerebellar nuclear neurons were recorded using an in vivo extracellular electrophysiology approach, and then after recording, a small amount of WGA-Alexa 555 tracer was injected into the recording site to mark the nucleus (Fig. 6). WGA-Alexa is particularly useful for this application because the tracer is taken up quickly (and thus the animal can be perfused at the end of the recording session), the recording site is marked brightly (Fig. 6a), and additional histological or immunohistochemical staining may be carried out on the same tissue section. Because of its rapid transport, robust signal, and reliability in labeling discrete regions, WGA-Alexa potentially offers new avenues for labeling tracts in alert behaving animals while simultaneously recording the neural activity of individual neurons [41, 42, 52], and perhaps because it travels transynaptically, one could conceivably track circuit firing across interlinked WGA-Alexa-mapped neurons. The goal in such studies would be to mark, map, and manipulate specific projections, their circuits, and the global networks they operate within.

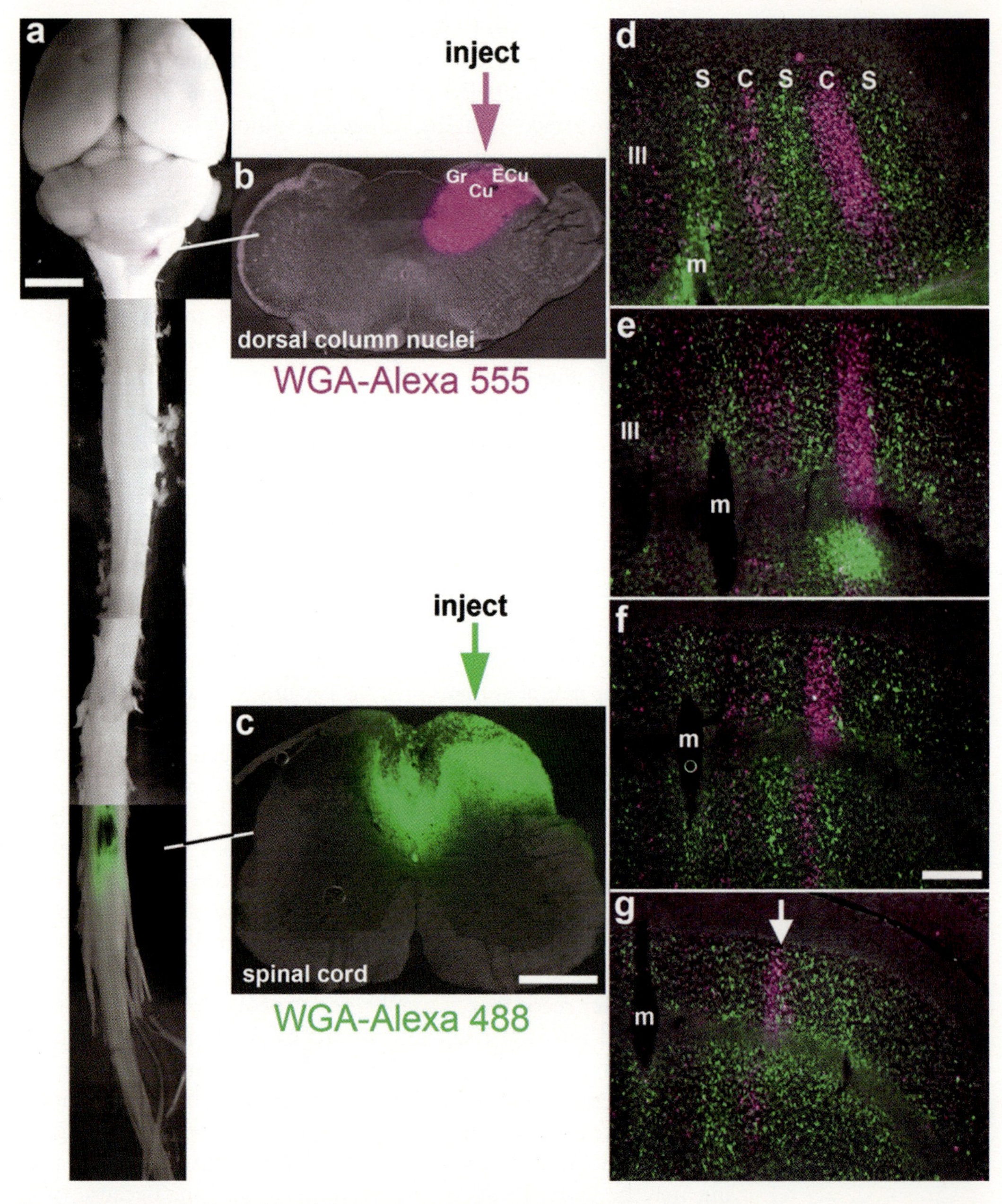

Fig. 3 WGA-Alexa tracing in the adult mouse cerebellar system. Whole brain and spinal cord dissected (**a**) after WGA-Alexa 555 was injected in the cuneate nucleus (**b**) and WGA-Alexa 488 injected into the lumbar spinal cord (**c**). The dual color injections reveal that the two tracts terminate in a striking complementary pattern of zones (alternating "C" and "S" for cuneocerebellar and spinocerebellar mossy fibers). The *arrow* is demonstrating how the use of the two WGA tracers can reveal domains of circuit overlap. *m* cerebellar midline, *Cu* cuneate nucleus, *Gr* gracile nucleus, *ECu* external cuneate nucleus. Scale bar = 2 mm in a, 500 μm in (**c**) (applies to **b**), 200 μm in **f** (applies to **d**–**g**). Adapted from Gebre et al. [19] Brain Structure and Function 217: 165–180

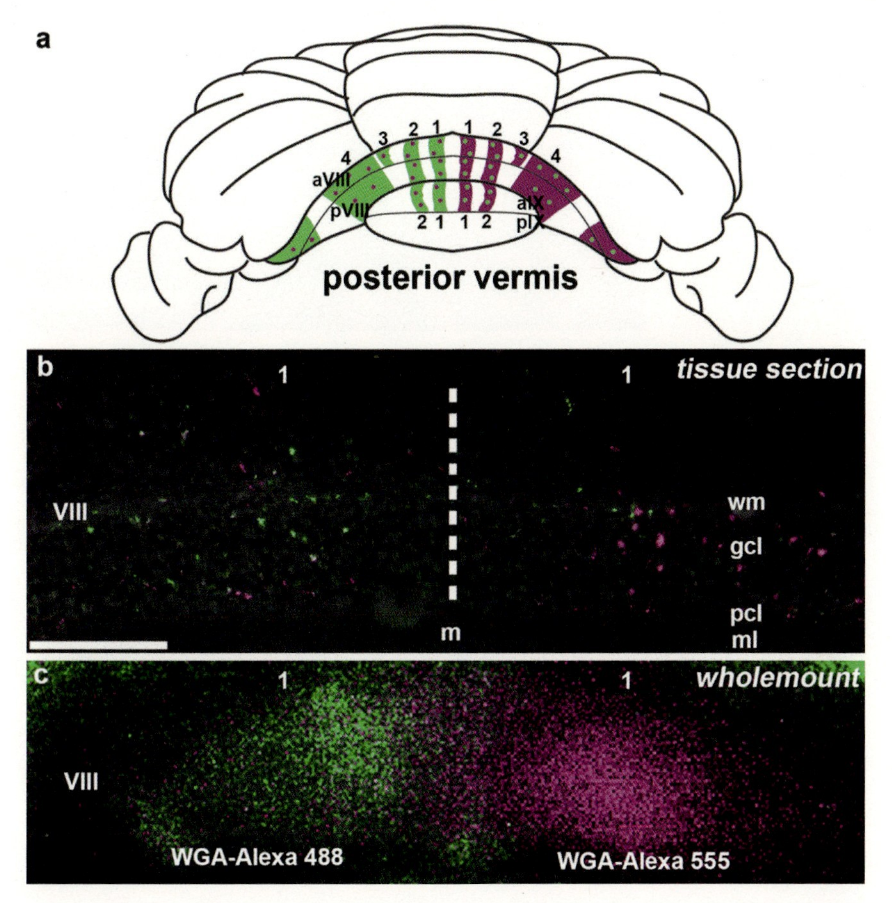

Fig. 4 Bilateral injections of WGA-Alexa 555 and WGA-Alexa 488 allow dual color, three-dimensional imaging of afferent projections. (**a**) Wholemount schematic of the mouse cerebellum summarizing the pattern of WGA-Alexa 555 and WGA-Alexa 488 labeled mossy fiber zones following bilateral injections into the same segment of the lumbar spinal cord. The *magenta* and *green dots* in the schematic represent contralateral projections of each tracer. (**b**) Injections of WGA-Alexa 555 and 488 into the spinocerebellar tract reveal ipsilateral projecting mossy fiber terminals with minor contralateral terminations in lobule VIII. (**c**) Dual labeling of ipsilateral and contralateral divisions of the spinocerebellar tract can be visualized in wholemount preparations. *ml* molecular layer, *pcl* Purkinje cell layer, *gcl* granule cell layer, *wm* white matter, *m* cerebellar midline, *aVIII* anterior lobule VIII, *pVIII* posterior lobule VIII, *aIX* anterior lobule IX, *pIX* posterior lobule IX. The *numbers 1–4* indicate cerebellar zonal modules (please refer to [57] for nomenclature). Scale bar in **b** = 100 μm (applies to **c**). Adapted from Reeber et al. [57] Brain Structure and Function 216: 159–169

8 Concluding Remarks

Neuroscience, as a discipline, is grounded on what we understand about brain structure. WGA and its derivatives have a firm place in this foundation, as they have provided several decades worth of exciting and important information about how the brain is wired for function and behavior. The utility of this remarkable tracer has spanned species ranging from invertebrates all the way through to vertebrate

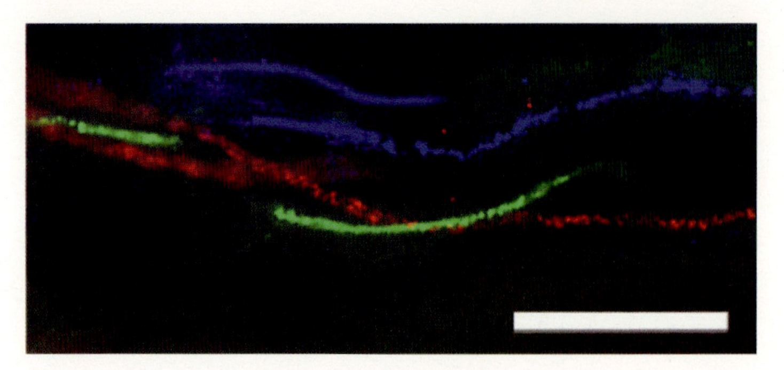

Fig. 5 Triple color tracing of axons in the cerebellar peduncles using WGA-Alexa 555 (*red*), WGA-Alexa 488 (*green*), and WGA-Alexa 350 (*blue*). The staining is very bright, axons are revealed with excellent clarity, and when used in combination, there is no "fibers of passage" problem. The scale bar = 20 μm

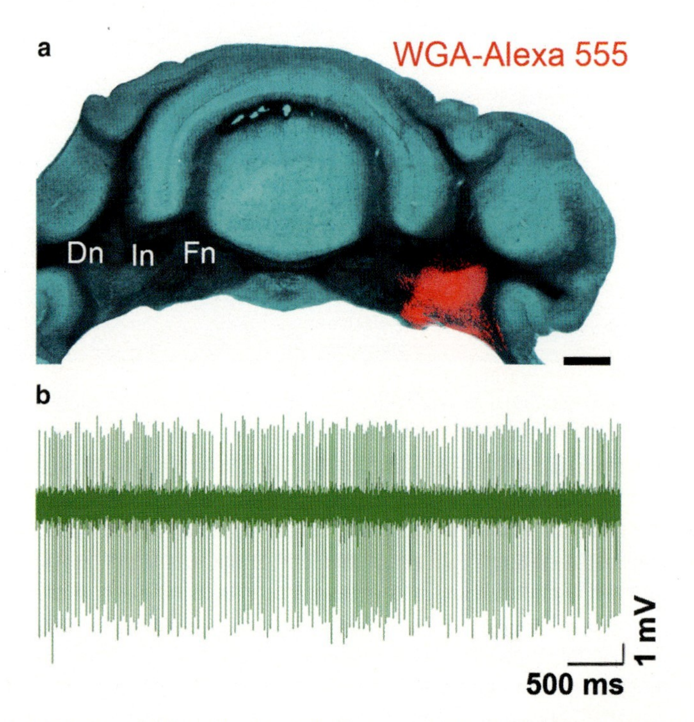

Fig. 6 WGA-Alexa 555 is ideal for labeling specific brain nuclei (**a**) after in vivo electrophysiology recording (**b**). The cerebellar nuclei were injected with a glass pipette that was also used for recording (7–10 MΩ impedance). In this example, 25 nl of WGA-Alexa 555 was delivered into the cerebellum. The spontaneous activity of cerebellar nuclear neurons was recorded using an ELC-03XS (NPI) universal extracellular amplifier and the signals digitized into Spike2 (CED, UK). The tissue was imaged under bright field (pseudo-colored blue) and then the WGA-Alexa 555 imaged with fluorescence (*red*). *Fn* fastigial nucleus, *In* interposed nucleus, *Dn* dentate nucleus. Scale bar in **a** = 1 mm

systems including humans. It has evolved from revealing the basic connectivity in the brain to unveiling the spectacular precision of how genetically distinct populations of cells are instructed to integrate in into specific circuits, and it has shed new light upon how the major freeways in the brain are topographically organized to allow sensorimotor integration, learning, memory, and many other cognitive and volitional behaviors to be processed with accuracy and speed. WGA has truly "marked" its spot as a revolutionary tool in the field, and its benefits will be appreciated for many generations to come—it is exciting to image what the next conjugate or modification will be!

Acknowledgments

This work was supported by funds from Baylor College of Medicine and Texas Children's Hospital (Houston, TX). R.V.S. was supported by the Caroline Wiess Law Fund for Research in Molecular Medicine, a BCM IDDRC Project Development Award, and by BCM IDDRC Grant Number 5P30HD024064 from the Eunice Kennedy Shriver National Institute of Child Health and Human Development, by Grant Number C06RR029965 from the National Center for Research Resources and by 1R01NS089664-01. Tissue work was performed in the BCM IDDRC Neuropathology Core. The content is solely the responsibility of the authors and does not necessarily represent the official views of the National Center for Research Resources or the National Institutes of Health.

References

1. Acharya S, Rayborn ME, Hollyfield JG (1998) Characterization of SPACR, a sialoprotein associated with cones and rods present in the interphotoreceptor matrix of the human retina: immunological and lectin binding analysis. Glycobiology 8(10):997–1006
2. Basbaum AI, Menetrey D (1987) Wheat germ agglutinin-apoHRP gold: a new retrograde tracer for light- and electron-microscopic single- and double-label studies. J Comp Neurol 261(2):306–318
3. Basbaum AI (1989) A rapid and simple silver enhancement procedure for ultrastructural localization of the retrograde tracer WGAapoHRP-Au and its use in double-label studies with post-embedding immunocytochemistry. J Histochem Cytochem 37(12): 1811–1815
4. Borges LF, Sidman RL (1982) Axonal transport of lectins in the peripheral nervous system. J Neurosci 2(5):647–653
5. Bráz JM, Basbaum AI (2009) Triggering genetically-expressed transneuronal tracers by peripheral axotomy reveals convergent and segregated sensory neuron-spinal cord connectivity. Neuroscience 163(4):1220–1232
6. Braz JM, Rico B, Basbaum AI (2002) Transneuronal tracing of diverse CNS circuits by Cre-mediated induction of wheat germ agglutinin in transgenic mice. Proc Natl Acad Sci U S A 99(23):15148–15153
7. Cajal SR (1995) Histologie du Système Nerveux de l'Homme et des Vertébrés (Maloine, Paris, 1909–1911). In: Swanson N, Swanson LW (eds) Santiago Ramón y Cajal: histology of the nervous system in man and vertebrates. Oxford University Press, New York
8. Ceccarelli B, Hurlbut WP, Mauro A (1973) Turnover of transmitter and synaptic vesicles at the frog neuromuscular junction. J Cell Biol 57(2):499–524
9. Collins CE, Stepniewska I, Kaas JH (2001) Topographic patterns of v2 cortical connections in a prosimian primate (Galago garnetti). J Comp Neurol 431(2):155–167

10. Coulter JD, Sullivan MC, Ruda MA (1980) Lectins as markers of neuronal connectivity (Abstract). Neuroscience 6:339
11. Cowan WM, Gottlieb DI, Hendrickson AE, Price JL, Woolsey TA (1972) The autoradiographic demonstration of axonal connections in the central nervous system. Brain Res 37(1): 21–51
12. DeOlmos J, Hardy H, Heimer L (1978) The afferent connections of the main and the accessory olfactory bulb formations in the rat: an experimental HRP-study. J Comp Neurol 181(2):213–244
13. Dumas M, Schwab ME, Thoenen H (1979) Retrograde axonal transport of specific macromolecules as a tool for characterizing nerve terminal membranes. J Neurobiol 10(2):179–197
14. Edwards SB (1972) The ascending and descending projections of the red nucleus in the cat: an experimental study using an autoradiographic tracing method. Brain Res 48:45–63
15. Erichsen JT, May PJ (2002) The pupillary and ciliary components of the cat Edinger-Westphal nucleus a transsynaptic transport investigation. Vis Neurosci 19:15–29
16. Fink RP, Heimer L (1967) Two methods for selective silver impregnation of degenerating axons and their synaptic endings in the central nervous system. Brain Res 4(4):369–374
17. Furue M, Uchida S, Shinozaki A, Imagawa T, Hosaka YZ, Uehara M (2011) Trajectories in the spinal cord and the mediolateral spread in the cerebellar cortex of spinocerebellar fibers from the unilateral lumbosacral enlargement in the chicken. Brain Behav Evol 77(1):45–54
18. Gebre SA, Reeber SL, Sillitoe RV (2012) Parasagittal compartmentation of cerebellar mossy fibers as revealed by the patterned expression of vesicular glutamate transporters VGLUT1 and VGLUT2. Brain Struct Funct 217:165–180
19. Gerfen CR, O'leary DD, Cowan WM (1982) A note on the transneuronal transport of wheat germ agglutinin-conjugated horseradish peroxidase in the avian and rodent visual systems. Exp Brain Res 48(3):443–448
20. Goldstein IJ, Hayes CE (1978) The lectins: carbohydrate-binding proteins of plants and animals. Adv Carbohydr Chem Biochem 35: 127–340
21. Gonatas NK, Harper C, Mizutani T, Gonatas JO (1979) Superior sensitivity of conjugates of horseradish peroxidase with wheat germ agglutinin for studies of retrograde axonal transport. J Histochem Cytochem 27(3):728–734
22. Goshgarian HG, Buttry JL (2014) The pattern and extent of retrograde transsynaptic transport of WGA-Alexa 488 in the phrenic motor system is dependent upon the site of application. J Neurosci Methods 222:156–164
23. Hamburger V, Hamilton HL (1951) A series of normal stages in the development of the chick embryo. J Morphol 88(1):49–92
24. Hendrickson A, Edwards SB (1978) The use of axonal transport for autoradiographic tracing of pathways in the central nervous system. Neuroanat Res Tech 1978:242–285
25. Herzog J, Kümmel H (2000) Fixation of transsynaptically transported WGA-HRP and fluorescent dyes used in combination. J Neurosci Methods 101(2):149–156
26. Hardy H, Heimer L (1977) A safer and more sensitive substitute for diamino-benzidine in the light microscopic demonstration of retrograde and anterograde axonal transport of HRP. Neurosci Lett 5(5):235–40
27. Heuser JE, Reese TS (1973) Evidence for recycling of synaptic vesicle membrane during transmitter release at the frog neuromuscular junction. J Cell Biol 57(2):315–344
28. Higo S, Udaka N, Tamamaki N (2007) Long-range GABAergic projection neurons in the cat neocortex. J Comp Neurol 503(3):421–431
29. Holtzman E, Teichberg S, Abrahams SJ et al (1973) Notes on synaptic vesicles and related structures, endoplasmic reticulum, lysosomes and peroxisomes in nervous tissue and the adrenal medulla. J Histochem Cytochem 21(4):349–385
30. Itaya SK, Van Hoesen GW, Barnes CL (1986) Anterograde transsynaptic transport of WGA-HRP in the limbic system of rat and monkey. Brain Res 398(2):397–402
31. Kim DS, Jeon SE, Park KC (2004) Oxidation of indole-3-acetic acid by horseradish peroxidase induces apoptosis in G361 human melanoma cells. Cell Signal 16(1):81–88
32. Klop EM, Mouton LJ, Holstege G (2005) Neurons in the lateral sacral cord of the cat project to periaqueductal grey, but not to thalamus. Eur J Neurosci 21(8):2159–2166
33. Klop EM, Mouton LJ, Holstege G (2002) Nucleus retroambiguus projections to the periaqueductal gray in the cat. J Comp Neurol 445(1):47–58
34. Lafuente J, Uriguen M, Cervós-navarro J (1999) Changes in glucidic radicals in contused human brains. Neuropathology 19(1):28–32
35. Lahiri D, Landers RA, Hollyfield JG (1995) Development of the interphotoreceptor matrix in Xenopus laevis. J Morphol 223(3): 325–339
36. Lee HS, Kim MA, Waterhouse BD (2005) Retrograde double-labeling study of common afferent projections to the dorsal raphe and the nuclear core of the locus coeruleus in the rat. J Comp Neurol 481(2):179–193

37. LeVine D, Kaplan MJ, Greenaway PJ (1972) The purification and characterization of wheat-germ agglutinin. Biochem J 129(4):847–856
38. Margolis TP, LaVail JH (1981) Rate of anterograde axonal transport of [125I]wheat germ agglutinin from retina to optic tectum in the chick. Brain Res 229(1):218–223
39. Margolis TP, Marchand CM, Kistler HB Jr, LaVail LH (1981) Uptake and anterograde axonal transport of wheat germ agglutinin from retina to optic tectum in the chick. J Cell Biol 89(1):152–156
40. Margrie TW, Brecht M, Sakmann B (2002) In vivo, low-resistance, whole-cell recordings from neurons in the anaesthetized and awake mammalian brain. Pflugers Arch 444(4): 491–498
41. Margrie TW, Meyer AH, Caputi A et al (2003) Targeted whole-cell recordings in the mammalian brain in vivo. Neuron 39(6):911–918
42. Mason CA, Gregory E (1984) Postnatal maturation of cerebellar mossy and climbing fibers: transient expression of dual features on single axons. J Neurosci 4(7):1715–1735
43. Mesulam M (1982) Tracing neural connections with horseradish peroxidase. John Wiley & Sons, Ed Marsel M Mesulam
44. Mesulam MM, Rosene DL (1977) Differential sensitivity between blue and brown reaction procedures for HRP neurohistochemistry. Neurosci Lett 5(1–2):7–14
45. Mesulam MM (1978) Tetramethyl benzidine for horseradish peroxidase neurohistochemistry: a non-carcinogenic blue reaction product with superior sensitivity for visualizing neural afferents and efferents. J Histochem Cytochem 26(2):106–117
46. Nicolson GL (1974) The interactions of lectins with animal cell surfaces. Int Rev Cytol 39: 89–190
47. O'leary DM, Gerfen CR, Cowan WM (1983) The development and restriction of the ipsilateral retinofugal projection in the chick. Brain Res 312(1):93–109
48. Oeltmann TN, Wiley RG (1986) Wheat germ agglutinin-ricin A-chain conjugate is neuronotoxic after vagal injection. Brain Res 377(2): 221–228
49. Peschanski M, Ralston HJ (1985) Light and electron microscopic evidence of transneuronal labeling with WGA-HRP to trace somatosensory pathways to the thalamus. J Comp Neurol 236(1):29–41
50. Phelan KD, Sacaan A, Gallagher JP (1996) Retrograde labeling of rat dorsolateral septal nucleus neurons following intraseptal injections of WGA-HRP. Synapse 22(3):261–268
51. Pinault D (1996) A novel single-cell staining procedure performed in vivo under electrophysiological control: morpho-functional features of juxtacellularly labeled thalamic cells and other central neurons with biocytin or Neurobiotin. J Neurosci Methods 65(2):113–136
52. Prochnow N, Lee P, Hall WC, Schmidt M (2007) In vitro properties of neurons in the rat pretectal nucleus of the optic tract. J Neurophysiol 97(5):3574–3584
53. Puelles L, Martinez-de-la-Torre M, Paxinos G, Watson C, Martinez S (eds) (2007) The chick brain in stereotaxic coordinates: an atlas correlating avian and mammalian neuroanatomy. Academic Press, San Diego, CA
54. Quigg M, Elfvin LG, Aldskogius H (1990) Anterograde transsynaptic transport of WGA-HRP from spinal afferents to postganglionic sympathetic cells of the stellate ganglion of the guinea pig. Brain Res 518(1–2):173–178
55. Reeber SL, Gebre SA, Filatova N, Sillitoe RV (2011) Revealing neural circuit topography in multi-color. J Vis Exp 57:pii:3371
56. Reeber SL, Gebre SA, Sillitoe RV (2011) Fluorescence mapping of afferent topography in three dimensions. Brain Struct Funct 216:159–169
57. Reeber SL, Sillitoe RV (2011) Patterned expression of a cocaine- and amphetamine-regulated transcript peptide reveals complex circuit topography in the rodent cerebellar cortex. J Comp Neurol 519:1781–1796
58. Sakai N, Insolera R, Sillitoe RV, Shi SH, Kaprielian Z (2012) Axon sorting within the spinal cord marginal zone via Robo-mediated inhibition of N-cadherin controls spinocerebellar tract formation. J Neurosci 32(44): 15377–15387
59. Sawczuk A, Covell DA (1999) Wheat germ agglutinin conjugated to TRITC: a novel approach for labeling primary projection neurons of peripheral afferent nerves. J Neurosci Methods 93(2):139–147
60. Schwab ME, Javoy-Agid F, Agid Y (1978) Labeled wheat germ agglutinin (WGA) as a new, highly sensitive retrograde tracer in the rat brain hippocampal system. Brain Res 152(1):145–150
61. Shen Y, Chen J, Liu Q et al (2011) Effect of wheat germ agglutinin density on cellular uptake and toxicity of wheat germ agglutinin conjugated PEG-PLA nanoparticles in Calu-3 cells. Int J Pharm 413(1–2):184–193
62. Sillitoe RV, Vogel MW, Joyner AL (2010) Engrailed homeobox genes regulate establishment of the cerebellar afferent circuit map. J Neurosci 30:10015–10024
63. Song Q, Yao L, Huang M et al (2012) Mechanisms of transcellular transport of wheat

germ agglutinin-functionalized polymeric nanoparticles in Caco-2 cells. Biomaterials 33(28):6769–6782

64. Steindler DA (1982) Differences in the labeling of axons of passage by wheat germ agglutinin after uptake by cut peripheral nerve versus injections within the central nervous system. Brain Res 250(1):159–167
65. Stöckel K, Dumas M, Thoenen H (1978) Uptake and subsequent retrograde axonal transport of nerve growth factor (NFG) are not influenced by neuronal activity. Neurosci Lett 10(1–2):61–64
66. Sugita M, Shiba Y (2005) Genetic tracing shows segregation of taste neuronal circuitries for bitter and sweet. Science 309(5735):781–785
67. Tabuchi K, Sawamoto K, Suzuki E et al (2000) GAL4/UAS-WGA system as a powerful tool for tracing Drosophila transsynaptic neural pathways. J Neurosci Res 59(1):94–99
68. Teune TM, Van der Burg J, De Zeeuw CI, Voogd J, Ruigrok TJ (1998) Single Purkinje cell can innervate multiple classes of projection neurons in the cerebellar nuclei of the rat: a light microscopic and ultrastructural triple-tracer study in the rat. J Comp Neurol 392(2): 164–178
69. Toonen M, Van Dijken H, Holstege JC et al (1998) Light microscopic and ultrastructural investigation of the dopaminergic innervation of the ventrolateral outgrowth of the rat inferior olive. Brain Res 802(1–2):267–273
70. Van der Want JJ, Klooster J, Cardozo BN, De Weerd H, Liem RS (1997) Tract-tracing in the nervous system of vertebrates using horseradish peroxidase and its conjugates: tracers, chromogens and stabilization for light and electron microscopy. Brain Res Brain Res Protoc 1(3):269–279
71. Van Rossum J (1969) Corticonuclear and corticovestibular projections from the cerebellum. Thesis, Van Gorcum, Assen
72. Vanderhorst VG, Terasawa E, Ralston HJ, Holstege G (2000) Monosynaptic projections from the lateral periaqueductal gray to the nucleus retroambiguus in the rhesus monkey: implications for vocalization and reproductive behavior. J Comp Neurol 424(2):251–268
73. Vanderhorst VG, Terasawa E, Ralston HJ (2001) Monosynaptic projections from the nucleus retroambiguus region to laryngeal motoneurons in the rhesus monkey. Neuroscience 107(1):117–125
74. Vogel MW, Prittie J (1994) Topographic spinocerebellar mossy fiber projections are maintained in the lurcher mutant. J Comp Neurol 343(2):341–351
75. Voogd J, Broere G, van Rossum J (1969) The medio-lateral distribution of the spinocerebellar projection in the anterior lobe and the simple lobule in the cat and a comparison with some other afferent fibre systems. Psychiatr Neurol Neurochir 72(1):137–151
76. Walling SG, Brown RA, Miyasaka N, Yoshihara Y, Harley CW (2012) Selective wheat germ agglutinin (WGA) uptake in the hippocampus from the locus coeruleus of dopamine-β-hydroxylase-WGA transgenic mice. Front Behav Neurosci 6:23
77. Wood JG, Byrd FI, Gurd JW (1981) Lectin cytochemistry of carbohydrates on cell membranes of rat cerebellum. J Neurocytol 10(1): 149–159
78. Yoshihara Y, Mizuno T, Nakahira M et al (1999) A genetic approach to visualization of multisynaptic neural pathways using plant lectin transgene. Neuron 22(1):33–41
79. Yoshihara Y (2002) Visualizing selective neural pathways with WGA transgene: combination of neuroanatomy with gene technology. Neurosci Res 44(2):133–140
80. Zaborszky L, Wouterlood FG, Lanciego JL (eds) (2010) Neuroanatomical tract-tracing, molecules, neurons, and systems. Springer, New York
81. Zhang J, Kleinschmidt J, Sun P, Witkovsky P (1994) Identification of cone classes in Xenopus retina by immunocytochemistry and staining with lectins and vital dyes. Vis Neurosci 11(6):1185–1192

Chapter 3

Retrograde Tract-Tracing "Plus": Adding Extra Value to Retrogradely Traced Neurons

José L. Lanciego

Abstract

Classical neuroanatomical tract-tracing methods have formed the basis for most of our current understanding of brain circuits. However, to obtain a deeper knowledge of the main operational principles of the brain, the simple delineation of brain connectivity is not sufficient. This particularly holds true in regard to the analysis of connections within the diseased brain, for instance, the study of a number of major neurological disorders through the use of animal models. In other words, the information gathered from tract-tracing techniques is often too static, and recent findings in the fields of neurophysiology, receptor mapping, and neuroimaging (among others) need to be integrated within the context of structural data of brain connectivity as seen with neuroanatomical tracing techniques. During the past few years, our laboratory has pioneered a number of combinations of retrograde tracers with in situ hybridization, analyzing the changes in mRNA expression levels within brain circuits of interest. More recently, we have succeeded in combining a number of tract-tracing methods with a newly introduced technique known as in situ proximity ligation assay (PLA). The PLA technique is particularly well suited for the analysis of protein-protein interactions. This combination of methods enabled us to elucidate unequivocally the presence of GPCR heteromers within identified brain circuits and we strongly believe that this will soon become a popular approach in the field. Here we provide readers with a landscape view of these approaches, together with step-by-step protocols so that these methods may be easily reproduced even by inexperienced users.

Key words Cholera toxin, Fluoro-Gold, Dextran amines, In situ hybridization, Proximity ligation assay, Neuroanatomical tracing

1 Introduction

Neuroanatomical tract-tracing methods employ a large number of substances that share a common principle: once injected into the brain, these substances are taken up by neuronal processes and transported either from the axon terminal backwards to the parent cell body (known as retrograde tracers) or from the neuronal soma and dendrites forward to the axon terminals (known as anterograde tracers). It is worth noting that a number of neuroanatomical tracers are indeed transported in both directions (anterograde and retrograde) and are therefore considered as bidirectional tracers.

Benjamin R. Arenkiel (ed.), *Neural Tracing Methods: Tracing Neurons and Their Connections*, Neuromethods, vol. 92, DOI 10.1007/978-1-4939-1963-5_3, © Springer Science+Business Media New York 2015

In the past few decades, many substances have been added to the technical arsenal of neuroanatomical tract-tracing. This review will only focus on the most popular ones. In order to gain a better historical perspective, a number of excellent books, reviews, and special issues have been made available. Key "bedside" books in the field are those published by Cowan and Cuénod [12], Heimer and Robards [18], Heimer and Zaborszky [19], Bolam [6], and [58]. Additional information can be obtained in two Special Issues of the Journal of Neuroscience Methods coedited by Lanciego and Wouterlood [29, 31] as well as in review contributions issued by Köbbert et al. [24] and by Lanciego and Wouterlood [32].

1.1 Most Popular Retrograde and Anterograde Tracers

Modern neuroanatomical tract-tracing methods form part of a rather short history that began in the 1970s with the introduction of horseradish peroxidase (HRP), the first substance transported retrogradely [25, 35, 39, 40]. Retrograde tracing with HRP was quickly adopted as the tracer of choice for retrograde tracing purposes and was widely used for more than two decades. A few years later, the introduction of a number of fluorescent dyes lead to a "neuroanatomical revolution." For the very first time, the implementation of multiple retrograde labeling, by combining several dyes in the same experimental paradigm [27], became a feasible option. Fluorescent retrograde tracers such as Fast Blue, Diamidine Yellow, propidium iodide, True Blue, and Lucifer Yellow were among the most popular ones during the 1980s.

The beginning of the 1980s also witnessed the arrival of a conjugated, nontoxic cholera toxin subunit B (choleragenoid, CTB-HRP). Initially proposed as a more sensitive tracer than native HRP [43, 44], it soon became clear that CTB possessed a number of advantageous properties when compared to HRP, such as (1) limited uptake by axons of passage; (2) improved neuronal filling; (3) receptor-mediated uptake; (4) less rapid elimination from labeled neurons, thus enabling longer survival times and therefore facilitating the combination of CTB with other neuroanatomical tracers; and (5) immunohistochemical detection.

Schmued and Fallon [50] introduced Fluoro-Gold (FG) as a "retrograde tracer with unique properties," and it quickly became the top ranked tracer among those most commonly used. Under UV illumination, FG-labeled neurons are visualized with intense white-yellow fluorescence. Furthermore, fluorescent emission is very resistant to bleaching; a common drawback that often arises when dealing with fluorescent dyes other than FG. Moreover, the availability of an antibody against FG raised in rabbit [8] boosted the use of FG due to some added methodological advantages such as (1) immunohistochemical detection followed by the use of secondary antibodies tagged with the fluorochrome of choice for detection under the confocal microscope without the need of using expensive UV lasers, (2) ultrastructural examination of

FG-labeled material, and (3) the easy combination of FG with other tracers or with the immunohistochemical detection of neuroactive substances to further elucidate the neurochemical profile of FG-labeled neurons.

Although all of the above tracers are mainly transported retrogradely, it soon became clear that there was an urgent need for tracers transported in the opposite direction, from cell bodies to axon terminals (anterograde tracers). The first approach to successfully fulfill this need was a plant lectin known as *Phaseolus vulgaris* leucoagglutinin (PHA-L). Initially introduced by Gerfen and Sawchenko [17], PHA-L is be taken up almost exclusively by neuronal cell bodies and is transported anterogradely toward axon terminals. The immunohistochemical detection of PHA-L allows for the proper visualization of axons, axon collaterals, and terminal boutons with an unprecedented level of detail. Similar to CTB and FG, the immunohistochemical detection of PHA-L (commercial antibodies against PHA-L raised in goat and rabbit have been made available) has opened the door for combinations with other tracing methods by means of multiple immunostaining. This allows for the ultrastructural examination of neuronal processes showing PHA-L labeling as well as for a whole spectrum of combinations under the confocal microscope. All these properties made PHA-L the anterograde tracer of choice for more than a decade. Although there are now a number of alternatives to PHA-L, it is still widely used in the field.

Biotinylated dextran amine (BDA) is the most relevant anterograde tracer of the dextran amine family and has been overwhelmingly applied since its introduction by Veenman et al. [57]. BDA is undoubtedly the anterograde tracer of choice and best exemplifies the broad success of modern neuroanatomical tract-tracing procedures. Readers interested in BDA tracing may gain a deeper insight into this technique by reading review manuscripts published by Reiner et al. [47] and Reiner and Honig [45]. Compared to PHA-L, the main advantage of BDA is the histochemical detection using streptavidin labeled with either HRP (for colorimetric detection) or with a full repertoire of fluorescent tags (for confocal examination). In other words, there are no antibodies engaged in BDA detection, and therefore, BDA is often the best option for experimental designs that try to combine multiple markers (reviewed in [33]; *see* also [28, 30, 32]). Although BDA 10 kDa is the most widely used, it is worth noting that BDA at a lower molecular weight (BDA 3 kDa) is also available. BDA 3 kDa performs better than BDA 10 kDa in retrograde tracing studies [36, 47].

The use of viruses as tracing agents presents a totally different approach. Viral strains such as herpes simplex, pseudorabies, and rabies virus (RV) are neurotropic organisms that upon entering a neuron replicate and cross the synapse to further infect a second-order neuron and so on. In other words, viruses are particularly well suited to the study of chains of interconnected neurons. Initially introduced by Krister Kristensson and colleagues [26], tract-tracing using viruses has been exploited by a number of leading neuroanatomists such as Hans Kuypers, Gabriella Ugolini, Arthur Loewy, Peter Strick, Patrick Card, and Gary Aston-Jones, among others. For those readers interested in viral tracing, a number of review manuscripts are available [1, 16, 23, 41, 56]. It is also worth noting that in collaboration with the group of Lydia Kerkerian-Le Goff (Université de la Méditerranée, Marseille, France), we have introduced an approach that consistently yields Golgi-like retrograde labeling of neurons traced with the rabies virus, resulting in sensitive visualization of even the finest postsynaptic elements in labeled neurons [37, 49].

1.2 Selection of the Most Suitable Tracer

Selecting which tracer is best suited to a given experimental design is often a difficult choice. In our experience, the tracer that yields the best performance for retrograde tracing purposes is Fluoro-Gold (FG). The crystalline solution that results from dissolving the FG yellow powder in 0.1 M cacodylate buffer pH 7.3 can be injected either by pressure or by iontophoresis (depending on the desired size of the injection site). Survival periods range from 1 week to up to a year and most of the common fixatives can be used. More importantly, there are two ways to visualize FG-labeled neurons: direct visualization under UV illumination or indirect immunofluorescent visualization can be achieved using a commercial antibody against FG raised in rabbit [8]. In addition to FG, another suitable option for retrograde tracing is cholera toxin subunit B (CTB). CTB is usually pressure injected and therefore better suited for tracing studies in monkeys, where larger injection sites are commonly used to generate consistent results as well as to minimize mistargeting. Nevertheless, it is worth noting that CTB also performs very well in rodents and indeed is a very good choice when used in combination with FG for double retrograde tract-tracing paradigms [14]. Moreover, BDA 3 kDa could be also considered as a good retrograde tracer (with some non-negligible anterograde labeling). Considering that BDA 3 kDa can be detected directly using dye-labeled streptavidin (i.e., there are no antibodies engaged in BDA visualization), this tracer is a good option when planning triple retrograde tracing experiments that combine BDA 3 kDa, FG, and CTB.

Transsynaptic retrograde tracing using the RV is also a feasible option. This method requires a different approach to the previously mentioned tracers, with some inherent safety considerations, as well as with some important advantages, as follows. Firstly, when the RV is detected using antibodies against a viral phosphoprotein, Golgi-like labeling is observed, and small neuronal processes such as thin dendrites and dendritic spines are often visualized with an unprecedented level of detail. Secondly, by taking advantage of the transsynaptic spread, second-order neurons are detected, and this feature provides an attractive way to identify interneurons innervating projection neurons. A more detailed explanation of this method can be found elsewhere [32, 37, 49].

Many of the same considerations apply when selecting the most suitable anterograde tracer. Among the available tools, our tracer of choice is undoubtedly BDA 10 kDa. Dissolved in 10 nM phosphate buffer pH 7.25, BDA 10 kDa can be injected both by pressure and iontophoresis (depending on the desired size of the injection site). BDA is compatible with a wide range of fixatives and survival times and appears to be homogeneously distributed within labeled axons and axon terminals. Some degree of retrograde BDA transport is always expected in brain areas innervating the injection site [46]. Furthermore, BDA labeling is well preserved when undergoing ultrastructural examination. In addition to BDA, PHA-L remains a very good second option for anterograde tract-tracing. PHA-L is immunohistochemically detected using a primary antibody raised either in goat or rabbit (we strongly recommend the use of the latter). PHA-L is almost exclusively transported in the anterograde direction and retrograde PHA-L tracing is often only observed in neighboring areas to the injection site. Survival times range from 1 week to 3 months and the immunohistochemical detection of PHA-L results in the visualization of fibers, axon collaterals, and terminal boutons with exquisite detail. Similar to BDA, PHA-L has shown proven efficacy in a wide range of mammalian and non-mammalian species. It is worth noting that PHA-L and BDA can easily be combined in experimental designs that require dual anterograde tract-tracing paradigms, such as those analyzing the potential convergence/divergence of two different afferent systems toward a given brain area (as shown by [28]).

1.3 Making Cocktails of Tracers

In addition to injecting two or even three tracers in the same experimental animal to analyze complex brain circuits, two tracers can be mixed together within the same injection to study of reciprocally interconnected brain areas. Mixing together an anterograde and a retrograde tracer enables the visualization of efferent axons arising from the injected brain region, as well as afferent neurons projecting to this region. This combination was pioneered by Coolen et al. [11] by preparing an injectable cocktail of BDA and CTB. Later on, Thompson and Swanson [54] reported the use of two different cocktails of tracers (BDA and FG as well as PHA-L and CTB) within the same experimental animal, therefore obtaining a large amount of sophisticated data on brain connectivity.

2 Retrograde Tracing Combined with In Situ Hybridization

Even the most sophisticated multi-tracing experiments are only capable of providing a "static" structural picture of brain circuits. If we consider the study of brain circuits in neurodegenerative animal models, labeling connections may not be sufficient. In other words, a "diseased" circuit often looks the same as the one observed in naïve animals. Some years ago, it became evident that we needed to develop tools that enabled us to analyze what was happening in a given circuit of interest beyond connectivity. To achieve this, we developed a protocol combining retrograde tract-tracing (using either FG or CTB) together with in situ hybridization (ISH). The combination of retrograde tracing with ISH made it possible to elucidate changes in gene expression patterns within a brain circuit of interest, comparing control and lesioned animals [9, 32, 42]. Briefly, efferent neurons are traced retrogradely using either FG or CTB, followed by the detection of an mRNA transcript of interest by means of in situ hybridization. Using these demanding techniques, we have succeeded in detecting changes in the mRNA coding the vesicular glutamate transporter isoforms 2 (vGlut2) within FG-labeled thalamostriatal-projecting neurons in hemiparkinsonian rats [2] as well as in CTB-labeled subthalamic neurons innervating the ventral thalamus [48]. Furthermore, two different mRNAs can be detected simultaneously in projection neurons using differently tagged riboprobes (biotin- and digoxigenin-labeled riboprobes). Accordingly, we have shown that FG-labeled neurons from the entopeduncular nucleus innervating the caudal intralaminar nuclei co-express mRNA coding the GABA precursors GAD65 or GAD67, together with vGlut2 mRNA [4]. Similar procedures were conducted to detect the mRNA coding vGlut isoforms 1 and 2 within the same thalamostriatal projection neurons that had been identified using CTB [5]. More recently, changes in GAD65 and GAD67 mRNA expression levels in CTB-labeled pallidothalamic neurons were reported by comparing expression levels in control, parkinsonian, and dyskinetic primates [10]. Furthermore, it is worth noting that ISH often affords a superior sensitivity than that of conventional immunohistochemistry, and this particularly holds true when trying to detect proteins expressed at very low levels. Cannabinoid receptors 1 and 2 (CB_1R and CB_2R) are a good example of this. These GPCRs are difficult to detect by immunohistochemistry, and therefore, ISH is a far more feasible approach. We have recently shown that CB_1R and CB_2R mRNA are co-expressed in CTB-labeled pallidothalamic neurons in monkeys, together with a marked decline in their expression levels in parkinsonian monkeys showing levodopa-induced dyskinesia [51]. Illustrative examples are provided in Fig. 1.

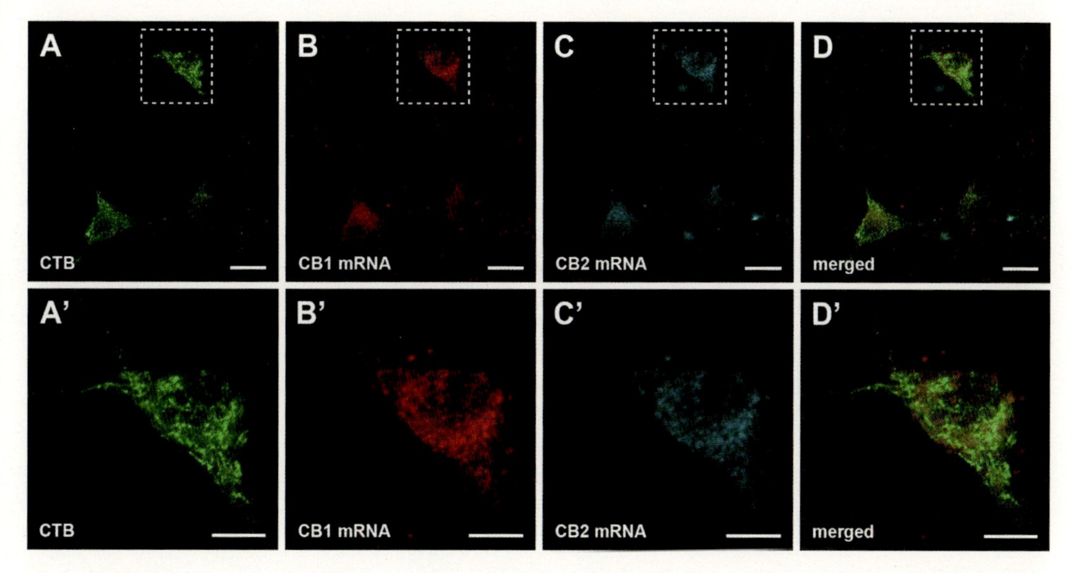

Fig. 1 Retrograde tracing with CTB combined with dual fluorescent in situ hybridization. (**a–d**) Neurons from the internal division of the globus pallidus projecting to thalamic ventral nuclei in monkeys are retrogradely labeled following the deposit of CTB in the ventral anterior/ventral lateral thalamic nuclei (*green channel*). All pallidothalamic projection neurons co-express CB_1R (*red channel*) and CB_2R mRNA (*blue channel*). (**a′–d′**) *Insets* taken from **a–d** at higher magnification. Scale bar is 20 μm in panels **a–d** and 10 μm in panels **a′–d′**

2.1 Protocol

In order to ensure the reproducibility of the method, a detailed procedure is provided below. The protocol uses a retrograde tracer (either FG or CTB), followed by the detection of two different transcripts (named "A" and "B") by means of dual fluorescent ISH. More detailed information on riboprobe preparation and alternative procedures can be found in Barroso-Chinea et al. [3].

Briefly, the detection of the biotin-labeled riboprobe is carried out first, followed by the digoxigenin-labeled riboprobe and finally the immunohistochemical procedure to visualize the tracer of choice. Although we routinely carry out the tracer detection once the dual fluorescent ISH is conducted, similar results were obtained by firstly visualizing the tracer, then later conducting the dual fluorescent ISH procedure.

2.1.1 Riboprobe Preparation

Sense and antisense riboprobes are transcribed from the respective plasmids. The plasmids are linearized and the sense or antisense probes are transcribed with the appropriate RNA polymerases (Boehringer Mannheim, Germany). The transcription mixture includes 1 μg template plasmid; 1 mM each of ATP, CTP, and GTP; 0.7 mM UTP and 0.3 mM DIG-UTP or biotin-UTP; 10 mM DTT; 50 U RNAse inhibitor; and 1 U of either T3, T7, or SP6 RNA polymerase in a volume of 50 μl. After 2 h at 37 °C, the template plasmid is digested with 2 U RNAse-free DNAse for 30 min at 37 °C. The riboprobes are then precipitated by the addition of 100 μl of 4 M ammonium acetate and 500 μl of ethanol and are recovered by centrifugation at 4 °C for 30 min. The quality of the synthesis is monitored in dot blots.

2.1.2 Histological Processing

(*Note: the following protocol is for the detection of a biotin-labeled riboprobe followed by a digoxigenin-labeled riboprobe. The detection of the retrograde tracer (FG or CTB) immediately follows the dual ISH steps.*)

1. 2 × 10 min washes with 0.1 % active PBS-DEPC (DEPC from Sigma).
2. Pre-equilibration: 5× SSC (0.75 M NaCl, 0.075 M Na-citrate) for 10 min.
3. Pre-hybridization: Incubate the sections for 2 h at 58 °C in the hybridization solution, prepared as follows (for a final volume of 5 ml):
   - 2.5 ml of deionized formamide (Sigma).
   - 1.25 ml of 20× SSC.
   - 10 μl of salmon DNA (20 mg/ml, Sigma).
   - 1.24 ml of H_2O-DEPC (inactive).
4. Hybridization:
   - Add 1 μl of each probe to 250 μl of hybridization solution (0.4 μg/ml).
   - Denature the probes for 5 min at 75 °C.
   - Add the denatured probes to the hybridization mix (400 ng/ml) and incubate the sections overnight at 58 °C.
5. Post-hybridization:
   - 1 × 30 min wash with 2× SSC.
   - 1 × 1 h wash with 2× SSC at 65 °C.
   - 1 × 1 h wash with 0.1× SSC at 65 °C.
6. Tyramide signal amplification (for the biotin-labeled riboprobe):
   - Add TNB blocking buffer to the sections with 0.5 % blocking reagent (TSA, kit NEN, Perkin-Elmer, Boston, MA) for 30 min. TN buffer comprises 100 mM Tris/HCl pH 7.5 and 150 mM NaCl pH 7.5.
   - Incubate the sections in peroxidase-conjugated streptavidin (Perkin-Elmer) diluted 1:100 in TSA for 30 min.
   - 3 × 5 min washes in TNT.
   - Incubate the sections in Alexa® Fluor633-conjugated ExtrAvidin (Molecular Probes-Invitrogen) for 90 min.

7. Digoxigenin signal amplification (for the digoxigenin-labeled riboprobe):
   - 1 × 10 min wash in TN (100 mM Tris/HCl, pH 7.5, and 150 mM NaCl).
   - Incubate the sections for 2 h (under gently shaking) with an alkaline phosphatase-conjugated anti-digoxigenin antibody raised in sheep (1:500 in 0.5 % blocking-TN; Roche Diagnostics GmbH).
   - 2 × 15 min washes in TN.
   - Equilibrate with buffer 4 (100 mM Tris/HCl pH 8.0, 100 mM NaCl, 10 nM $MgCl_2$) for 10 min.
   - HNPP/Fast Red TR substrate (Roche Diagnostics): add 10 μl of HNPP/10 μl of Fast Red to 1 ml of buffer 4 and leave the sections for 1 h with gentle shaking (emission range of HNPP/Fast Red TR is between 540 and 590 nm, with a peak at 562 nm).
   - 3 × 5 min washes in PBS pH 7.6.
8. Detection of the retrograde tracer of choice (CTB or FG):
   - Incubate the sections in a rabbit anti-FG antibody (Chemicon) overnight at 4 °C (1:2,000 in PBS pH 7.6) or in a rabbit anti-CTB antibody (GenWay) overnight a 4 °C (1:1,000 in PBS pH 7.6). Please note that a goat anti-CTB antibody is also available from List Biological Labs and can be then used as an alternative to the rabbit anti-CTB antibody.
   - 3 × 10 min washes in PBS pH 7.6.
   - Incubate the sections for 2 h at room temperature in an Alexa® Fluor488-conjugated donkey anti-rabbit IgG (1:200 in PBS pH 7.6; antibody from Molecular Probes-Invitrogen). *Note*: when using a goat anti-CTB antibody, then the required bridge antibody will be an Alexa® Fluor488-conjugated donkey anti-goat IgG (also from Molecular Probes-Invitrogen, to be used at the same dilution as the former one).
   - 3 × 10 min washes in PBS pH 7.6.
   - Mount the section on glass slides, air-dry in the dark, and dehydrate rapidly in toluene and coverslip with DPX (BDH Chemicals, UK). Note: in our experience, 2 × 5 min washes in toluene afford a similar preservation of the fluorescent signal compared to commercially available anti-fading mounting solutions.

3 Retrograde Tracing Combined with In Situ Proximity Ligation Assay

Detecting one or two mRNA transcripts within projection neurons provides important functional information on its own. Using this approach one can measure changes in the patterns of gene expression for a given mRNA of interest as seen under different experimental conditions. However, these techniques have their own inherent limitations, particularly because very little is known about the intracellular processing of mRNAs. Although the most conventional destination for any given mRNA is to be linked to translating polysomes, synthesized mRNAs could also be directed to processing bodies, sites specialized in both the storage of repressed mRNA as well as in mRNA degradation [53]. In other words, the presence of a particular mRNA does not necessarily imply that the related protein is going to be finally synthesized. This issue is of paramount importance in regard to the expression of G-protein-coupled receptors (GPCRs) within retrogradely traced neurons. The detection of GPCR heteromers (e.g., two different GPCRs assembled together) is currently a hot topic in the field of neuropharmacology. GPCR heteromers are functionally distinct molecular entities and do not merely represent the aggregation of two receptors with independent functions [15]. This is exemplified by GPCR heteromers made up of CB_1R and CB_2R. We have recently shown that the mRNA coding CB_1R and CB_2R are co-expressed within projection neurons in the primate globus pallidus [34, 51]. Whereas previous studies have shown that differences in mRNA expression levels of CB_1R are matched by differences in CB_1R immunoreactivity and CB_1R ligand binding [13, 20–22, 38], definitive proof that these mRNA are translated into the related proteins was still lacking. Going one step further, in addition to detecting the protein, we needed a way of determining whether these receptors, once synthesized, ultimately formed heteromers.

This methodological gap has recently been filled by the arrival of the so-called in situ proximity ligation assay (PLA). Initially introduced by Söderberg et al. [52], the PLA technique has been successfully employed to detect GPCR heteromers in the striatum [55] as well as in the globus pallidus [7]. The method is based on a clever and simple principle: two primary antibodies (each directed against each target receptor) are firstly used, followed by respective bridge antibodies. Bridge antibodies are modified by covalently coupling a pair of affinity oligonucleotide probes (a plus and a minus probe). Only when the target proteins are in close proximity (<17 nm) do the probes ligate and form templates for rolling circle amplification (1,000-fold amplification of the DNA molecule; *see* [52, 55]). Hybridization of complementary fluorescently labeled oligonucleotides with the amplified DNA is then visualized as a red dot with fluorescent microscopy, representing a single protein-protein interaction, i.e., a GPCR heteromer. In the case that primary antibodies against each receptor are raised in the same species, the primary antibodies can be tagged directly with a pair of affinity oligonucleotides probes and then detected using the standard procedure, i.e., without the use of bridge antibodies.

Combining the PLA technique with retrograde tract-tracing using CTB represents a feasible choice to unequivocally demonstrate the presence of GPCR heteromers within identified projection neurons. We have succeeded in detecting CB_1R-CB_2R heteromers in pallidothalamic projection neurons (retrogradely labeled with CTB) in monkeys, as well as the detection of adenosine 2A receptors ($A_{2A}R$) forming heteromers with either CB_1R or CB_2R. Detection of GPCR heteromers in these samples was carried out using the Duolink II in situ PLA detection kit (Olink, Bioscience, Sweden). Since primary antibodies against CB_1R and CB_2R are both raised in the same species (from Thermo Scientific, Rockford, USA, and from Cayman Chemical, Ann Arbor, USA, respectively), the rabbit anti-CB_1R antibody was linked to a plus PLA probe and the rabbit anti-CB_2R was linked to a minus PLA probe following the instructions supplied by the manufacturer. A different strategy was used for the detection of $A_{2A}R$-CB_1R and $A_{2A}R$-CB_2R heteromers. In this case, the anti-$A_{2A}R$ antibody is raised in mouse (Upstate, Millipore, USA), whereas the anti-CB_1R and the anti-CB_2R antibodies are raised in rabbit. Accordingly, the bridge antibodies linked to a pair of affinity probes (plus or minus) were used after the primary antibodies. Illustrative examples are provided in Fig. 2.

3.1 Protocol

(*Note: the following protocol is for the detection of CB_1R-CB_2R heteromers using primary antibodies coupled to a plus and minus PLA probe, respectively, followed by the detection of the retrograde tracer CTB.*)

1. 3×5 min washes in PBS (free-floating, gently shaking).
2. 1×12 min wash with 0.01 % Triton in PBS.
3. 2×5 min washed in PBS, then mount the sections in slides and air-dry.
4. Incubation for 1 h at 37 °C with the blocking solution in a preheated humidity chamber.
5. Incubate the sections overnight at 4 °C with PLA probe-linked primary antibodies. Antibodies are diluted by using the buffer provided labeled as *Antibody diluent 1*×. Recommended dilution is twice more concentrated as the dilution used in standard immunohistochemistry.
6. Wash the sections twice with buffer A at room temperature.

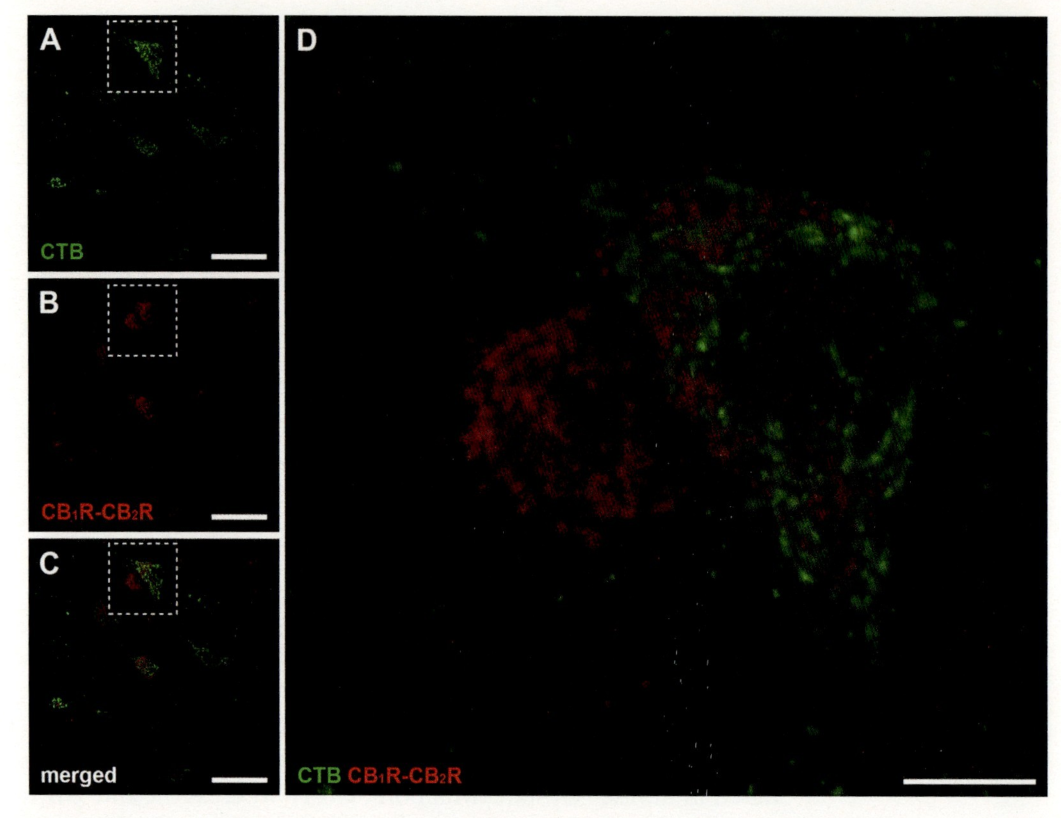

Fig. 2 Retrograde tracing with CTB combined with in situ proximity ligation assay (PLA). The use of the PLA technique allows the unequivocal elucidation of whether two given mRNA transcripts are ultimately translated into the related proteins. More importantly, the PLA technique demonstrates that CB_1R and CB_2R form GPCR receptor heteromer complexes. (**a–c**) CB_1R-CB_2R heteromers (*red channel*) are located postsynaptically in CTB-labeled pallidothalamic projection neurons in monkeys (*green channel*). Each *red dot* represents one CB_1R-CB_2R receptor heteromer. (**d**) *Inset* taken from **a–c** at higher magnification showing cannabinoid receptor heteromers within two pallidal neurons, one labeled with CTB (therefore identified as a pallidothalamic projection neuron) and the other one lacking CTB labeling. Scale bar is 50 μm in panels **a–c** and 10 μm in panel **d**

7. Incubate the sections with the ligation solution for 30 min to 1 h at 37 °C in a humidity chamber. The ligation solution is prepared as follows:
   - 8 μl of *Ligation* 5×.
   - 31 μl of H_2O (miliQ).
   - 1 μl of *Ligase.*
8. 3 × 5 min washes with buffer A at room temperature.

   (*Note: from now on, all kit elements and slides should be preserved from light exposure.*)
9. Incubate the sections with the amplification solution (100 min) at 37 °C in a humidity chamber. The amplification solution is prepared as follows:

  - 8 μl of *Amplification* 5×.
  - 31.5 μl of H_2O (miliQ).
  - 0.5 μl of *Polymerase (diluted 1:80)*.

10. 2×10 min washes with buffer B 1×.
11. 1×1 min wash with buffer B ×0.01.
12. Air-dry at room temperature in the dark.
13. Mount the sections with a minimal amount of *Duolink II Mounting Medium with DAPI*.

4 Some Advice to Novel Users

Prior to planning a tract-tracing experiment, there are few issues that need to be considered. Firstly, one should select the type of tracer that is best suited to your needs. In our experience, FG delivered by iontophoresis in rodents is the first choice in retrograde tracing. Alternatively, pressure-injected CTB can also be used when larger injections sites are required. In monkeys, CTB performs better than FG. For anterograde tract-tracing purposes, BDA 10 kDa is undoubtedly the best choice, for both iontophoretic and pressure delivery. PHA-L only works when injected by iontophoresis and is a very good alternative to BDA. Secondly, when considering multiple neuroanatomical tracing, a number of issues need to be addressed. These considerations or "golden rules" have been summarized (for details, *see* [32]). The best choice for dual retrograde tracing combined with anterograde tract-tracing is the combination of FG, CTB, and BDA. This requires the use of a rabbit anti-FG antibody, a goat anti-CTB antibody and a streptavidin for the BDA protocol. Detection can be achieved both by immunofluorescence (using Alexa®-tagged secondary antibodies together with an Alexa®-labeled streptavidin) and by immunohistochemistry using three different colored peroxidase substrates such as nickel-enhanced DAB (for BDA detection), a regular DAB solution (for CTB detection), and Vector VIP chromogen (for visualizing FG-containing neurons). It is important to use these substrates in this order, i.e., DAB-Ni first, then DAB and finally V-VIP. Moreover, the use of cocktails of tracers is also a feasible option as summarized above in Sect. 1.3. Finally, combining tract-tracing with in situ hybridization for single- or dual-mRNA detection might provide an important added value to the structural data gathered from tracing experiments, as shown in Sect. 3. Furthermore, detecting protein-protein interactions within traced elements is an approach of paramount importance when trying to demonstrate that two different mRNAs are ultimately translated into the related proteins and are interacting to each other. In this regard, combining retrograde (or anterograde) tracing with PLA (as explained in Sect. 4) is the technique of choice and possesses an extraordinary added value to the field of GPCR heteromer complexes.

5 Appendix: Some Tract-Tracing Books You Need to Have in Your Library

As pointed out by Laszlo Zaborszky and colleagues in the preface to the third edition of Neuroanatomical Tract-Tracing Methods (2006), we still feel that books hold an important place in this field despite the current thinking that *what is not on the Web does not exist*. In the last five decades, a vast list of bibliographical references dealing with neuroanatomical tracing has been generated, and thus, it is almost impossible to deal with every single technique that has been described. Nevertheless, making some basic choices to plan a tract-tracing experiment is not as difficult as it may be initially seen. A number of books and book series have been published at different times in the last couple of decades. Together, they cover most of the existing knowledge in relation to tract-tracing methods.

One of the most successful book series has been the three books entitled "Neuroanatomical Tract-Tracing Methods." The first edition was published in 1981 and was edited by Heimer and Robards, entitled *Neuroanatomical Tract-Tracing Methods* [18]. In 1991, Heimer and Zaborszky [19] published an updated book entitled *Neuroanatomical Tract-Tracing Methods 2: Recent Progress*. One decade later, we had the honor and responsibility of preparing the third book of this successful series, entitled *Neuroanatomical Tract-Tracing Methods 3: Molecules, Neurons, and Systems* [58]. The third edition was the product of a joint effort of my former mentor Floris Wouterlood, our close friend Laszlo Zaborszky and myself. At present, I believe that it is too early to anticipate whether or not we will go ahead with the preparation of a fourth version of this series of books. Although I recognize that we are currently living in an era of molecular neurobiology, my feeling is that we are on the verge of a revival of structural neuroanatomy, mainly due to the need to integrate data gathered using molecular biology tools within a structural context. Finally, in addition to these series of books, it is also worth acknowledging another successful book published by Paul Bolam [6] in 1992, entitled *Experimental Neuroanatomy*. These four seminal contributions provide potential readers an overall and comprehensive historical perspective of neuroanatomical tracing. More importantly, all the techniques covered in these books are explained in a very detailed manner. Most of the techniques are explained using step-by-step recipes, as well as a number of very helpful "tips and tricks" to facilitate the reproduction of these methods even by the most inexperienced users. The same philosophy has been applied to the present book you are currently reading.

In addition to books, some insight can also be gained from two Special Issues coedited by myself together with Floris Wouterlood in the Journal of Neuroscience Methods. The first of thesetwoissuesappearedonNovember15,2000("Neuroanatomical Tracing Methods at the Millennium"), the second on December 15, 2010 ("Neuroanatomical Tracing and Systems Neuroscience: The State of the Art"). I wish to express my deep gratitude to all the scientists that contributed to both Special Issues, as well as to both the former editor in chief (N.G. Bowery) and current coeditors-in-chief (V. Crunelli and G. Gerhardt).

All the books and Special Issues listed above, as well as the current book you have in your hands, have been written with graduate students in mind since they are the main expected audience for all these protocols. My students from the Master in Neurosciences and Cognition program have received instruction based on the contents of this chapter, and I hope that many students will benefit this in the future.

Acknowledgments

It is a great pleasure to acknowledge the continuous support received from colleagues, laboratory team members, and technicians. To mention just a few team members: Alberto J. Rico, Iria González-Dopeso, Salvador Sierra, Mónica Pérez-Manso, Pedro Barroso-Chinea, Natasha Luquin, Virginia Gómez-Bautista, Iciar P. López, Lorena Conte-Perales, Nancy Gonzalo, María Castle, and Elena Erro. Furthermore, I am also particularly indebted to my technician Elvira Roda who joined the laboratory more than 10 years ago and who still contributes continuously with new ideas and improvements. Moreover, it is also worth recognizing the extensive training that I received in the Department of Neuroanatomy and Neuropharmacology at the Amsterdam Vrije Universiteit under the guidance and mentorship of Floris G. Wouterlood, as well as from the laboratory technicians Barbara Jorritsma-Byham, Annaatje Pattiselanno, and Peter Goede. They all made my stay in Amsterdam (1992 and 1996) an easy, pleasant, and fruitful experience. Supported by grants from the Ministerio de Economía y Competitividad (BFU2012-37907, SAF2008-03118-E and SAF39875-C02-01), Eranet-Neuron (Heteropark), CiberNed (CB06/05/0006), Departamento de Salud, Gobierno de Navarra, and UTE project/ Foundation for Applied Medical Research (FIMA). Salary for S.S. is partially supported by a grant from Mutual Médica.

References

1. Aston-Jones G, Card JP (2000) Use of pseudorabies virus to delineate multi-synaptic circuits in brain: opportunities and limitations. J Neurosci Meth 103:51–61
2. Aymerich MS, Barroso-Chinea P, Pérez-Manso M, Muñoz-Patiño AM, Moreno-Igoa M, González-Hernández T, Lanciego JL (2006) Consequences of unilateral nigrostriatal denervation on the thalamostriatal pathway in rats. Eur J Neurosci 23:2099–2108
3. Barroso-Chinea P, Castle M, Aymerich MS, Pérez-Manso M, Erro E, Tuñón T, Lanciego JL

(2007) Expression of the mRNAs encoding for the vesicular glutamate transporters 1 and 2 in the rat thalamus. J Comp Neurol 501:703–715
4. Barroso-Chinea P, Rico AJ, Pérez-Manso M, Roda E, López IE, Luis-Ravelo D, Lanciego JL (2008) Glutamatergic pallidothalamic projections and their implications in the pathophysiology of Parkinson's disease. Neurobiol Dis 31:422–432
5. Barroso-Chinea P, Castle M, Aymerich MS, Lanciego JL (2008) Expression of vesicular glutamate transporters 1 and 2 in the cells of origin of the rat thalamostriatal pathway. J Chem Neuroanat 35:101–107
6. Bolam JP (1992) Experimental neuroanatomy. Oxford University Press, Oxford, UK
7. Callén L, Moreno E, Barroso-Chinea P, Moreno-Delgado D, Cortés A, Mallol J, Casadó V, Lanciego JL, Franco R, Lluis C, Canela EI, McCormick PJ (2012) Cannabinoid receptors CB1 and CB2 form functional heteromers in brain. J Biol Chem 287:20851–20865
8. Chang HT, Kuo H, Whittaker JA, Cooper NGF (1990) Light and electron microscopic analysis of projection neurons retrogradely labelled with Fluoro-Gold: notes on the application of antibodies to Fluoro-Gold. J Neurosci Meth 35:31–37
9. Conte-Perales L, Barroso-Chinea P, Rico AJ, Gómez-Bautista V, López IP, Roda E, Wouterlood FG, Lanciego JL (2010) Neuroanatomical tracing combined with in situ hybridization: analysis of gene expression patterns within brain circuits of interest. J Neurosci Meth 194:28–33
10. Conte-Perales L, Rico AJ, Barroso-Chinea P, Gómez-Bautista V, Roda E, Luquin N, Sierra S, Lanciego JL (2011) Pallidothalamic-projecting neurons in *Macaca fascicularis* co-express GABAergic and glutamatergic markers as seen in control, MPTP-treated and dyskinetic monkeys. Brain Struct Funct 216:371–386
11. Coolen LM, Jansen HT, Goodman RL, Wood RI, Lehman MN (1999) A new method for simultaneous demonstration of anterograde and retrograde connections in the brain: co-injections of biotinylated dextran amine and the beta subunit of cholera toxin. J Neurosci Meth 91:1–8
12. Cowan WM, Cuénod M (1975) The use of axonal transport for studies of neuronal connectivity. Elsevier, Amsterdam
13. Egertova M, Elphick MR (2000) Localisation of cannabinoid receptors in the rat brain using antibodies to the intracellular C-terminal tail of CB. J Comp Neurol 422:159–171
14. Erro E, Lanciego JL, Giménez-Amaya JM (2002) Re-examination of the thalamostriatal projections in the rat with retrograde tracers. Neurosci Res 42:45–55
15. Ferré S, Baler R, Bouvier M, Caron MG, Devi LA, Durroux T, Fuxe K, George SR, Javitch JA, Lohse MJ, Mackie K, Milligan G, Pfleger KD, Volkow ND, Waldoher M, Wood AS, Franco R (2009) Building a new conceptual framework for receptor heteromers. Nat Chem Biol 5:131–134
16. Geerling JC, Mettenleiter TC, Loewy AD (2006) Viral tracers for the analysis of neural circuits. In: Zaborszky L, Wouterlood FG, Lanciego JL (eds) Neuroanatomical tract-tracing 3: molecules, neurons, and systems. Springer, New York, NY, pp 262–303
17. Gerfen CR, Sawchenko PE (1984) A method for anterograde axonal tracing of chemically specified circuits in the central nervous system: combined *Phaseolus vulgaris*-leucoagglutinin (PHA-L) tract tracing and immunohistochemistry. Brain Res 343:144–150
18. Heimer L, Robards M (1981) Neuroanatomical tract-tracing methods. Plenum, New York, NY
19. Heimer L, Zaborszky L (1991) Neuroanatomical tract-tracing methods 2. Recent progress. Plenum, New York, NY
20. Herkenham M, Lynn AB, Johnson MR, Melvin LS, de Costa BR, Rice KC (1991) Neuronal localization of cannabinoid receptors in the basal ganglia of the rat. Brain Res 547: 267–274
21. Herkenham M, Lynn AB, Johnson MR, Melvin LS, de Costa BR, Rice KC (1991) Characterization and localization of cannabinoid receptors in rat brain: a quantitative in vitro autoradiographic study. J Neurosci 11: 563–583
22. Julian MD, Martin AB, Cuellar B, Rodriguez De Fonzeca F, Navarro M, Moratalla R, Garcia-Segura LM (2003) Neuroanatomical relationship between type 1 of cannabinoid receptors and dopaminergic systems in the rat basal ganglia. Neuroscience 119:309–318
23. Kelly RM, Strick PL (2000) Rabies as a trans-neuronal tracer of circuits in the central nervous system. J Neurosci Meth 103:63–72
24. Köbbert C, Apps R, Bechmann I, Lanciego JL, Mey J, Thanos S (2000) Current concepts in neuroanatomical tracing. Prog Neurobiol 62: 327–351
25. Kristensson K, Olsson Y, Sjöstrand J (1971) Axonal uptake and retrograde transport of exogenous proteins in the hypoglossal nerve. Brain Res 32:399–406
26. Kristensson K, Ghetti B, Wisniewski HM (1974) Study on the propagation of herpes simplex virus (type 2) into the brain after intraocular injection. Brain Res 69:189–201

27. Kuypers HG, Bentivoglio M, van der Kooy D, Catsman-Berrevoets CE (1980) Double retrograde neuronal labeling through divergent axon collaterals, using two fluorescent tracers with the same excitation wavelength which label different features of the cell. Exp Brain Res 40:383–392
28. Lanciego JL, Wouterlood FG (1994) Dual anterograde axonal tracing with *Phaseolus vulgaris leucoagglutinin* (PHA-L) and biotinylated dextran amine (BDA). Neurosci Protoc 94-050-06-01-13
29. Lanciego JL, Wouterlood FG (2000) Neuroanatomical tract-tracing methods beyond 2000: what's now and next. J Neurosci Meth 130:1–2
30. Lanciego JL, Wouterlood FG (2006) Multiple neuroanatomical tract-tracing: approaches for multiple tract-tracing. In: Zaborszky L, Wouterlood FG, Lanciego JL (eds) Neuroanatomical tract-tracing 3: molecules, neurons, and systems. Springer, New York, NY, pp 336–365
31. Lanciego JL, Wouterlood FG (2010) Proceedings of the workshop "neuroanatomical tracing and systems neuroscience: the state of the art", 7th FENS meeting, Amsterdam, The Netherlands, 3 Jul 2010. J Neurosci Meth 194:1
32. Lanciego JL, Wouterlood FG (2011) A half century of experimental neuroanatomical tracing. J Chem Neuroanat 42:157–183
33. Lanciego JL, Wouterlood FG, Erro E, Arribas J, Gonzalo N, Urra X, Cervantes S, Giménez-Amaya JM (2000) Complex brain circuits studied via simultaneous and permanent detection of three transported neuroanatomical tracers in the same histological section. J Neurosci Meth 103:127–135
34. Lanciego JL, Barroso-Chinea P, Rico AJ, Conte-Perales L, Callen L, Roda E, Gómez-Bautista V, Lopez IP, Lluis C, Labandeira-Garcia JL, Franco R (2011) Expression of the mRNA coding the cannabinoid receptor 2 in the pallidal complex of *Macaca fascicularis.* J Psychopharmacol 25:97–104
35. LaVail JH, LaVail MM (1972) Retrograde axonal transport in the central nervous system. Science 176:1416–1417
36. Lei W, Jiao Y, Del Mar N, Reiner A (2004) Evidence for differential cortical input to direct pathway versus indirect pathway striatal projection neurons in rats. J Neurosci 24:8289–8299
37. López IP, Salin P, Kachidian P, Barroso-Chinea P, Rico AJ, Gómez-Bautista V, Conte-Perales L, Coulon P, Kerkerian-Le Goff L, Lanciego JL (2010) The added value of rabies virus as a retrograde tracer when combined with dual anterograde tract-tracing. J Neurosci Meth 194:21–27
38. Mailleux P, Vanderhaeghen JJ (1992) Distribution of neuronal cannabinoid receptor in the adult brain: a comparative receptor binding autoradiography and in situ hybridization histochemistry. Neuroscience 48:655–668
39. Mesulam MM (1976) The blue reaction product in horseradish peroxidase neurohistochemistry. J Histochem Cytochem 24:1273–1280
40. Mesulam MM (1978) Tetramethylbenzidine for horseradish peroxidase neurohistochemistry: a non-carcinogenic blue reaction product with superior sensitivity for visualizing neural afferents and efferents. J Histochem Cytochem 26:106–117
41. Morecraft RJ, Ugolini G, Lanciego JL, Wouterlood FG, Pandya DN (2009) Classic and contemporary neural tract tracing techniques. In: Johansen-Berg H, Behrens T (eds) Diffusion MRI: from quantitative measurement to in-vivo neuroanatomy. Oxford University Press, Oxford, pp 273–308
42. Pérez-Manso M, Barroso-Chinea P, Aymerich MS, Lanciego JL (2006) 'Functional' neuroanatomical tract-tracing: analysis of changes in gene expression of brain circuits of interest. Brain Res 1072:91–98
43. Trojanowski JQ, Gonatas JO, Gonatas NK (1981) Conjugates of horseradish peroxidase (HRP) with cholera toxin and wheat germ agglutinin are superior to free HRP as orthogradely transported markers. Brain Res 223: 381–385
44. Trojanowski JQ, Gonatas JO, Steiber A, Gonatas NK (1982) Horseradish peroxidase (HRP) conjugates of cholera toxin and lectins are more sensitive retrograde transported markers than free HRP. Brain Res 231:33–50
45. Reiner A, Honig MG (2006) Dextran amines: versatile tools for anterograde and retrograde studies of nervous system connectivity. In: Zaborszky L, Wouterlood FG, Lanciego JL (eds) Neuroanatomical tract-tracing 3: molecules, neurons, and systems. Springer, New York, NY, pp 304–335
46. Reiner A, Veenman CL, Honig MG (1993) Anterograde tracing using biotinylated dextran amine. Neurosci Protoc 93-050-14
47. Reiner A, Veenman CL, Medina Y, Jiao N, Del Mar N, Honig MG (2000) Pathway tracing using biotinylated dextran amines. J Neurosci Meth 103:11–22
48. Rico AJ, Barroso-Chinea P, Conte-Perales L, Roda E, Gómez-Bautista V, Gendive M, Obeso JA, Lanciego JL (2010) A direct projection

from the subthalamic nucleus to the ventral thalamus in monkeys. Neurobiol Dis 39: 381–392

49. Salin P, Castle M, Kachidian P, Barroso-Chinea P, López IP, Rico AJ, Kerkerian-Le Goff L, Coulon P, Lanciego JL (2008) High-resolution neuroanatomical tract-tracing for the analysis of striatal microcircuits. Brain Res 1221:49–58
50. Schmued LC, Fallon JH (1986) Fluoro-Gold: a new fluorescent retrograde axonal tracer with numerous unique properties. Brain Res 377: 147–154
51. Sierra S, Luquin N, Rico AJ, Gómez-Bautista V, Roda E, Dopeso-Reyes IG, Vásquez A, Martínez-Pinilla E, Labandeira-García JL, Franco R, Lanciego JL (2014) Detection of cannabinoid receptors CB_1 and CB_2 within basal ganglia output neurons in macaques: changes following experimental parkinsonism. Brain Struct Funct (in press, DOI: 10.1007/ s00429-014-0823-8)
52. Söderberg O, Leuchowius KJ, Gullberg M, Jarvius M, Weibrecht I, Larsson LG, Landegren U (2008) Characterizing proteins and their interaction in cells and tissues using the in situ proximity ligation assay. Methods 45:227–232
53. Sossin WS, DesGroseillers L (2006) Intracellular trafficking of RNA in neurons. Traffic 7: 1581–1589
54. Thompson RH, Swanson LW (2010) Hypothesis-driven structural connectivity: analysis supports network over hierarchical model of brain architecture. Proc Natl Acad Sci U S A 34:15235–15239
55. Trifilieff P, Rives ML, Urizar E, Piskorowski RA, Vishwasrao HD, Castrillon J, Schmauss C, Slattman M, Gullberg M, Javitch JA (2011) Detection of antigen interactions ex vivo by proximity ligation assay: endogenous dopamine D_2-adenosine A_{2A} receptor complexes in the striatum. Biotechniques 51:111–118
56. Ugolini G (2010) Advances in viral transneuronal tracing. J Neurosci Meth 194:2–20
57. Veenman CL, Reiner A, Honig MG (1992) Biotinylated dextran amine as an anterograde tracer for single- and double-label studies. J Neurosci Meth 41:239–254
58. Zaborszky L, Wouterlood FG, Lanciego JL (2006) Neuroanatomical tract-tracing methods 3: molecules, neurons, and systems. Springer, New York, NY

Chapter 4

Simultaneous Collection of In Vivo Functional and Anatomical Data from Individual Neurons in Awake Mice

Brittany N. Cazakoff and Stephen D. Shea

Abstract

Ideally, to trace neural circuits, one often desires access to functional data that may be linked to anatomical attributes such as neuron type or projection target. Here we describe methods used for this purpose in our laboratory. We aim for this chapter to serve as a practical guide to applying "loose patch" or "cell-attached patch" electrophysiology techniques in vivo to simultaneously obtain information about neuronal firing (e.g., responses to sensory stimuli) and detailed anatomical information including dendritic morphology and axonal targeting. Since data on neuronal circuit function are often most useful when gathered during wakeful behavior, we pay special attention here to the use of these techniques in awake, head-fixed mice. However, the same methods may easily be applied to anesthetized animals.

Key words Electrophysiology, In vivo, Awake, Loose patch, Extracellular, Neuroanatomy

1 Introduction

Ultimately, to better understand neural circuits, neuroscientists must be able to associate the detailed wiring diagrams obtained via anatomical tracing methods with data characterizing the activity of circuits in terms of neuronal firing [1]. The most direct approach to making this connection is to simultaneously obtain electrophysiological and anatomical data from the same neurons. Moreover, assuming a major objective of modern neuroscience is to understand how neural circuitry gives rise to behavior, it is perhaps optimal to obtain functional data on neural firing patterns from animals that are consciously perceiving stimuli and behaviorally responding [4]. These electrophysiology data, when corresponded with anatomical morphology and connectivity, are invaluable for revealing what information is carried in the "wires" traced using anatomical methods.

A number of existing techniques allow measurement of electrical activity of isolated neurons in vivo. Here we focus on a

Benjamin R. Arenkiel (ed.), *Neural Tracing Methods: Tracing Neurons and Their Connections*, Neuromethods, vol. 92, DOI 10.1007/978-1-4939-1963-5_4, © Springer Science+Business Media New York 2015

technique ("loose patch" recording) that is compatible with acquisition of anatomical information about the recorded cell. We find the loose patch method is a particularly mechanically stable recording configuration, so we also describe procedures specifically for making recordings in awake animals. In our laboratory, we use these techniques in mice, although in principle they could be used in rats or other small animals. As a sample application, here we demonstrate the fidelity and stability of these methods by describing their use to record from granule cells of the mouse main olfactory bulb, a class of small cells that are otherwise difficult to electrophysiologically target.

We have included general protocols and notes for performing the following procedures:

1. Construction of a head-fixed recording rig.
2. Implantation of a head fixation bar.
3. Loose patch recording/cell-labeling techniques.

2 Materials

2.1 Construction of a Head-Fixed Recording Rig

In this section, we describe each of the major components of our head-fixed recording rig and explain some of our design choices. The reader may make different choices or assemble them differently depending on the demands of a particular experiment.

2.1.1 Head Fixation System

In our laboratory, we use a custom-machined head bar and clamp system (Fig. 1). The head bar (20 mm × 3 mm) is laser cut from a 1.27 mm-thick titanium sheet and is tapered on the ends where it is grasped on each side by a clamp. The clamps are held in place by custom rails and optical hardware (Thorlabs). The tapered ends of the bar fit into the mouths of the clamps, and custom thumbscrews allow quick fixation and release of the head bar. In Sect. 2.2, we describe procedures for stably affixing these head bars to the skull of a mouse.

In choosing this design, we had a number of considerations. First, we wanted to achieve maximal stability of the preparation; thus, we chose to grasp the bar in two locations on either side of the head. Second, we wanted the implant to be as lightweight as possible so that it minimally interferes with the animal's behavior. Third, we wanted the implant to be as low profile as possible so that it could be placed very close to the craniotomy. As we will describe in Sect. 2.2, secure placement close to the target brain region is critical for achieving stable recordings. We have used these head bars effectively in recordings from the main olfactory bulb, the auditory cortex, and the somatosensory cortex. It is important to consider optimal placement of the head bar for different target regions.

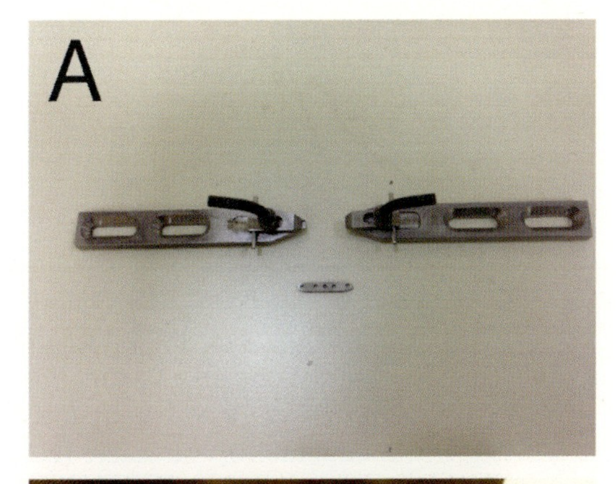

Fig. 1 Photos of the head fixation system. (**a**) Head fixation clamps and a head bar. (**b**) Clamping system positioned over the wheel

2.1.2 Animal Support Structure

In our laboratory, the head clamps are positioned so that the mouse rests on a freely rotating cylindrical wheel, cut from a foam pillow (Exervo TeraNova EVA foam roller) (Fig. 1). This particular product has the dual advantages of (1) having a textured surface that facilitates running and (2) being easy to clean. The wheel has a rod through the center that is held on either side by optical posts. The design of our wheel is simple and also somewhat arbitrary. Any wheel or ball can suffice as long as it rotates easily with minimal drag. Optional enhancements can include spring mounts for the axle or a ball that moves in two dimensions as opposed to axially [3].

Unless the experiment explicitly requires or uses a ball that rotates with more than one degree of freedom, we recommend avoiding this as the animal may slide laterally. Another feature we have included in our set up is the microcontroller-based velocity sensor that computes and outputs an analog signal proportional to linear velocity of the wheel surface (see below).

As alternatives to a running wheel or ball, one may use a static platform or tube. Many laboratories have reported success using these structures to hold the animal; however, in our experience, when the animal is unable to run, there is a tendency for it to grab an edge of the structure and apply force. This could lead to movement that will be inhibitory to stable recordings or, in the worst case, detachment of the head bar. Therefore, the choice of a wheel has two benefits. First, it keeps the animal alert, active, and relatively comfortable. Second, when the animal does move, the wheel diffuses the kinetic energy, and the animal cannot push or pull on the head bar.

2.1.3 Amplifier

This choice is mostly up to personal preference, as essentially any current clamp or "bridge" amplifier will work. Features that are important include (1) an ability to gate and/or command current pulses flexibly, (2) a head stage capable of making positive and negative current pulses of up to 1 nA through a micropipette of up to 50 **MΩ** resistance, (3) a "buzz capacitance" button with adjustable duration for transiently overcompensating the electrode capacitance, and (4) an adjustable high-pass filter ranging from DC to at least 100 Hz. Virtually all amplifiers are compatible with a wide range of pipette holders. Select a pipette holder that fits 1.5 mm OD glass pipettes, includes a silver wire that reaches down close to the pipette tip, and also features a side pressure port for applying positive and negative pressure.

2.1.4 Wheel Velocity Sensor

Depending on your experimental needs, you may optionally equip the running wheel with a rotary encoder that outputs a signal that can be used to calculate running velocity. Here we provide a simple inexpensive method for building this component and use a microcontroller to produce an analog signal proportional to velocity.

2.1.5 Micromanipulator

To stably and flexibly position your recording pipette above and into the brain, you will first need a coarse manipulator with three or, ideally, four axes. As an example, we recommend the MX110 from Siskiyou Design. Second, you will need one axis of fine manipulation used to advance the pipette once it is in the brain. The full array of choices for this task is beyond the scope of this paper; however, in our laboratory we use either a hydraulic microdrive from Soma Scientific or a Luigs and Neumann piezoelectric drive. Each one has a different user feel, so the choice is somewhat a matter of personal preference. You should also of course consider the range in the depth axis to be sure it is appropriate for your target.

Parts List

Amplifier (we use NPI ELC-03 or BA-0X).

Desktop computer with data acquisition system (we use Spike2 equipped with a Power1401 ADC/DAC board).

Oscilloscope.

Appropriate BNC cables to connect devices to the computer.

Head bar clamp and frame (constructed from custom-machined hardware and optical posts and rails from Thorlabs).

Coarse four-axis micromanipulator and mounting post.

Foam wheel cylinder (Exervo TeraNova EVA foam roller; cut into 6″ cylinder).

Axial mount for wheel (constructed from optical posts and rails from Thorlabs).

Rotary encoder (model COM-11102 available from SparkFun).

Arduino Uno microcontroller.

MAX500 DAC integrated circuit chip.

12 V DC power supply.

Miscellaneous wire, connectors, and a power source for assembling the velocity sensor.

2.2 Implantation of a Head Fixation Bar

Parts List

Absorbent points (Henry Schein, part no. 9004703).

Air hose.

Anesthetic for recovery surgery (e.g., ketamine and xylazine).

C&B-Metabond Quick! cement system (Parkell Inc., part no. S380).

Cotton swabs.

Dental cement powder (A-M Systems, part no. 525000).

Dental cement solvent (A-M Systems, part no. 526000).

Gelfoam absorbable gelatin dental sponge (Pfizer).

Head bar (2 cm × 0.3 cm × 0.1 cm) with tapered ends.

Microdrill or dental drill.

Stainless steel, flat-head machine screws (Amazon Supply, part no. B002SG89OI).

Surgical tools (fine forceps, scalpel or spring-loaded scissors, needle driver).

Sutures.

2.3 Making Loose Patch Recordings in Awake Mice

Parts List

Borosilicate glass with filament (O.D. 1.5 mm, I.D. 0.86 mm; BF150-86-10, Sutter Instrument).

Micropipette puller with box filament.

1.5 % Neurobiotin dissolved in patch solution.

Patch solution.

- 10 mM HEPES.
- 2 mM magnesium chloride.
- 10 mM potassium chloride.
- 125 mM potassium gluconate.
- Filter and freeze in 10–20 μl aliquots.

Pipette holder for patch pipettes.

Silver chloride pellet for ground electrode.

Silver wire.

Surgical tools (blade breaker, fine forceps).

Wire to connect ground to head stage.

3 Methods

3.1 Construction of a Head-Fixed Recording Rig

There is no fixed protocol or set procedure for constructing an electrophysiology rig because the needs of every experiment are different. However, careful thought should be given to the placement of all components so that there is easy access to the animal, plenty of room to see the preparation under a surgical microscope, and clearance to easily and quickly withdraw and replace your pipette. We strongly recommend that you perform your experiments inside a Faraday cage and carefully ground all the components of your rig to a common point that has a secure and heavy gauge path to AC ground. It is also important to keep AC-powered devices outside of the Faraday cage. Finally, you may wish to perform the experiments on an air table to dampen vibrations, although this is obviously less important when the animal will be awake and moving on its own.

One special feature we describe here is the attachment of a rotary encoder to the wheel in order to measure the direction and speed of running. Figure 2 is a schematic of how to wire together such a device. The axle of the encoder is placed in the center of the running wheel, and its output wires connect to an Arduino Uno microcontroller. In Appendix, we include software code that can be loaded onto the microcontroller so that it counts clicks from the encoder to integrate position change per unit time. At every time step, the microcontroller serially communicates with a DAC chip

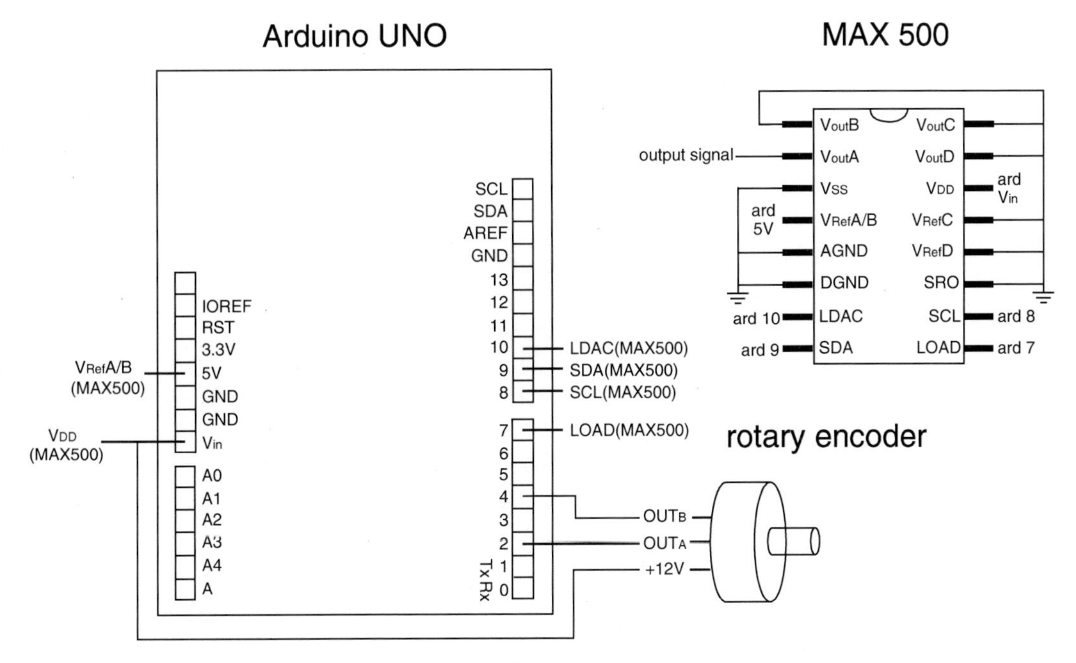

Fig. 2 Schematic for building a running wheel velocity sensor

to update the velocity reading, which the DAC chip outputs as an analog voltage that can be acquired along with your neural data and other physiological measures.

3.2 Implantation of a Head Fixation Bar

The procedure described here is for implantation of a head bar on an adult (approx. 6–10 weeks old) mouse but may be adapted for use in other animals. We secure head bars to the animal's skull using a combination of small machine screws and dental cement. This increases both the stability and longevity of the implant with minimal disruption to the animal's behavior. Surgery should be completed at least 3 days in advance of recording to allow the animal sufficient time to recover.

3.2.1 Surgery

The animal is first anesthetized, positioned in the stereotaxic frame, and the skull is exposed using a scalpel or spring-loaded scissors. You should completely clear the membranous tissue on the surface of the skull using fine forceps or by rubbing the skull surface with a cotton swab.

Using the microdrill, we make burr holes in which to position machine screws. Using a screwdriver and a pair of fine forceps, we then fasten these screws to the skull. Screws should be secured only halfway into the skull, taking care not to damage the underlying brain. While positioning of the screws will depend on the recording area of interest, we typically position 3–4 screws in a square pattern with two screws positioned far from the intended recording site and 1–2 screws positioned just next to the recording site.

Where possible, we position the caudal and rostral screws far enough apart from each other so the head bar may be positioned between them. Any bleeding from the burr holes should be cleaned using the absorbent tips or Gelfoam.

The head bar is secured to the skull using dental cement. While this cement is adequate for stabilizing the head bar, it does not itself adhere well to the skull. Instead, we first coat the skull surface in a thin layer of Metabond (prepared according to the manufacturer's instructions). We find that Metabond adheres tightly to the skull and provides a better surface on which the dental cement can be applied. Both the Metabond and the dental cement should only be applied to dry surfaces as fluid prevents adherence and thus compromises the stability of the head cap. The skull surface can be dried using a combination of cotton swabs, Gelfoam, or absorbent tips placed under the skin and pressurized air blown onto the skull surface. Once the Metabond is dry, the head bar can be placed on top of the Metabond and secured with ample dental cement, ensuring that the cement does not stick to the tapered ends of the head bar. Dental cement should completely cover the screws and the middle section of the head bar. Using this method, we find the head caps are stable across multiple recording sessions and continue to stick to the skull surface weeks after the initial surgery.

3.3 Making Loose Patch Recordings in Awake Mice

3.3.1 Craniotomy

For awake recordings, we first anesthetize the mouse with isoflurane anesthesia (1 %) and make a craniotomy. Individual mice are used for multiple recordings, and a new craniotomy is made each day. This better ensures that Neurobiotin fills can be resolved postmortem. Mice tolerate well the brief period of anesthesia, and recordings are not made until at least one half hour after the animal wakes up. Local lidocaine anesthetic (4 % Topicaine) is used on the skin surrounding the area of interest to minimize discomfort to the animal during recording.

Small (200 μm × 200 μm) clean craniotomies are essential for achieving stable loose patch recordings in awake animals. Large burr holes or cranial windows provide less support to the patch pipette and tend to bleed more often, inhibiting recording success. To achieve small craniotomies, we first shave down several layers of skull without breaking through skull to brain. When the skull is sufficiently thinned (pliable), we use a blade breaker to cut a small opening in the skull and expose brain. We keep the craniotomy moist and devoid of blood by covering it with Gelfoam soaked in 0.9 % saline until we are ready to record. For most of our recordings, we do not find it necessary to clear the dura mater as the patch pipettes are sharp and strong enough to pierce through this tissue.

3.3.2 Ground Electrode

We use a silver pellet ground wire positioned under the skin near the recording area of interest. Throughout the recording, we find it necessary to keep the tip of the ground wire wet using saline-soaked dental foam. This increases the surface area in contact with the preparation and reduces noise.

Placement of the ground wire is critical for loose patch recordings. One should not position the ground wire near muscles that may displace the ground when the animal moves or breaths. This results in additional noise and/or breathing artifact in the recording. We find positioning the ground underneath the skin on one side of the top of the skull adequate and superior to positioning the ground underneath the skin near the neck muscles at the back of the head.

3.3.3 Loose Patch Recording Pipettes

Patch pipettes are pulled from standard wall borosilicate tubing using a Flaming/Brown type micropipette puller. While pulling parameters may differ with puller as well as ambient temperature and humidity, pipettes should be 15–25 MΩ resistance and have a relatively long taper. Depending on the depth of the target region, you will want the first few millimeter of the tip to flare minimally to prevent damage superficial to the recording site. We find that heating the glass capillary over three steps and pulling on the last step give us better resistance than pulling slowly over more steps or pulling in one step.

Using a 10 μl Hamilton Syringe, we fill loose patch pipettes with 1.5 % Neurobiotin in loose patch solution (3–4 μl). In some cases, Neurobiotin can spill out into the brain area of interest and label more cells than intended. It may then be necessary to backfill the pipette tip with label-free loose patch solution (~0.5 μl) and then fill the rest of the pipette with Neurobiotin patch solution. Patch solution should be aliquoted and kept frozen until ready for daily use.

3.3.4 Cell Search

Here we describe procedures similar to those pioneered by Pinault [5]. In a blind loose patch recording, you can determine whether you have encountered a cell by observing changes in voltage while applying fixed current (current clamp mode). Therefore, you should begin by using the computer or timing device such as a Master-8 (AMPI) to command repetitive negative current pulses from the amplifier. We use -200 pA square current pulses of 0.2 s duration delivered at ~2.5 Hz.

To begin searching for cells, first load the pipette onto the pipette holder and apply positive pressure to the pipette through the tubing attached to a 10 ml syringe. Generally, depressing the syringe plunger about 2–3 ml provides enough pressure to keep the pipette tip clear of debris as the pipette is advanced in the tissue. The positive pressure can be maintained by closing a three-way valve attached to the syringe on one end and the tubing on the other. While applying positive pressure, lower the pipette until it touches the surface of the brain. This can be observed through the microscope, or it may also be heard if the voltage output of the amplifier is connected to an audio monitor. Once the pipette touches the surface, zero the fine manipulator reading and the amplifier potential, and then check the resistance of the patch pipette. Many amplifiers will have a test button for checking the resistance, but it can also be calculated using Ohm's law ($V = IR$). While applying 200 pA negative current steps, the voltage drop on the oscilloscope or recording software should be 3–5 mV for a 15–25 MΩ resistance.

You will be looking for abrupt transient increases in the size of this voltage step that result from increased resistance when approaching a cell. Therefore, you can make them easier to detect by compensating for most of the pipette resistance. Advance the pipette quickly (25–50 μm per s) to ~100 μm above the region of interest. While lowering the pipette, one should periodically "buzz" the capacitance to keep the pipette free of debris. Once you reach the area of interest, you should advance the pipette more slowly (2–3 μm per s) while checking for increases in voltage indicating the presence of a cell.

With experience, it will become evident that contact with a cell is marked by an abrupt and dramatic increase in the size of the voltage response to the current pulses. Slower drifting changes are almost always due to clogging of the pipette with debris. If such an abrupt increase is observed and maintained over several 2–3 μm steps, immediately vent away the positive pressure. Frequently, at this point, the pipette will increase further the tightness of its seal to the cell by two or more fold. Exactly how far you should advance the pipette before removing positive pressure, and whether you should advance pipette further after that, depends on the type of cell you are trying to record. You should experiment with different approaches. You may also find that applying a small amount of negative pressure accelerates the sealing process, but this is often not necessary. During good recordings, the series resistance measured through the pipette is usually one hundred to several hundred MOhm, but there is no specific target as long as there is adequate signal-to-noise ratio to allow good single unit isolation. Figure 3 depicts some raw traces obtained during the sealing process of a recording from a granule cell in the mouse main olfactory bulb.

3.3.5 Labeling Cells

Cells are filled with Neurobiotin using steps of positive current injection (+700–1,000 pA; 500 ms square pulse, 1 Hz). You should continue applying current injections for at least 20 min although, depending on the cell of interest, more or less time may be required. The cell may continue to spike during dye labeling, an indication that the seal is still present; however, loss of spiking is not always an indication that one has lost the seal. Instead, cell

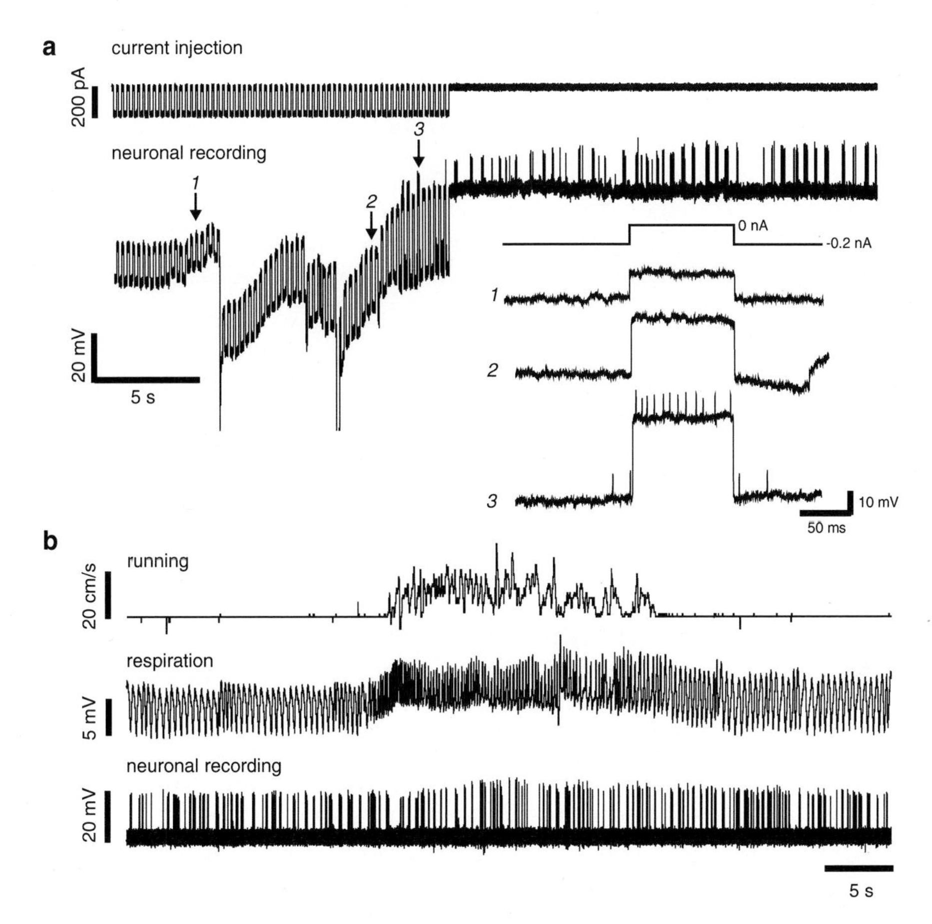

Fig. 3 Example data gathered during isolation and recording of a mouse main olfactory bulb granule cell in an awake, running mouse. (**a**) Raw physiology traces of current (*top*) and voltage (*bottom*) during acquisition and isolation of a granule cell. For the first half of this recording, 200 pA negative current pulses were used to monitor pipette resistance while searching for a cell. The numbered traces show the voltage response at the labeled time points. Note the increase in the voltage deflection. Near the halfway point, the pulses are turned off, and an isolated single unit emerges. (**b**) In our setup, unit recordings are stable during free running. The *top trace* shows the signal output of our wheel velocity sensor, reflecting a 20–25 s episode of running. The *middle trace* depicts the signal from a nasally implanted thermistor to sense respiration. The *bottom trace* is the voltage trace from the same neuron depicted in (**a**)

resistance can be checked periodically throughout the fill, and in cases where the seal is lost, filling should be terminated. The animal can then be anesthetized, and the skin over the recording area can be sutured, or the animal can be perfused with PBS followed by 4 % paraformaldehyde. Tissue is then sectioned and treated with a streptavidin-conjugated fluorophore to identify labeled cells as described [2].

4 Notes

Helpful Tips for Stable Head Caps

1. The skull surface should be kept as dry as possible throughout the surgery and especially before the application of dental cement. Both careful drilling of burr holes so as to minimize bleeding as well as drying the skull with pressurized air assist in this goal.
2. Screws should not be inserted so deep that the head of the screw is flush with the skull. In such a case, screws may spin in the burr holes and fail to provide a secure hold for the dental cement. Further, deep screws have the potential to damage the underlying brain.
3. Dental cement should extend adjacent to the recording area of interest. Securing the head bar and dental cement as close as possible to the recording site better ensures the target region will have little movement during recording.
4. Dental cement should not stick to nearby skin or muscle tissue. These surfaces are generally not stable or dry enough for the cement adherence. Further, glue on the skin is irritating to the animal.
5. Apply a final layer of superglue on the head cap to harden the cap and provide additional stability.

Helpful Tips for Loose Patch Recording

1. While recordings are generally attainable in mice of any age, young adult mice (5–7 weeks) are best.
2. Habituating the animal to the recording apparatus for at least 2 days (1 h each day) ensures the animal is calm and minimizes the loss of seals due to excessive running or struggling by the animal. Where the experiment allows, it is also helpful to water deprive the animal 24 h prior to the recording and provide water to the animal on the ball through a lick port.
3. Brain pulsation disrupts the stability of recording over long periods. Pulsation can arise with a messy or bloody craniotomy for several hours into an experiment. Covering the craniotomy with saline-soaked dental foam for several minutes may help, but we find it best to abandon the craniotomy and make a new one in the other hemisphere.
4. On occasion, one may encounter breathing artifact in the voltage recording. This can often be remedied by repositioning the ground wire.
5. Upon finding a cell, one can often advanced the electrode several more steps (10 μm) in order to improve the quality of the recording and achieve better seals.

Acknowledgments

The authors wish to thank R. Eifert for custom machining and J. Sanders for technical advice on design and construction of the velocity sensor.

Appendix: Arduino Code

```
#include<digitalWriteFast.h>  //include this library available from  //https://code.google.com/p/digitalwritefast/
   // Define Arduino inputs
   #define c_EncoderInterrupt 0
   #define c_EncoderPinA 2
   #define c_EncoderPinB 4
   //Define Arduino serial outputs
   #define LOAD 7
   #define CLOCK 8
   #define DATA 9
   #define LDAC 10
   //Define variables
   volatile bool _EncoderBSet;
   volatile long _EncoderTicks=0;
   volatile long tmpdata=0;
   volatile long velocity=0;
   //set clock timing
   #define HALF_CLOCK_PERIOD 10
   //initialization routine
   void setup()
    {
   //Assign functions to inputs
   pinMode(c_EncoderPinA, INPUT);
   digitalWrite(c_EncoderPinA, LOW);
   pinMode(c_EncoderPinB, INPUT);
   digitalWrite(c_EncoderPinB, LOW);
   attachInterrupt(c_EncoderInterrupt,
HandleMotorInterruptA, RISING);
   //Assign functions to outputs
   pinMode(DATA, OUTPUT);
   pinMode(CLOCK,OUTPUT);
   pinMode(LOAD,OUTPUT);
   pinMode(LDAC,OUTPUT);
   //initialize serial outputs
   digitalWriteFast(DATA,LOW);
   digitalWriteFast(CLOCK,HIGH);
   digitalWriteFast(LOAD,HIGH);
```

```
digitalWriteFast(LDAC,HIGH);
Serial.begin (9600);
}
//Running loop function
void loop()
{
tmpdata=_EncoderTicks;
delay(40);//bin size in ms. if changed you
need to change denominator in //next line
velocity = ((((_EncoderTicks-
tmpdata))/0.04)*0.047)+128; //0.047 scal-
ing //term assumes a 6" diameter wheel
writeValue(velocity);
Serial.print(velocity-128);
Serial.print("\n");
}
// Interrupt function triggered by a click on
encoder OutA
void HandleMotorInterruptA()
{
// Reading the state of encoder OutB deter-
mines the direction of rotation
if (digitalReadFast(c_EncoderPinB) == LOW) {
_EncoderTicks=_EncoderTicks-1;
} else {
_EncoderTicks=_EncoderTicks+1;
}
}
//write value to serial connection to MAX500
void writeValue(uint8_t value)
{
//start of sequence
digitalWriteFast(DATA,LOW);
delayMicroseconds(HALF_CLOCK_PERIOD);
digitalWriteFast(CLOCK,LOW);
delayMicroseconds(HALF_CLOCK_PERIOD);
digitalWriteFast(CLOCK,HIGH);
digitalWriteFast(DATA,LOW);
delayMicroseconds(HALF_CLOCK_PERIOD);
digitalWriteFast(CLOCK,LOW);
delayMicroseconds(HALF_CLOCK_PERIOD);
digitalWriteFast(CLOCK,HIGH);
//send the 8 bit sample data
for(int i=7;i>=0;i--){
digitalWriteFast(DATA,((value&(1<<i)))>>i);
delayMicroseconds(HALF_CLOCK_PERIOD);
digitalWriteFast(CLOCK,LOW);
delayMicroseconds(HALF_CLOCK_PERIOD);
```

```
digitalWriteFast(CLOCK,HIGH);
 }
//latch enable, DAC output is set
digitalWriteFast(DATA,LOW);
delayMicroseconds(HALF_CLOCK_PERIOD);
digitalWriteFast(LOAD,LOW);
delayMicroseconds(HALF_CLOCK_PERIOD);
delayMicroseconds(HALF_CLOCK_PERIOD);
digitalWriteFast(LOAD,HIGH);
delayMicroseconds(HALF_CLOCK_PERIOD);
digitalWriteFast(LDAC,LOW);
delayMicroseconds(HALF_CLOCK_PERIOD);
delayMicroseconds(HALF_CLOCK_PERIOD);
digitalWriteFast(LDAC,HIGH);
}
```

References

1. Bock DD, Lee WC, Kerlin AM, Andermann ML, Hood G, Wetzel AW, Yurgenson S, Soucy ER, Kim HS, Reid RC (2011) Network anatomy and in vivo physiology of visual cortical neurons. Nature 471:177–182
2. Cazakoff BN, Lau BY, Crump KL, Demmer HS, Shea SD (2014) Broadly tuned and respiration-independent inhibition in the olfactory bulb of awake mice. Nat Neurosci 17:569–576
3. Dombeck DA, Khabbaz AN, Collman F, Adelman TL, Tank DW (2007) Imaging large-scale neural activity with cellular resolution in awake, mobile mice. Neuron 56:43–57
4. Petersen CC (2009) Genetic manipulation, whole-cell recordings and functional imaging of the sensorimotor cortex of behaving mice. Acta Physiol (Oxf) 195:91–99
5. Pinault D (1996) A novel single-cell staining procedure performed in vivo under electrophysiological control: morpho-functional features of juxtacellularly labeled thalamic cells and other central neurons with biocytin or Neurobiotin. J Neurosci Methods 65:113–136

Chapter 5

Single-Cell Electroporation for In Vivo Imaging of Neuronal Morphology and Growth Dynamics

Sharmin Hossain, Kaspar Podgorski, and Kurt Haas

Abstract

Single-cell electroporation (SCE) is a technique for acutely transfecting or dye-labeling individual neurons within intact living tissues. In addition to fluorescently labeling neurons, SCE can be used to conduct cell-autonomous studies of protein function by co-delivering fluorophores with DNA, RNA, antisense constructs, peptides, proteins, or drugs. SCE involves inserting a thin glass pipette into neural tissue to restrict an electric field and exposure to a solution of delivery compounds to an individual neuron at the pipette tip. Application of a brief electric pulse induces transient pores in the target cell and iontophoretic transfer of delivery compounds only to that cell. SCE is not limited to specific cell types and leaves no residual delivery agents. SCE has proven to be useful for in vivo fluorescent imaging of neuronal morphology and connectivity and for conducting time-lapse imaging of structural changes due to growth and plasticity. Furthermore, "targeted SCE" allows selecting neurons based on connectivity, protein expression, activity patterns, or receptive field properties. Overall, SCE offers a relatively simple and highly versatile alternative to transgenic approaches for acutely labeling or transfecting post-differentiated neurons.

Key words Single-cell electroporation, Electroporation, In vivo imaging, Morphology, Dendritogenesis, Time lapse, Dendrite, Transfection, Dynamic morphometrics, Two-photon microscopy

1 Introduction

The primary goal of neuronal tracing is to accurately characterize neuronal structure in order to better elucidate underlying circuit function. To this end, large-scale connectomics projects have been undertaken to precisely map network structures, typically using high-volume rendering of electron microscopy serial sections of neural tissue [1, 2]. This approach will provide invaluable data on precise circuit morphology and connectivity, in the form of a precise 3D snapshot of neural circuits at one point frozen in time. However, much of neural circuit function occurs in the missing 4th dimension. Neuronal signals, synaptic strength, and neuronal morphology all vary dynamically in time and these changes are intrinsically tied to circuit function. Dynamic, time-varying changes are most pronounced

Benjamin R. Arenkiel (ed.), *Neural Tracing Methods: Tracing Neurons and Their Connections*, Neuromethods, vol. 92, DOI 10.1007/978-1-4939-1963-5_5, © Springer Science+Business Media New York 2015

during early brain circuit development, but continue in mature neural circuits as they respond to experience [3–7]. Capturing and analyzing time-varying neural circuit changes, or *dynamic morphometrics*, have proven exceptionally powerful for understanding how neural circuits are constructed and continue to change in response to external stimuli [6, 8–10]. High-resolution 3D imaging of neuronal morphology and its changes over time (4D) within the intact and unanesthetized brain offers our best opportunity to understand the relationship between neural circuit structure and function.

Imaging neuronal morphology within the living brain requires sparse fluorescent labeling to distinguish one cell from the surrounding tissue. Single-neuron labeling is optimal, since limitations in resolution of fluorescently labeled structures make the task of separating the intertwined axonal or dendritic arbors of multiple cells extremely challenging. Therefore, a number of strategies have been developed to label individual neurons or to sparsely label neurons throughout a tissue. These methods, including transgenics, virus, biolistics, and lipofection, each have their own advantages and disadvantages in the number and type of cells labeled, ease of application, potential damage, and versatility [11]. *Single-cell electroporation* (*SCE*) is another strategy for labeling cells within intact tissues with attractive properties (Fig. 1) [12–14]. SCE employs a fine glass pipette inserted into intact tissue with the tip touching the target cell to restrict an electroporative field and delivery compounds to that cell (see Fig. 1a, b). SCE can be used to fluorescently label neurons by transfection of plasmid DNA for expression of protein fluorophores (Fig. 1c) or by loading with dyes (Fig. 1d). The setup and application of SCE is relatively inexpensive, fast, and straightforward. While the brief stimulation (10–300 ms) required for SCE induces transient pores in the plasma membrane of target neurons, cells rapidly recover leading to no detectable lasting deficits and no residual delivery compounds. Moreover, SCE is highly versatile in its lack of cell-type specificity and in the ability to co-deliver multiple distinct plasmids encoding different proteins and of combination of dyes and peptides, RNA constructs, or drugs.

Since its first description in 2001 [12], SCE has proven to be a powerful tool for labeling individual neurons for high-resolution imaging of morphology in vivo (Fig. 1c, d). The most profound results from application of SCE, however, have been its use in the rapid time-lapse imaging of neuronal structural changes over time (Fig. 2) [6, 8–10, 15–21]. Such imaging reveals that neurons within the intact and awake developing brain exhibit rapid changes in axonal and dendritic arbor morphology during growth and in response to experience [8, 9, 15, 21, 22]. A critical further advance has been the development of computer software to identify, track, and measure all processes in 4D. Imaging and analyses of 4D data sets are collectively called *dynamic morphometrics*, which provides a comprehensive quantification of morphological changes required

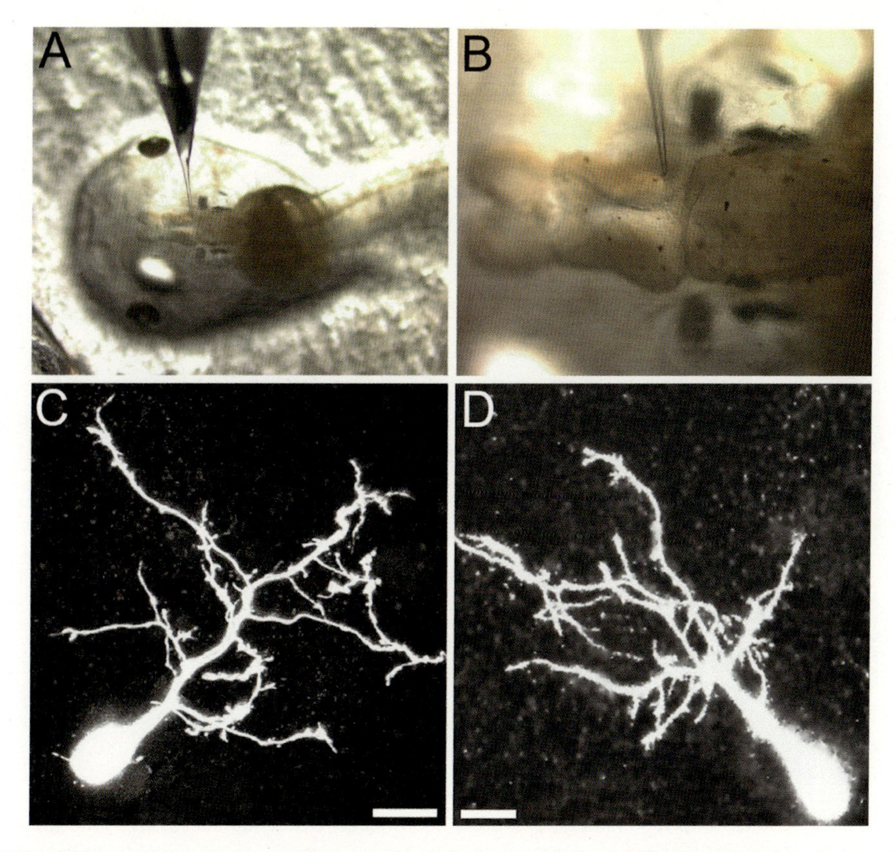

Fig. 1 SCE of DNA and dye for in vivo neuronal imaging in transparent *Xenopus laevis* tadpoles. Accurate in vivo imaging of entire dendritic morphology of brain neurons requires methods to label individual cells. (**a**) Anesthetized tadpole on moistened tissue with SCE pipette inserted into the brain. (**b**) Magnification of the tadpole brain showing the SCE pipette inserted into the right optic tectum. (**c**) Individual tectal neuron expressing GFP 24 h following SCE of plasmid DNA imaged in vivo using two-photon microscopy. Measure bar = 20 μm. (**d**) In vivo two-photon image of tadpole tectal neuron 1 h following SCE-mediated loading of the fluorescent dye Alexa Fluor 488 dextran. Measure bar = 20 μm

for realistic modeling of growth and a thorough understanding of how changes at short timescales contribute to larger changes in morphological patterning over longer periods [6, 8–10]. Such analyses provide an important dimension to the understanding of neural circuit behavior and function absent from connectomics approaches using fixed tissue. Here, we discuss the basics of SCE theory and implementation and its applications for neuronal tracing, dynamic imaging, targeted labeling of specific neuronal populations, as well as advanced uses for protein localization, biosensors, and cell-autonomous studies of protein function.

1.1 Background on Electroporation

Electroporation has been used for many decades for the transfer of DNA and other molecules into dissociated cells in solution [23]. More recently, electroporation has been applied to the bulk transfection of large numbers of cells within intact tissues in vivo [24, 25].

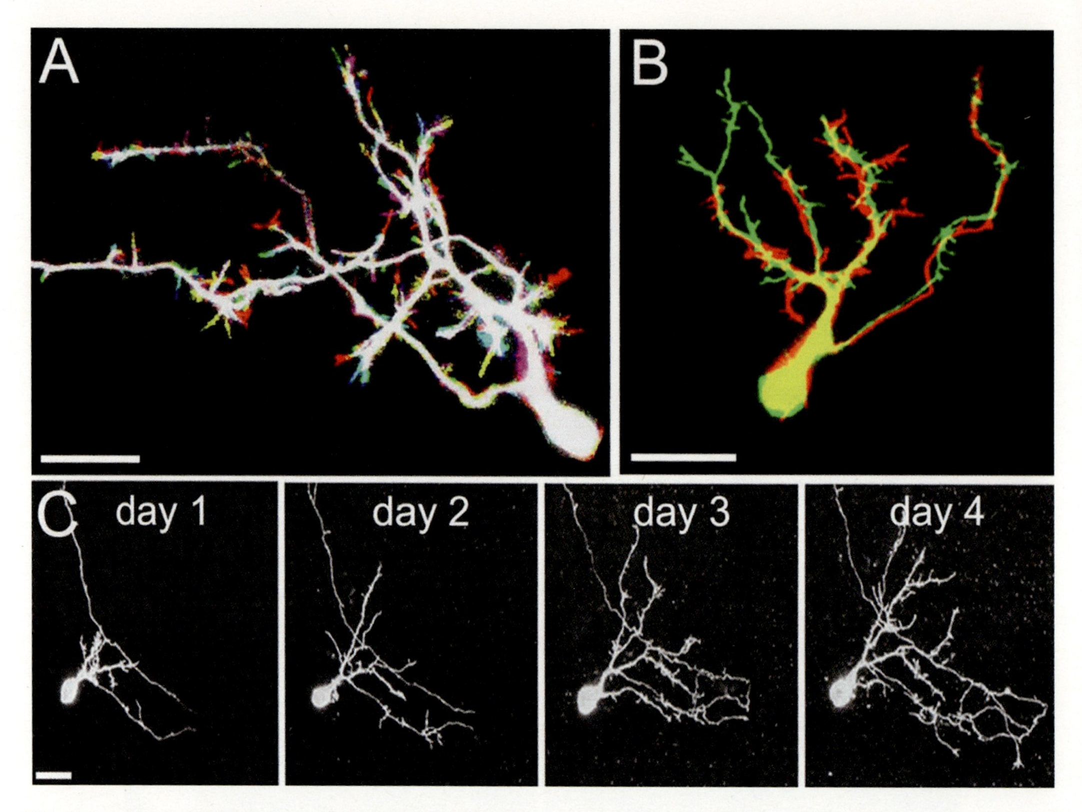

Fig. 2 In vivo two-photon time-lapse imaging of brain neuronal growth dynamics. In vivo time-lapse imaging reveals the dynamic morphological plasticity of growing brain neurons. (**a**) Individual tadpole tectal neuron expressing GFP via SCE imaged in vivo at 5-min intervals. Six successive images are overlaid, with each time point a different color and overlap = *white*. Colored processes indicate turnover. (**b**) Overlay of two images of a GFP-expressing tectal neuron at a 3-h interval. *Red* = initial image, *green* = 3 h later, *yellow* = overlap. (**c**) The same tectal neuron labeled by SCE-mediated transfection imaged at 24-h intervals. All measure bars = 20 μm

These techniques all employ exposure of target cells to a solution of the delivery compound and application of an external electric field, typically with large anode and cathode plate electrodes. There are two sequential components of electroporation: pore formation induced by an applied electric field and the iontophoretic transfer of charged compounds through these pores. When an external electric field is applied, charged molecules within the solution and inside cells are attracted to the oppositely charged electrode. Intracellular charged ions are trapped by the plasma membrane and build up at sites adjacent to each electrode, creating local increases in membrane potential. When the membrane potential reaches a threshold, estimated to be around 0.5 V, the electrostatic forces maintaining the lipid bilayer break down and the lipids reconfigure to form pores [23, 26–29]. The size of pores is influenced by the strength of the applied field [26]. Once formed, however, the lifetime of pores is independent of the applied field. If pores are excessively large, they will continue to grow and lyse the cell.

Smaller pores will shrink and collapse over a period of hundreds of milliseconds [27]. While the electric field is applied, charged molecules in the external solution can pass through open pores to enter cells. When the field is terminated and the pores collapse, these compounds are trapped within the cells. Electric field parameters are optimized for high-efficiency compound delivery with minimal cell lyses and rapid recovery. Electroporation induces transient osmotic shock by temporary pore formation, producing swelling from which cells typically recover within minutes without lasting detectable effects.

SCE makes use of the same phenomena, but restricts the electric field and delivery compounds to individual cells with a fine-tipped glass pipette [12, 30]. The pipette is filled with a solution of the delivery compounds and inserted into the preparation with the tip directly adjacent to or touching the target cell (a high resistant seal is not required). A brief electrical pulse is delivered between an electrode within the pipette and an external ground, which evokes an electric field with peak strength in close proximity to the pipette tip. This field induces pores in the plasma membrane of the target cell. Simultaneously, charged delivery compounds within the pipette solution move from the pipette through the pores and into the target cell as an electrical current. Following termination of the brief electrical stimulation, the pores reseal, the pipette is removed, and the delivery compounds have been delivered only to the target cell. The same pipette can then be reused to electroporate other cells.

SCE is a relatively simple technique requiring inexpensive and common laboratory equipment. An electrode puller is needed to produce sharp-tipped pipettes with shallow shanks, similar to pipettes used for patch clamp electrophysiology. SCE success is sensitive to the tip diameter, since small tips tend to clog or break and larger tips fail to focus sufficient field strength to generate pore formation. We find that optimal pipettes have tip diameters between 0.6 and 1 μm and relatively shallow angle shanks similar to patch pipettes to avoid clogging and electrostatic interactions between the pipette walls at the tip. Pipette glass with an inner filament should be used to allow easy backfilling of solution and tip filling via capillary action. A simple square-pulse electrical stimulator able to generate 10–100 V and 1–2 μA is sufficient. We find that trains of short duration square pulses are effective, such as those produced by common Grass Technologies SD9 stimulators or the Axoporator 800a from Molecular Devices. An oscilloscope is useful to monitor pipette tip resistance. Pipette holders, manipulators, and microscopes are also required. A silver wire is inserted into the glass pipettes and a second ground wire is placed anywhere in electric contact with the preparation. These wires are connected to the two poles of the stimulator. The tissue or organism to be labeled is placed on the stage of a microscope and the pipette held in the pipette holder mounted on a manipulator. The pipette is then inserted into the tissue with the tip touching, but not puncturing the target cell. Following brief stimulation, the pipette can be reused for additional sites within the same tissue or another preparation.

1.2 SCE-Mediated Transfection

SCE-mediated delivery of DNA for neuronal transfection is a powerful method to fluorescently label cells and to regulate protein function [9, 10, 21]. SCE of expression plasmids encoding protein fluorophores produces brightly labeled cells for in vivo imaging and morphological reconstructions (Fig. 1c). Since plasmids are stable within neurons, the continuous fluorophore production allows replenishing signal lost to fluorescent imaging-induced bleaching. Furthermore, combinations of different plasmids can be co-electroporated with near 100 % efficiency to simultaneously express multiple exogenous proteins (Fig. 3a, b) [12]. This allows expression of a space-filling fluorophore for morphometrics along with additional fluorescently tagged proteins targeting

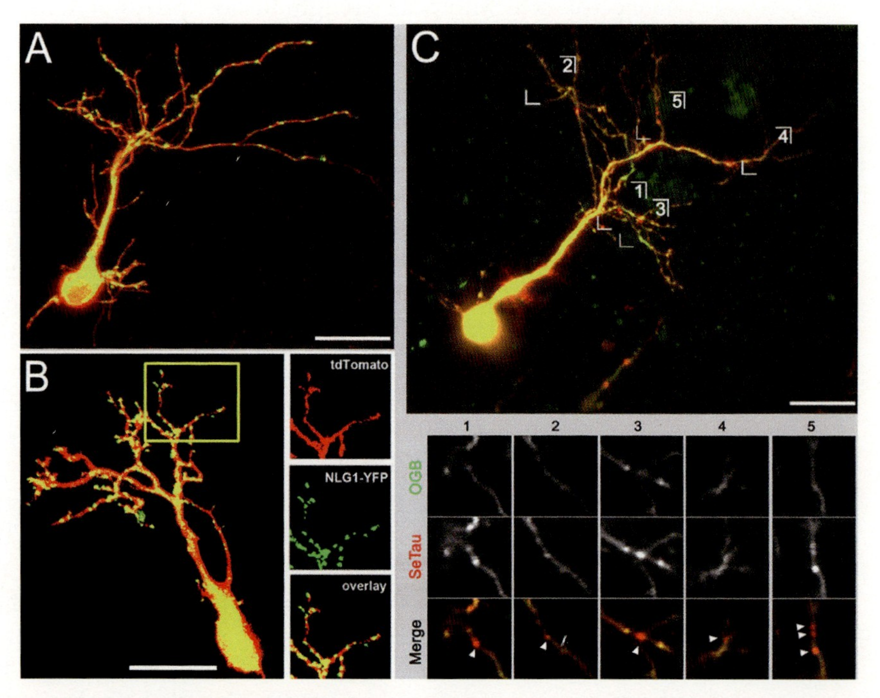

Fig. 3 SCE-mediated co-delivery of multiple compounds for protein localization. SCE is highly versatile in the types and combinations of molecules that can be simultaneously delivered to cells. (**a**) In vivo imaging of tectal neuron co-electroporated with plasmids encoding a red space-filling protein fluorophore and the postsynaptic protein PSD-95 fused to GFP. (**b**) SCE-mediated co-electroporation of plasmids encoding the red protein fluorophore tdTomato and the cell adhesion protein neuroligin 1 (NLG1) fused to YFP is used for protein localization studies. (**c**) SCE-mediated co-delivery of small molecules is used to label synapses. Synapses are labeled using a peptide. The space-filling green dye Oregon Green BAPTA (OGB) is delivered along with a peptide that binds to endogenous PSD-95. The PSD-95-binding peptide is labeled with the red squaraine rotaxane dye SeTau. *Bottom panel* shows enlargement of corresponding regions marked above, separating the *green* (OGB) and *red* (SeTau) channels to demonstrate punctate synaptic labeling. All measure bars = 20 μm

subcellular domains or constructs to enhance or interfere with specific protein function. SCE-mediated transfection for altered protein function has the advantage over transgenic approaches in that SCE is typically applied to post-differentiated neurons, thereby limiting compensatory effects.

Electroporation-mediated transfection is a multistep process involving DNA-membrane interactions, DNA delivery to the nucleus, transcription, and translation [31]. The efficiency of SCE transfection varies between preparations, potentially due to differences in the ability to directly deliver DNA to the nucleus in different cell types. Different electrical stimulation parameters are required for SCE-mediated delivery of different delivery compounds. Likely due to the relatively large size of plasmid DNA molecules, long trains are required, such as 300 ms trains of 200 μs—1 ms square pulses. A challenge with SCE of DNA (typically used at a concentration of 2 μg/μl) is that there is no immediate indicator for successful cell or nuclear loading, since one must wait at least 12 h for protein fluorophore expression. Therefore, it is time-consuming to troubleshoot parameters if no expression is seen. Two approaches can minimize this problem. Since SCE failures may be due to altered tip resistance due to breaking or clogging, it is useful to monitor the tip current using an oscilloscope and replace the pipette or adjust the voltage accordingly. Moreover, since SCE-mediated dye loading typically has a much higher efficiency than DNA transfection, dye loading can be used to test for appropriate tip geometry. Tip parameters found to be successful for dye loading can then be used for DNA transfection.

1.3 Fluorescently Tagged Proteins and Cell-Autonomous Studies

SCE-mediated transfection for expression of multiple proteins has been useful for labeling subcellular structures and for cell-autonomous studies of protein function. For example, SCE-mediated co-delivery of plasmids encoding a space-filling protein fluorophore with plasmid encoding a protein fluorophore fused to the synaptic protein PSD-95 allows identification of putative synapses throughout a dendritic arbor (Fig. 3a) [9]. In experiments designed to test the roles of the cell adhesion molecule neuroligin in dendritogenesis, different dominant negative mutations lacking distinct protein-protein binding domains have been expressed (Fig. 3b) [9]. Since these experiments employed SCE to restrict exogenous gene expression to individual neurons, results could confidently be attributed to cell-autonomous effects of altered target protein function, rather than secondary effects from larger circuit changes.

1.4 SCE for Loading Dyes into Individual Neurons

SCE-mediated loading of neurons with fluorescent dyes offers a much simpler methods to label cells and provides immediate feedback when optimizing pipette tip parameters (Fig. 1d) [10, 12]. Unlike DNA, dye loading only requires delivery across the plasma membrane. Therefore, SCE dye loading is very effective for all cell types, including mammalian neurons. Since dye molecules are smaller than plasmid DNA, shorter duration electroporative stimuli can be used: single pulses or trains of pulses lasting 15–30 ms. Typically, high concentrations of dye (100 μM to 1 mM) are used in order to rapidly transfer a bolus of dye to a single neuron. Longer duration stimuli often cause loading of neighboring cells. When using dyes for SCE, one must consider their charge and distribution throughout neurons. Dextran conjugates are used to aid distribution throughout axonal and dendritic processes and avoid sequestration in intracellular vacuoles. It is important to empirically test for the correct polarity of pulses to apply. Fluorescent dyes conjugated to dextrans can vary in their charge, and in some cases, the effective polarity of a dye changes with repeated SCE stimulation, potentially due to exchange of ions with the external solution. This is easily tested by directly observing fluorophore movement during SCE stimuli using fluorescence microscopy. One must take care, however, to limit fluorescent illumination of labeled cells since dyes rapidly bleach and their fluorescence is coupled with formation of free radicals, which cause phototoxicity. It is recommended that direct observation of SCE loading using fluorescence illumination be used to optimize stimulation parameters (polarity, amplitude, and duration) for each pipette tip on a sample preparation. Then, the same parameters can be used without fluorescent illumination to load cells to be used for experimentation employing less damaging confocal or two-photon imaging. Even when the ultimate goal is to deliver DNA, we recommend optimizing pipette tip geometries using dyes. While it would seem ideal to combine DNA with dye to overcome this problem, this is usually not possible because the differences in sizes of these compounds result in dye labeling of multiple cells before DNA is delivered to one.

1.5 Dyes Versus Protein Fluorophores

When considering using SCE for imaging neuronal morphology, one should balance the advantages and disadvantages of loading dyes versus DNA. DNA transfection has the strengths of continuous production of protein fluorophores to counter bleaching and expression of fluorescently tagged proteins and constructs to alter protein function. However, SCE-mediated DNA transfection efficiency is relatively low and highly dependent on the cells targeted. While SCE of DNA yields high success rates in *Xenopus* tadpoles and zebrafish embryos, rates are low for mature mammalian pyramidal neurons. These differences may be due to the requirement to directly deliver DNA to the nucleus by SCE, which is easier in immature fish and frog neurons with small somata and relatively large nuclei. SCE-mediated dye loading, in contrast, is highly efficient in all cells since delivery compounds need only traverse the plasma membrane. Although bleaching can critically interfere with long-term imaging of dye-filled cells since dye is not replenished, we have found that photostable dyes can be used for long duration rapid time-lapse imaging. We find that dextran conjugates of Alexa Fluor dyes can completely and brightly fill neurons for days to weeks and are amenable to 5-min imaging for 1–2 or 24-h interval imaging over 5 days [10].

1.6 SCE of Small Compounds for Regulating Protein Function and for Protein Localization

SCE-mediated small-molecule loading can also be used for controlling protein function and expression and expression of fluorescently labeled proteins. Similar to dye loading, SCE is also useful for delivering other small molecules which only need to cross the plasma membrane, including RNA constructs, peptides, proteins, and drugs [8, 10, 32, 33]. A space-filling dye for morphometrics can be combined with mRNA for expression of exogenous protein, or RNA interference strategies for decreased target protein expression, including Morpholinos, RNAi, and shRNA [8, 10, 33, 34]. Combining space-filler dye with peptides can be used to alter protein function and interfere with protein-protein interactions. Fluorescently tagged peptides can also be used to identify intracellular localization of target endogenous proteins. If possible, all delivery compounds (Morpholinos, peptides, etc.) should be tagged with fluorophores distinguishable from the space-filling dye in order to assure delivery and to determine the duration of detectable presence within cells. Fluorescently tagged peptides can also be used to localize proteins. For example, co-electroporation of a space-filling dye along with a fluorescently tagged peptide that binds to endogenous PSD-95 can be used to label synapses (Fig. 3c) [32].

1.7 Dynamic Morphometrics: SCE for In Vivo Imaging of Neuronal Growth

SCE labeling allows for high-resolution 3D images of living neurons within their native environments, including within intact brain circuits. SCE has proven particularly effective for conducting time-lapse imaging of growth (Fig. 2). In developing brains of *Xenopus*, SCE has been used to fluorescently label neurons within the optic tectum for imaging of dendritic arbor maturation [6, 8–10, 15, 21]. Rapid, 5-min interval imaging over hours captures fast dynamic growth behavior (Fig. 2a), while longer 1–24 h interval imaging shows how rapid events culminate to produce lasting patterning of mature neuronal morphology (Fig. 2b, c). Both SCE-mediated DNA transfection and dye loading have been effective for these studies. The utility of SCE for studies of morphogenesis has been significantly aided by the development of advanced imaging technologies to capture full 3D morphologies of labeled cells and computer software to reconstruct neuronal structures for quantification [6, 8–10]. Two-photon microscopy allows imaging deep within intact tissues with minimal damage for high-resolution 3D imaging and for repeated time-lapse imaging of growth. Numerous commercial softwares and freewares are available to render 3D morphologies. However, a significant advance has been the development of a freely available Matlab-based program called *Dynamo*, for comprehensive quantification of morphological changes over time due to growth and experience-driven structural plasticity [6, 8–10]. Dynamo employs a semiautomatic alignment and the identification and measurement of all processes in 3D across all time points. Rapid time-lapse imaging that captures all morphological changes over time and Dynamo software yield accurate measures of rates of process addition and retraction, motility, and lifetime. Such values are critical for fully charactering structural plasticity in order to understand underlying molecular mechanisms. For example, rapid time-lapse imaging of growing *Xenopus* tectal neurons reveals the direct effects of altered sensory experience on dynamic growth. Visual stimuli that induce lasting potentiation of visual-evoked firing produce stabilization of dynamic dendritic processes [8]. This stabilization was detected as a decrease in process motility and increased lifetimes. By using SCE to co-express a space-filling protein fluorophore along with fluorescently tagged PSD-95, it could be determined that experience-driven dendrite stabilization is mediated through synapse formation. These studies support a model in which neural activity driven by sensory experience directs brain neuronal growth through the selective stabilization of processes connected to activated circuits.

1.8 SCE of Biosensors

Another powerful application of SCE is for delivery or expression of biosensors, like calcium-sensitive fluorescent dyes and proteins (Fig. 3c) [35, 36]. While bulk-loading sensors to all neurons within a tissue is useful for sampling action potential-associated calcium transients in cell bodies of large numbers of neurons, restricting calcium signals to individual neurons allows discerning transients localized to dendrites or synapses, which are occluded by bulk-loading strategies. Since multiple dyes can be co-electroporated if they share the same polarity, a second non-calcium sensor can be used as a reference fluorescent signal to control for concentration.

1.9 Targeted SCE

There are two major approaches to conducting SCE—"blind" or "targeted." In the blind approach, the SCE pipette is inserted into the neural tissue into a region containing cell bodies without direct visualization of tip-target cell interaction (Fig. 1a, b). When low magnification microscopy is employed such as with a stereomicroscope, the long working distance affords rapid setup and movement between stimulation sites. Post-SCE screening using higher magnification can then be used to identify optimally labeled neurons. In contrast, targeted SCE involves high magnification, low working distance imaging to directly visualize pipette tip contact with target cells. This approach benefits from increased efficiency for loading specific cell types, but suffers from the increased time

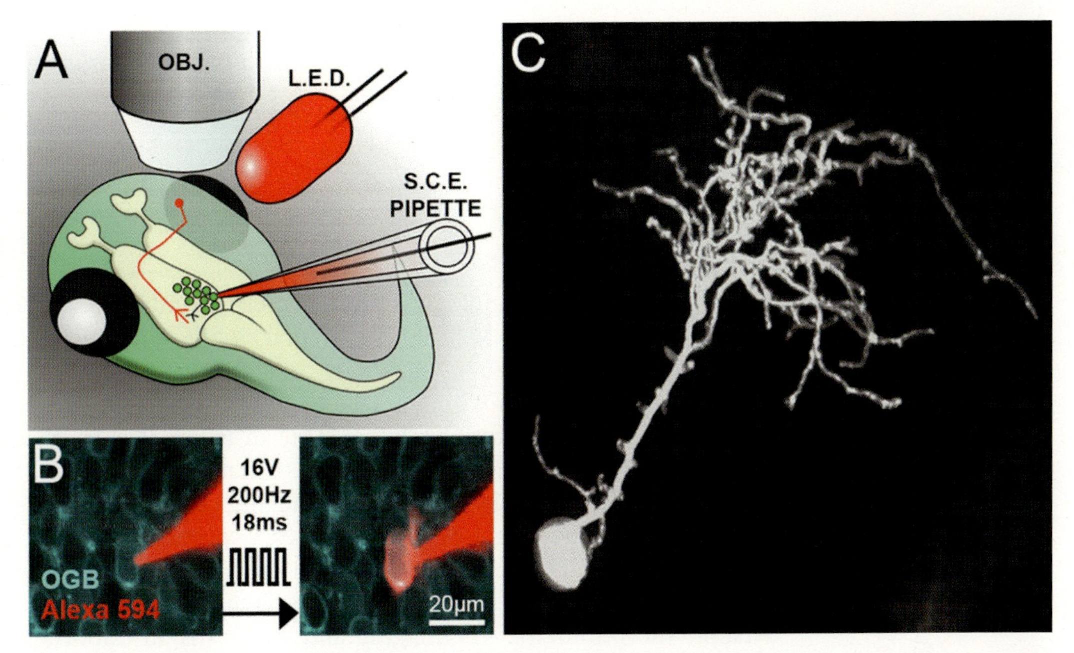

Fig. 4 Targeted SCE. SCE can be used to label neurons based on specific properties, such as their receptive field responses. (**a**) Schematic of setup to identify visually responsive brain neurons for targeted SCE. The tadpole optic tectum is first loaded with the calcium-sensitive dye OGB-1-AM in order to detect calcium transients associated with action potential firing evoked by visual stimuli from an LED light. An SCE pipette filled with the space-filling dye Alexa Fluor 594 dextran is then inserted into the brain to target visually responsive neurons. (**b**) *Left*: In vivo two-photon image of a field of tectal neurons loaded with OGB1 and the tip of the SCE pipette filled with Alexa Fluor 549 dextran. The tip is positioned so that it is touching a neuron with robust visual-evoked calcium transients. *Right*: Image directly following delivering a short train of electrical pulses to the pipette, showing SCE-mediated delivery of Alexa Fluor 549 to the target neuron. (**c**) Full in vivo two-photon 3D imaging of the target neuron dendritic arbor. Alternating between calcium imaging of visual-evoked responses and morphology allows correlation of activity and structural changes of brain neurons selected based on their response properties

needed for setup. One type of targeted SCE is "shadow SCE" in which an extracellular dye is first injected to make non-stained cell bodies stand out as shadows [37, 38]. Alternatively, all cells within a tissue can be bulk loaded with a fluorescent calcium-sensitive dye, such as OGB-1-AM (Fig. 4). In vivo imaging can then be used to select specific targets for SCE based on their morphology or action potential firing patterns reflected in calcium sensor fluorescent transients. A similar nonvisual technique involves recording electrical signals from the SCE pipette to identify target cells by the electrical firing properties [39, 40]. Other strategies target cells based on connectivity. For example, transgenics in which fluorophores are selectively expressed in distinct populations off neurons (such as GFP expression in interneurons), SCE can be used to dye-fill downstream neurons outlined by the transgenically labeled axons. Alternatively, in nontransgenic preparations upstream neurons may be first labeled by SCE. Such methods can facilitate identification interconnected neurons for paired electrophysiological recording. Another method to identify interconnected neurons is to use SCE to express genetically modified rabies virus which functions as a monosynaptic retrograde label [41].

2 Conclusions

The rapid pace of technological advancement in imaging brain structure has provided new insights to neural circuit function. SCE contributes to these ongoing efforts by providing a relatively easy means to label neurons within intact tissues for live imaging of morphology. SCE is exceptionally versatile in the molecules that can be delivered and the types of neurons that can be targeted. SCE is particularly useful for time-lapse imaging of individual neurons to capture the morphological changes associated with functional plasticity. Such continuous structural and functional plasticity are interrelated and are a fundamental to circuit function. Thus, dynamic morphometric 4D imaging and analysis is an important complement to connectomics approaches.

3 Materials

1. Pipette glass (with filament).
2. Silver wires for internal pipette electrode and ground.
3. Leads to connect silver wires to voltage stimulator.
4. Pipette holder.
5. Coarse manipulator with associated posts and stand.
6. Fluorescent stereomicroscope.
7. Horizontal pipette puller (same used for patch clamp electrophysiology pipettes).
8. Voltage stimulator capable of generating square pulses and trains of pulses at 1 μA and 10–70 V. Examples include SD9 Square Pulse Stimulator from Grass Technologies and the Axoporator 800A from Molecular Devices.
9. Delivery solution:

   For transfection: plasmid DNA at 2 μg/μl, endotoxin free, diluted in dH_2O.

   For dye loading: dye diluted in dH2O at a concentration of 100 μM to 1 mM (for Alexa Fluor 488 dextran 3000, adjust as needed for each dye).
10. Pipette loader to fill pipette tip with solution of the delivery compounds. Pipetter tips for gel loading are sufficient.

4 Methods

4.1 Experimental Method for SCE of Xenopus Laevis Tadpole Tectal Brain Neurons

The retinotectal system of *Xenopus laevis* has proven to be a useful model for the study of vertebrate brain circuit development. Brain neurons within the optic tectum are easily accessible for SCE and can be loaded with DNA or dye with high efficiency. The protocol below can be readily modified for other preparations.

4.1.1 Setup Preparation (10 min)

1. Pull single-cell electroporation glass pipettes.
   Pull SCE pipettes on a horizontal Sutter puller with a tip diameter between 0.6 and 1 μm and a tip taper angle of greater than 10°.
2. Fill pipette tip with solution of the delivery compound (dye, plasmid DNA, peptides, RNA constructs, etc.).
   Make a high concentration of the molecules to be delivered to cells: for DNA, use 2 μg/μl; for dye, use 100 μM to 1 mM. Fill pipette with solution either by backfilling, or by direct injection using a thin tube loader, such as a gel-loading tip.
3. Mount pipette on holder attached to a coarse manipulator.
   Insert pipette into holder mounted on a 3-axis coarse micromanipulator. Ensure that the silver wire electrode is inserted into micropipette in contact with the filling solution.

4.1.2 Electroporation (2 min per Tadpole)

1. Anesthetize tadpole.
   Place tadpoles in 0.02 % MS-222 solution for 5 min or until fully anesthetized.
2. Place tadpole onto petri dish under a stereomicroscope (see Fig. 1a).
   Transfer an anesthetized tadpole using a plastic transfer pipette, carefully positioning the tadpole dorsal side up using a moistened paintbrush. The tadpole can be placed on top of a moistened tissue or Kimwipe. Excess fluid can be removed so that the tadpole remains in place during insertion of the SCE pipette. A ground silver wire can be inserted under the tissue, in electrotonic contact with the tadpole via the bath salt solution.
3. Insert SCE pipette tip into tadpole brain (see Fig. 1b).
   Position the pipette tip above the target region in the tadpole brain and slowly advance the pipette until it dimples the skin. Apply brief, rapid forward force to puncture the skin and enter the superficial layers of the tadpole brain.
4. Apply SCE electrical stimulation.
   Apply brief electric stimulation using either a Grass Technologies SD9 stimulator or an Axoporator 800A. For delivery of plasmid DNA, use half-second trains of short square pulses and a current of 1 μA (e.g., 300 ms trains of 200 μs square pulses, at 500 Hz). For SCE-mediated dye

loading, short square pulses or trains of short pulses can be used (total duration ~15 ms) (see Fig. 4).

5. Retract pipette.
   Withdraw pipette off the brain, move to another appropriate site, and repeat steps 3 and 4.
6. Return tadpole to rearing solution for recovery.
7. Repeat steps 2–6 for additional tadpoles.

4.2 Screening

1. *Screening* (1 min per tadpole).
   Screen tadpoles for appropriate fluorescent labeling 15–30 min following SCE of dye and 12–24 h after SCE of DNA.
2. Anesthetize tadpoles in 0.02 % MS-222 solution for 5 min.
3. Mount anesthetized tadpoles in an imaging chamber and view with a fluorescence stereomicroscope or upright microscope.
4. Rapidly screen for successful labeling using appropriate excitation and emission filter sets.

5 Notes

There are a number of considerations or "tricks" that impact the ease and efficiency of SC. Prior to loading, briefly centrifuge solutions at high speed to reduce the amount of particulate debris, which may lead to clogging of the pipette. Use pipette glass with filaments to allow backfilling by dipping the non-tip end into the delivery solution and filling by capillary action. Only small volumes (0.5–1 μl) are required to fill the pipette tip, sufficient to contact the silver wire in pipette. Use of small volumes will limit waste of high-concentration delivery solutions. It is important to avoid bubbles in the tip since they can break electrical connectivity and limit iontophoresis. If bubbles occur, dislodge by vigorously flicking the pipette tip with your finger.

For labeling brain neurons in *Xenopus*, multiple tadpoles can be electroporated in a brief period one after the other. This is facilitated by simultaneously anesthetizing many tadpoles. Tadpoles can remain in MS-222 for 30 min. Brain neurons in the optic tectum are located directly under the skin, so care must be given to targeting superficial layers. To improve overall yield, multiple sites can be electroporated throughout the tadpole brain. The same SCE pipette can be reused for multiple sites and tadpoles, until it clogs or breaks. Tadpoles should recover from anesthesia and SCE within 5 min.

Use the correct polarity—DNA is negatively charged, but dyes and other compounds can vary. Fluorescent compounds should be tested empirically. Since DNA delivery cannot be directly monitored, it is convenient to monitor SCE current using an oscilloscope to make sure the pipette tip is not clogged or broken. Direct visualization of dye iontophoresis and cell loading via SCE with fluorescence microscopy should be used to optimize tip geometry and stimulation parameters. Using dye to optimize tips is recommended prior to SCE of DNA. However, it is important to remember that exposure of labeled cells to epifluorescent excitation should be limited due to rapid bleaching of fluorophores and phototoxicity. Thus, screening should be performed as quickly as possible.

If no neurons are labeled by SCE of plasmid DNA encoding protein fluorophores, like GFP, try SCE delivery of fluorescent dyes while directly viewing labeling. Direct feedback from dye loading will facilitate optimizing tip geometry (tip diameter and shank angle). If neurons are labeled using SCE-mediated dye loading, but they are dim, increase dye concentration, increase stimulation parameters (voltage and duration), and limit exposure to epifluorescent illumination. If neurons are dim following SCE-mediated delivery of plasmid-encoding protein fluorophores, increase DNA concentration, make sure that endotoxin-free DNA preparations are used, increase train duration, and wait longer following SCE for increased protein production.

References

1. Morgan JL, Lichtman JW (2013) Why not connectomics? Nat Methods 10:494–500
2. Helmstaedter M, Briggman KL, Turaga SC, Jain V, Seung HS, Denk W (2013) Connectomic reconstruction of the inner plexiform layer in the mouse retina. Nature 500:168–174
3. Trachtenberg JT, Chen BE, Knott GW, Feng G, Sanes JR, Welker E et al (2002) Long-term in vivo imaging of experience-dependent synaptic plasticity in adult cortex. Nature 420:788–794
4. Wu GY, Cline HT (2003) Time-lapse in vivo imaging of the morphological development of Xenopus optic tectal interneurons. J Comp Neurol 459:392–406
5. Wu GY, Zou DJ, Rajan I, Cline H (1999) Dendritic dynamics in vivo change during neuronal maturation. J Neurosci 19:4472–4483
6. Hossain S, Hewapathirane DS, Haas K (2012) Dynamic morphometrics reveals contributions of dendritic growth cones and filopodia to dendritogenesis in the intact and awake embryonic brain. Dev Neurobiol 72:615–627
7. Chen BE, Lendvai B, Nimchinsky EA, Burbach B, Fox K, Svoboda K (2000) Imaging high-resolution structure of GFP-expressing neurons in neocortex in vivo. Learn Mem 7:433–441
8. Chen SX, Cherry A, Tari PK, Podgorski K, Kwong YK, Haas K (2012) The transcription factor MEF2 directs developmental visually driven functional and structural metaplasticity. Cell 151:41–55
9. Chen SX, Tari PK, She K, Haas K (2010) Neurexin-neuroligin cell adhesion complexes contribute to synaptotropic dendritogenesis via growth stabilization mechanisms in vivo. Neuron 67:967–983
10. Liu XF, Tari PK, Haas K (2009) PKM zeta restricts dendritic arbor growth by filopodial and branch stabilization within the intact and awake developing brain. J Neurosci 29:12229–12235
11. Karra D, Dahm R (2010) Transfection techniques for neuronal cells. J Neurosci 30:6171–6177
12. Haas K, Sin WC, Javaherian A, Li Z, Cline HT (2001) Single-cell electroporation for gene transfer in vivo. Neuron 29:583–591
13. Hewapathirane DS, Haas K (2008) Single cell electroporation in vivo within the intact developing brain. Journal of Visualized Experiments 17:705
14. Liu XF, Haas K (2011) Single-cell electroporation in Xenopus. Cold Spring Harb Protoc 9:pii:pdb.top065607
15. Sin WC, Haas K, Ruthazer ES, Cline HT (2002) Dendrite growth increased by visual activity requires NMDA receptor and Rho GTPases. Nature 419:475–480
16. Sorensen SA, Rubel EW (2006) The level and integrity of synaptic input regulates dendrite structure. J Neurosci 26:1539–1550

17. Bestman JE, Cline HT (2009) The relationship between dendritic branch dynamics and CPEB-labeled RNP granules captured in vivo. Front Neural Circuit 3:10
18. Shen W, Da Silva JS, He H, Cline HT (2009) Type A GABA-receptor-dependent synaptic transmission sculpts dendritic arbor structure in Xenopus tadpoles in vivo. J Neurosci 29:5032–5043
19. Ewald RC, Van Keuren-Jensen KR, Aizenman CD, Cline HT (2008) Roles of NR2A and NR2B in the development of dendritic arbor morphology in vivo. J Neurosci 28:850–861
20. Sorensen SA, Rubel EW (2011) Relative input strength rapidly regulates dendritic structure of chick auditory brainstem neurons. J Comp Neurol 519:2838–2851
21. Haas K, Li J, Cline HT (2006) AMPA receptors regulate experience-dependent dendritic arbor growth in vivo. Proc Natl Acad Sci U S A 103:12127–12131
22. Poulain FE, Gaynes JA, Stacher Horndli C, Law MY, Chien CB (2010) Analyzing retinal axon guidance in zebrafish. Methods Cell Biol 100:3–26
23. Ho SY, Mittal GS (1996) Electroporation of cell membranes: a review. Crit Rev Biotechnol 16:349–362
24. Tabata H, Nakajima K (2001) Efficient in utero gene transfer system to the developing mouse brain using electroporation: visualization of neuronal migration in the developing cortex. Neuroscience 103:865–872
25. Nakamura H, Funahashi J (2001) Introduction of DNA into chick embryos by in ovo electroporation. Methods 24:43–48
26. Bilska AO, DeBruin KA, Krassowska W (2000) Theoretical modeling of the effects of shock duration, frequency, and strength on the degree of electroporation. Bioelectrochemistry 51:133–143
27. DeBruin KA, Krassowska W (1999) Modeling electroporation in a single cell. I. Effects of field strength and rest potential. Biophys J 77:1213–1224
28. Freeman SA, Wang MA, Weaver JC (1994) Theory of electroporation of planar bilayer membranes: predictions of the aqueous area, change in capacitance, and pore-pore separation. Biophys J 67:42–56
29. Neumann E, Kakorin S, Toensing K (1999) Fundamentals of electroporative delivery of drugs and genes. Bioelectrochem Bioenerg 48:3–16
30. Haas K, Jensen K, Sin WC, Foa L, Cline HT (2002) Targeted electroporation in Xenopus tadpoles in vivo – from single cells to the entire brain. Differ Res Biol Divers 70:148–154
31. Neumann E, Kakorin S, Tsoneva I, Nikolova B, Tomov T (1996) Calcium-mediated DNA adsorption to yeast cells and kinetics of cell transformation by electroporation. Biophys J 71:868–877
32. Podgorski K, Terpetschnig E, Klochko OP, Obukhova OM, Haas K (2012) Ultra-bright and -stable red and near-infrared squaraine fluorophores for in vivo two-photon imaging. PLoS One 7:e51980
33. Boudes M, Pieraut S, Valmier J, Carroll P, Scamps F (2008) Single-cell electroporation of adult sensory neurons for gene screening with RNA interference mechanism. J Neurosci Methods 170:204–211
34. Tanaka M, Yanagawa Y, Hirashima N (2009) Transfer of small interfering RNA by single-cell electroporation in cerebellar cell cultures. J Neurosci Methods 178:80–86
35. Nevian T, Helmchen F (2007) Calcium indicator loading of neurons using single-cell electroporation. Pflug Arch Eur J Physiol 454: 675–688
36. Kassing V, Engelmann J, Kurtz R (2013) Monitoring of single-cell responses in the optic tectum of adult zebrafish with dextran-coupled calcium dyes delivered via local electroporation. PLoS One 8:e62846
37. Judkewitz B, Rizzi M, Kitamura K, Hausser M (2009) Targeted single-cell electroporation of mammalian neurons in vivo. Nat Protoc 4: 862–869
38. Kitamura K, Judkewitz B, Kano M, Denk W, Hausser M (2008) Targeted patch-clamp recordings and single-cell electroporation of unlabeled neurons in vivo. Nat Methods 5: 61–67
39. Graham LJ, Del Abajo R, Gener T, Fernandez E (2007) A method of combined single-cell electrophysiology and electroporation. J Neurosci Methods 160:69–74
40. Rathenberg J, Nevian T, Witzemann V (2003) High-efficiency transfection of individual neurons using modified electrophysiology techniques. J Neurosci Methods 126:91–98
41. Marshel JH, Mori T, Nielsen KJ, Callaway EM (2010) Targeting single neuronal networks for gene expression and cell labeling in vivo. Neuron 67:562–574

Chapter 6

Practical Methods for In Vivo Cortical Physiology with 2-Photon Microscopy and Bulk Loading of Fluorescent Calcium Indicator Dyes

Stephen D. Van Hooser, Elizabeth N. Johnson, Ye Li, Mark Mazurek, Julie H. Culp, Arani Roy, Rishabh Kasliwal, and Kelly Flavahan

Abstract

In vivo 2-photon imaging of neurons that have been bulk-loaded with fluorescent calcium indicator dyes is permitting many fundamental principles of neural circuit organization and development to be uncovered. In this article, we describe the materials and procedures that we have used in our investigations of ferrets, tree shrews, and mice. Special attention is given to the design and construction of custom stereotaxic devices and the prevention of stray light from entering the 2-photon microscope during vision experiments.

Key words Head plate, Headplate, Two-photon, 2-photon, AM dyes, AM calcium dyes, Oregon Green BAPTA-1, Light block, Optical chamber, Epifluorescence

1 Introduction

In 1990, Winfried Denk and Watt Webb developed 2-photon laser scanning microscopy, which has revolutionized the ability of scientists to observe and manipulate biological processes in living animals and tissues [1]. The critical insight underlying 2-photon microscopy, which was first described in 1931 by Maria Goeppert-Mayer [2], is that photochemical reactions that can be initiated by a single photon with a certain energy can also be initiated by the nearly simultaneous arrival of two photons with lower energy. By using lower energy photons, one can achieve very precise control over the location of the photochemical reaction, as 2-photon excitation events will occur in a very small region around the focal point of the microscope lens, and stray excitation events will not occur throughout the entire light cone, as in traditional microscopy. In addition, the lower energy light that is typically employed in biological experiments is in the

Benjamin R. Arenkiel (ed.), *Neural Tracing Methods: Tracing Neurons and Their Connections*, Neuromethods, vol. 92, DOI 10.1007/978-1-4939-1963-5_6,

infrared spectrum, which is transmitted much farther into living tissues as compared to visible light. Therefore, by a single insight, 2-photon microscopy allows the precise observation and manipulation of photochemical events deep inside intact, living biological tissues, which is exactly where it is most revealing to study the processes of the nervous system [3, 4].

By the mid-1990s, 2-photon microscopy was combined with fluorescent calcium indicator dyes [5] to reveal calcium dynamics in single dendrites [6, 7] and dendritic spines [8, 9]. Imaging of calcium indicator dyes with the 2-photon has allowed transformative advances in our understanding of the nervous system at many scales, from individual synaptic connections to dendritic computation, circuitry, and networks [10].

In the early 2000s, Arthur Konnerth's group demonstrated that hundreds of individual cells within the living brain could be bulk-loaded with calcium indicator dyes conjugated to acetoxymethyl (AM) ester groups [11]. These AM-conjugated dyes [12] can be injected directly into the extracellular space, where they are taken up by neurons and glial cells. When these molecules pass through cell membranes into a cell, the AM group is cleaved by endogenous esterases, and the indicator dye can then exhibit calcium-dependent fluorescence. Once free of the AM group, the indicator dye is also relatively confined to the cell, leaking out only slowly over a period of 6–18 h [13]. When combined with 2-photon imaging, AM dye loading enables one to monitor the activity of hundreds of neighboring neurons that are part of the same neural circuit.

Subsequent 2-photon calcium imaging studies have revealed many basic principles of cortical development and organization. Network studies have revealed that visual cortical orientation, direction, and binocular disparity maps are precise at the level of single neurons [14–16]. In addition, because 2-photon microscopy provides images of the cell bodies that are being recorded, the responses of the same identified neurons can be followed over time as responses change. This has enabled biologists to monitor the response properties of single neurons within the developing brain and to watch as these response properties are altered by sensory experience [17–19]. By combining 2-photon calcium imaging with other fluorescent markers, one can examine the response properties of neurons that have been labeled by genetic [20–23], viral [24], or anatomical [25] methods. Further, in a particularly powerful set of approaches, one can combine characterization of functional response properties in vivo with subsequent characterization of the connectivity among these cells with slice physiology [26] or electron microscopy [27, 28]. Finally, Konnerth and colleagues have recently employed 2-photon calcium imaging to characterize all of the functional inputs (thousands of spines) and the functional output (spiking activity) of single neurons, which will greatly enhance our ability to understand the computations performed by neurons within networks [29–32].

1.1 Scope of the Article

In this article, we will discuss the materials and procedures needed to perform bulk loading of calcium indicator dyes and 2-photon imaging in the cortex of living mammals. There are already many published protocol papers and videos that linearly explain the basic procedures for performing 2-photon calcium imaging [11, 33–42]. However, the technique is quite challenging and depends on the success of many small but important steps. In particular, visual experiments require that one completely blocks the light from the stimulus monitor from entering the 2-photon microscope. Here, our goal is to provide two additional services beyond a linear protocol. First, we aim to describe the design choices we made as we developed materials and procedures for our particular animal preparations and experiments in visual cortex, so that an interested reader can quickly adapt these procedures for their own specific needs. Second, we aim to describe the most difficult procedures so that a novice experimentalist can quickly learn to perform them. We focus on practical considerations and share our own mythology about procedures that have uncertain outcomes.

Neuroscience is a highly interdisciplinary science, and this article assumes that the reader is already familiar with the basic concepts underlying fluorescence imaging, 2-photon microscopy, and in vivo surgery for acute, anesthetized experiments. That is, the reader should be able to understand how to go to an optics catalog and choose a fluorescent filter cube that is appropriate for imaging a particular fluorescent molecule and understand why the filter cube will work. Further, the reader should understand conceptually how to perform 2-photon or confocal imaging of fluorescent material on a slide [1, 3, 43–45]. Finally, we assume that the reader has some familiarity with performing in vivo surgeries for other procedures, such as microelectrode recordings or intrinsic imaging, and we will not describe methods of anesthesia and initial surgical procedures such as tracheostomy.

1.2 Outline of a Basic Experiment

A basic in vivo experiment that employs bulk loading of calcium indicator dyes involves the following steps: (1) surgical preparation of the animal, including mounting the animal in a suitable stereotaxic device for 2-photon imaging, (2) preparation of the calcium indicator dye solution, (3) injection of the calcium indicator dye into the cortex, and (4) 2-photon imaging of the cells and the responses. We will first discuss at length the materials and equipment needed for these experiments (Sect. 2), followed by step-by-step procedures (Sect. 3), and Sect. 4.

2 Materials

2.1 Choice of Research Animals

The first decision point for materials is to select the appropriate species and age of research animals for 2-photon calcium imaging. Experience suggests that younger animals are easier to image, as the brain tissue is more transparent so it is easier to image deeper into the cortex. Some investigators think that the loading process itself works better in younger animals. We generally use ferrets that are between postnatal days 30 and 70 (P30–P70), tree shrews that are P40–P65, and mice that are younger than P60. Using a conventional 2-photon setup, we expect to be able to observe activity in cell bodies to tissue depths of approximately 200–400 μm.

2.2 2-Photon Microscope and Optics

We use an Ultima IV in vivo 2-photon microscope from Prairie Technologies (Madison, WI). We have used lasers from both major manufacturers, including Chameleon lasers from Coherent and Mai Tai lasers from Newport-Spectra-Physics. When selecting a microscope for physiology and dye injection, we recommend choosing a microscope that allows bright field imaging and epifluorescence imaging through oculars in order to ease the process of finding the tissue to be imaged with 2-photon mode as well as to allow visualization of the dye injection without using 2-photon mode. Further, we recommend choosing a microscope that has the ability to provide digital timing triggers when each frame is recorded in order to ease the process of synchronizing the imaging to other events, such as external stimulation.

Objectives for in vivo 2-photon imaging must easily pass both infrared and visible light and have a long working distance of several millimeters to allow room for an optical interface on the brain and, optionally, simultaneous pipette access. There are several very high-quality 2-photon objectives, including objectives from Olympus and Nikon that have numerical apertures greater than 1.0 and adjustable collars. However, these high numerical aperture lenses cost in excess of $25,000 and are cost prohibitive for many labs. We recommend an excellent and affordable 16× Nikon objective (model CFI175) that offers a 500 μm × 500 μm field of view, a numerical aperture of 0.8, and a long working distance of 3.0 mm. We also use the LUMPLFL40XW/IR 40× objective from Olympus (water immersion), but it offers a smaller field of view (240 μm × 240 μm) compared to the CFI175.

It is important to calibrate the radiant flux (i.e., laser power) that is delivered to the tissue through the objective so as not to overheat the neural tissue. We use a PM100D optical power meter with S121C sensor (sensitive to wavelengths in 400–1,100 nm) from ThorLabs. At the beginning of the experimental day (or at least once per week), we set the power meter on top of a flashlight (it's just the right height to make a perfect stand), power up the laser, and put on the 16× objective. We lower the 16× objective into the meter but do not actually touch the meter. On our microscope, the laser power is adjusted by a Pockels cell, and we measure how many "Pockels cell" units correspond to 10, 20, 30, 40, 50, 60, 70, and 100 mW. When imaging, we try to keep the laser power as low as is feasible and in particular try not to image with a radiant flux above 40–60 mW for extended periods of time.

In order to visualize calcium dye injection, we use two different filter cubes from Chroma Technologies in our epifluorescence light path, mounted in the turret of our microscope. One filter cube is appropriate for viewing Alexa Fluor 488 (exciter HC 480/40, emitter HQ535/50, beamsplitter Q505lp), and the other is for viewing Alexa Fluor 594 (exciter HQ575/50x, emitter HQ640/50, beamsplitter Q610lp).

Finally, we use a filter cube that consists of a dichroic mirror and filters to split the emitted visible light from the 2-photon to 2 photomultiplier tubes (PMTs). Again, these filters are chosen for viewing Alexa Fluor 488 and Alexa Fluor 594. The cube is from Chroma Technologies (exciter HQ525/70: emitter HQ607/45, beamsplitter 575dcxr). The greenish light of Alexa Fluor 488 is directed to one PMT, while the reddish light of Alexa Fluor 594 is directed to the other PMT.

2.3 Stereotaxic Devices

Typically, surgical procedures such as skull exposure and craniotomy are difficult to perform underneath the 2-photon microscope, so most in vivo 2-photon labs have developed small stereotaxic frames that can be moved in and out of the microscope. The surgery is performed in a nearby open space, and then the animal is moved underneath the 2-photon microscope. The stereotaxic frame is secured in place by attaching it to the optical table with screws or a magnetic clamp. We use gas anesthesia provided through a tracheostomy (ferret and tree shrew) or a face mask (mouse). We have found it very helpful to attach, directly to the stereotaxic frame, an arm that holds a clamp for the anesthesia tube. This setup allows the animal and frame to be moved without disrupting the delicate anesthesia tubing.

There are three important requirements for the stereotaxic equipment that will hold the animal during 2-photon calcium imaging. First, the animal must be held very still, as 2-photon optical sections are only about 2 μm thick, and movement from frame to frame must be minimized: even a few microns of vertical movement per frame is unacceptable. Thus, the stability requirements for 2-photon imaging are more stringent than for single-unit recording and perhaps even whole-cell recording [46]. Second, it is critical that the mounting system allows the user to maintain a pool of sterile saline or ACSF so that a water-immersion objective may be used for 2-photon imaging. Third, in experiments that involve externally applied light such as visual stimulation, it is useful if the choice of mounting system can aid in blocking out light from the external light source.

We have developed several solutions for mounting ferrets, tree shrews, and mice on the 2-photon. In all of these species, we use custom stainless steel head plates that are adhered to the skull with dental caulk or dental cement (*see* also [37, 40, 42, 47, 48]). Each head plate holds the animal firmly from above and provides a more rigid attachment to the stereotaxic frame as compared to ear bars, particularly in young animals. The head plate is mounted onto metal bars that fit into the normal ear bar slots on our custom stereotaxic frame. Head plate specifications for different species are shown in Fig. 1. Depending upon the structure of the species's skull and the area of the brain to be imaged, one can opt for a small opening (Fig. 1a) or a longer slotted opening (Fig. 1b). In ferret, where visual cortex is located on the temporal slope of the skull, in a place where the skull is not particularly flat, we use a small, circular opening that facilitates both a secure mechanical attachment and the formation of a watertight seal around the head plate opening with the cement. Note that when the ferret is mounted in the stereotaxic via the head plate, the head is at approximately a 45° angle, so the surface of the cortex is normal to the microscope. In tree shrew, visual cortex is located close to the midline in a relatively flat part of the skull; we took advantage of this structure by designing a head plate with an elongated opening, which enables us to expose and image visual cortex in both hemispheres with the same head plate.

In our experiments, we provide visual stimulation to the animal, and we must prevent the light from the stimulus monitor from entering the objective. Our head plates are designed to assist in this process. In our ferret and tree shrew head plates, we have added a cylindrical metal well that helps prevent stray light from entering the objective. The diameter of the well was chosen to be wide enough so that the objective can easily be lowered into the well, and the height of the well was chosen to be as high as possible while still permitting pipettes to enter at shallow angles (30–45°) and have access to most locations within the head plate hole (example in Fig. 1d). The mouse head plates must be quite small in the anterior/posterior dimension so as not to block the animal's upper visual field when mounted on the skull, and there is insufficient room to mount a well that is wide enough to accommodate the objective (Fig. 2a). Our solution was to fashion a well out of a standard medium weigh boat that is spray painted black (Fig. 2b). A hole is cut in the bottom of the boat with a #11 scalpel blade to permit access to the head plate and skull, and the weigh boat is attached to the head plate with cyanoacrylate glue (Fig. 2c, d). The weigh boat's walls are sloped in a way that allows more space for the objective to be close to the site of the craniotomy while simultaneously blocking less of the mouse's visual field. Further, the weigh boat is plastic and flexible, which helps it conform to the surface of the head plate and skull or bend slightly if the objective presses against it.

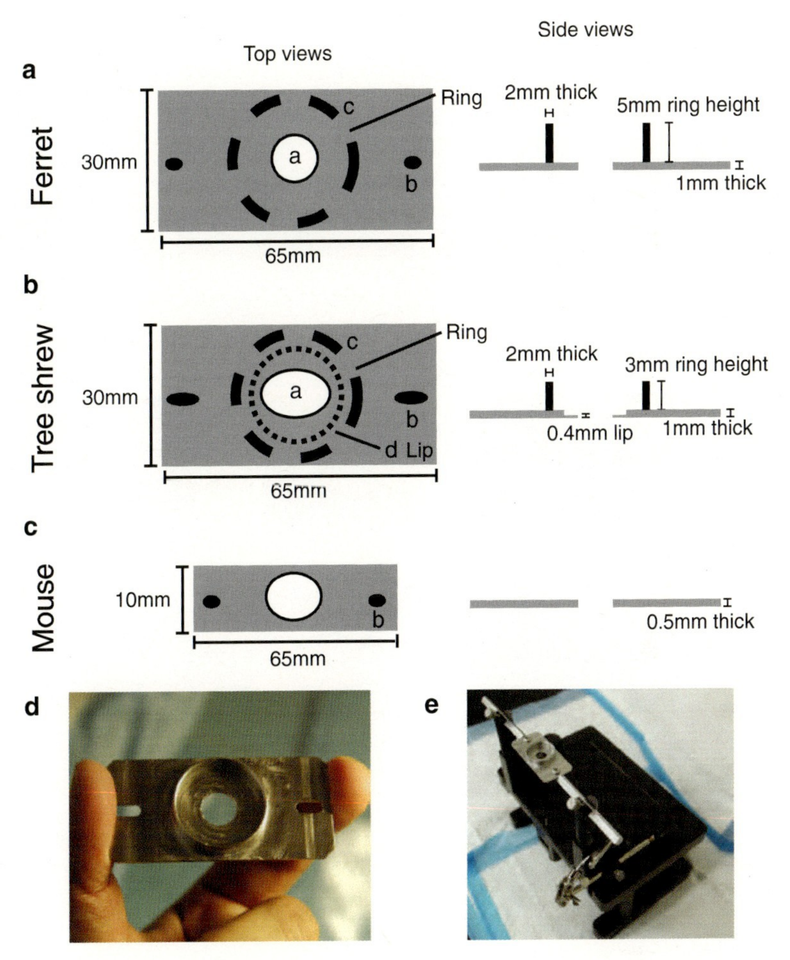

Fig. 1 Head plate designs for acute experiments in anesthetized (**a**) ferret, (**b**) tree shrew, and (**c**) mouse. All head plates are made from 1 mm stainless steel and include a center opening (*a*) and mounting holes (*b*) for attachment to the stereotaxic frame. The opening in the ferret head plate is circular, 10 mm in diameter, which helps ensure adequate cement contact on the skull above visual cortex, which is sloped. The opening in the tree shrew head plate is elongated (12 mm × 8 mm), which allows access to visual cortex in both hemispheres at the same time. Holes for mounting screws (*b*) are elongated across the long dimension of the head plate (4 mm × 8 mm) to allow flexibility in attachment to the stereotaxic frame. Ferret and tree shrew head plates also include a raised cylindrical well (*c*) with a thickness of 2 mm and a diameter of 27 mm; this structure is 5 mm high in the ferret head plate and 3 mm high in the tree shrew head plate. The tree shrew head plate also includes a thinned central region, 0.4 mm thick, 13 mm in diameter, that allows easier pipette access because the top of the head plate is even closer to the skull (0.4 mm vs. 1.0 mm). (**d**) Photograph of the ferret head plate. (**e**) Photograph of the ferret head plate mounted on the stereotaxic frame. The head plate is screwed into metal bars that attach to the stereotaxic frame using the clamps that normally hold the ear bars

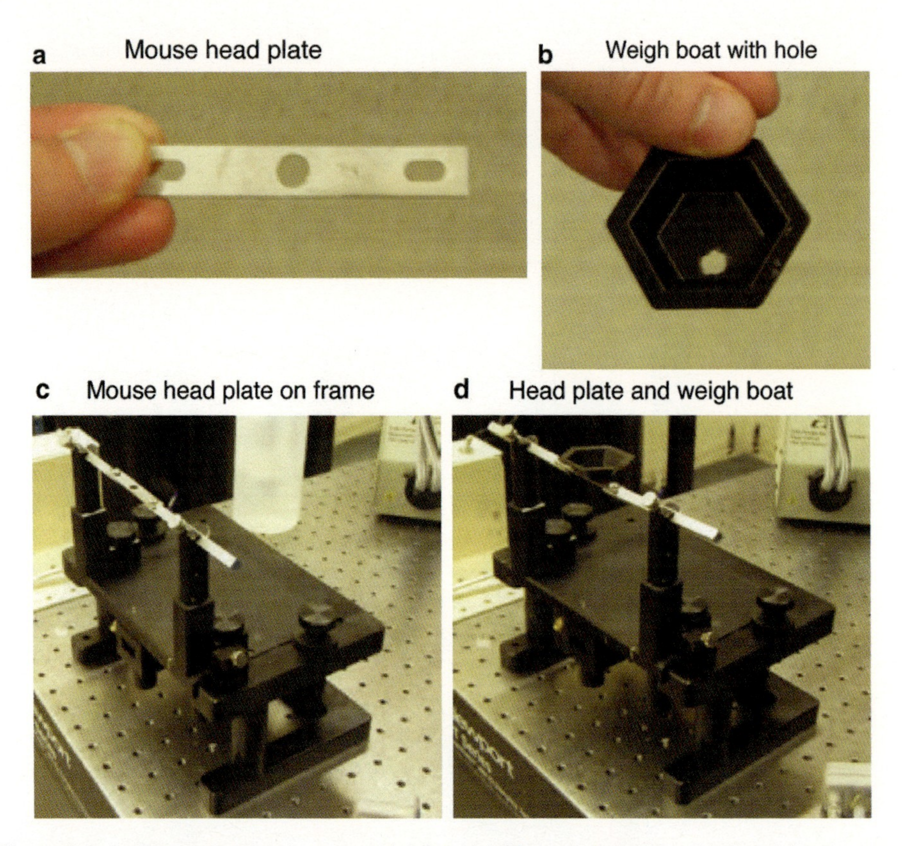

Fig. 2 Mouse head plate and weigh boat combination. (**a**) Mouse head plate. (**b**) Weigh boat spray painted black with an N × N cut out near the edge that will be placed anteriorly. (**c**) Mouse head plate mounted in the stereotaxic frame. (**d**) Mouse head plate with weigh boat glued in place

2.4 Light-Blocking Instruments for Experiments That Involve Light

The signal from our visual stimulus monitor is much brighter than the fluorescence coming from the molecules excited by the 2-photon microscope, so we must take great care to block all of this light from reaching our PMTs.

The first place that stray light can reach the PMTs is through the microscope housing. Our microscope was designed to be used in complete darkness inside a large, light-tight enclosure that includes a door that is intended to be shut during imaging (enclosure from Prairie Technologies, Fig. 3a). The microscope is therefore not designed to be used in a light environment, and like many microscopes, there are small openings that permit light to enter the PMTs. We cover all of these openings with high-quality light-tight tape (ThorLabs, Newton, NJ). To identify the openings, we turn on the microscope without an objective in place and tape over the objective tube to block light from entering where the objective is normally located. Next, we turn off the lights and run the microscope in scanning mode in order to record through the PMTs. We do not enable the infrared laser during this work because it is not needed and could possibly burn a hole in the tape

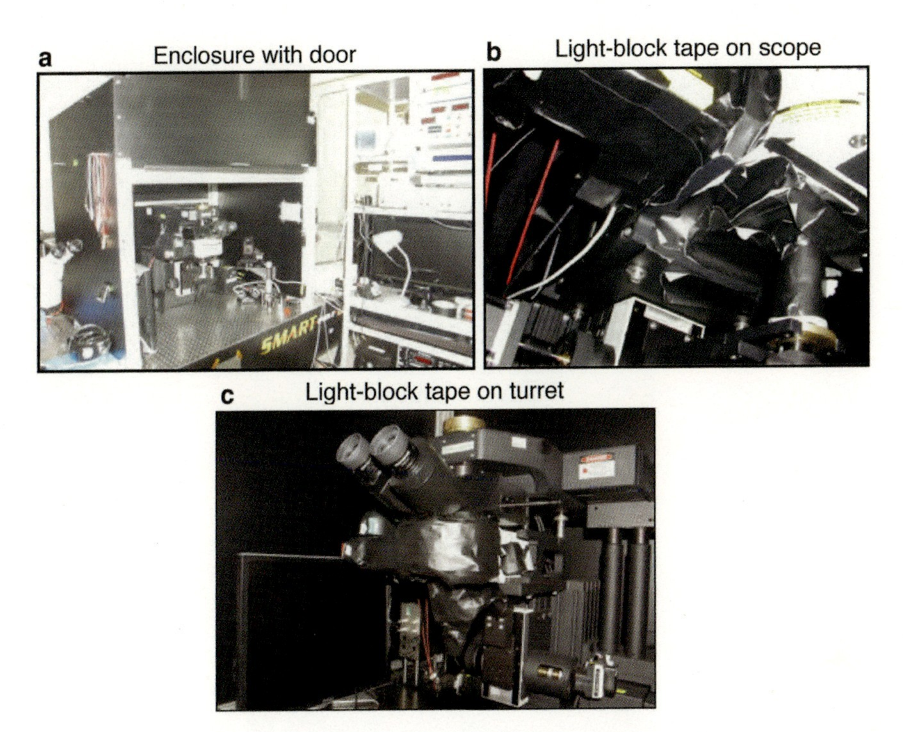

Fig. 3 Blocking stray light from entering the microscope. (**a**) The *black enclosure box* that surrounds our microscope, with front access door open. (**b**) Light-tight tape applied to the underbody of our first 2-photon microscope. Our newer microscope does not require so much tape, but it is important to sweep over the surface with a laser pointer or flashlight to identify all openings and tape them. (**c**) An example of the turret controls taped for in vivo imaging with visual stimulation. This tape is not left in place permanently but rather is added only at the time of imaging

over the objective tube. Then, we shine either a laser pointer or a bright flashlight at every point on the microscope, from every angle, and identify locations that produce signal on the PMTs. We then divide these locations into two groups: those that we can cover permanently without blocking any microscope controls (Fig. 3b) and those that we will cover only during the imaging phase of the experiment (Fig. 3c). We block all of the openings with tape until we can shine the laser pointer or flashlight all over the scope without causing activation of the PMTs. At the conclusion of this process, the locations that can be covered permanently remain taped, and tape is removed from areas that cover microscope controls (like the filter cube selection turret).

The second major pathway for stray light is through the objective itself. To prevent stray light from entering the objective, we constructed an objective cloak (Fig. 4) made out of light-tight fabric (ThorLabs). Strips of braided elastic (Michaels craft store) are sewn into the top and bottom of the cloak so that it grips the objective and the head plate. Often, it is necessary to add a couple pieces of light-block tape on the front of the head plate or the animal's ears to fully prevent light from entering.

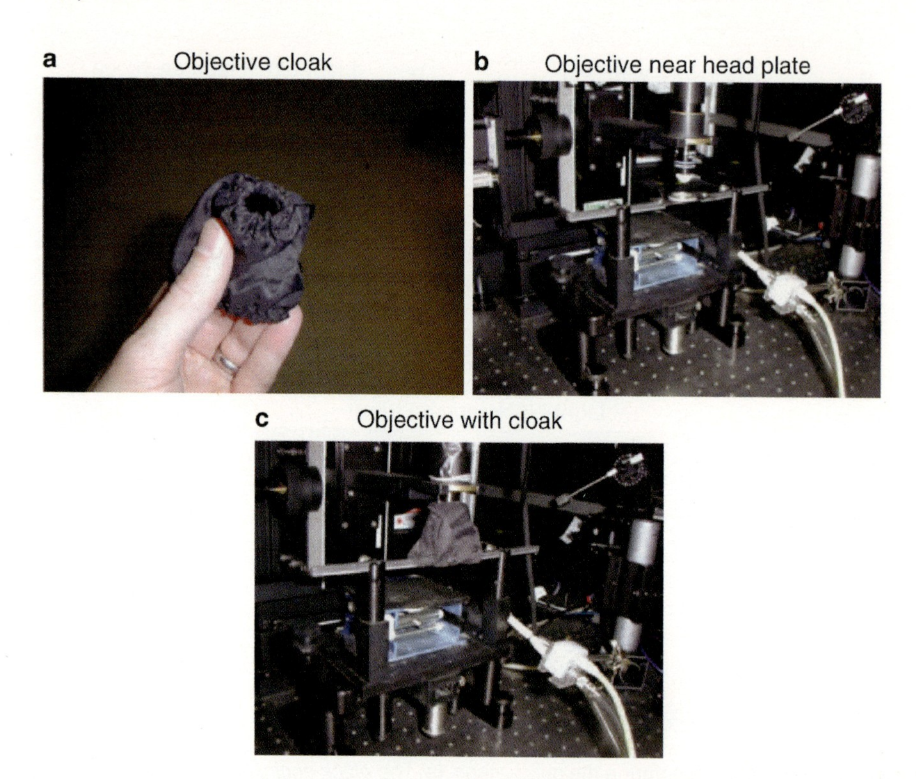

Fig. 4 Blocking stray light from entering the objective. (**a**) The objective cloak with elastic top and bottom. (**b**) The stereotaxic frame and head plate under the 2-photon microscope, with objective attached. (**c**) The objective cloak on the objective and pulled down around the head plate to block external light from entering the optical chamber

2.5 Selection of Dye for First Experiments

We use the calcium indicator dye Oregon Green BAPTA-1AM (OGB-1). We recommend that new users start with this dye because it seems to work in many species, such as mice [11], rats [15], cats [14–16], ferrets [18, 19, 49, 50], and monkeys [51], and to try other dyes only after gaining some experience with OGB-1. There are many other calcium dyes available in AM form, but many do not commonly appear in in vivo papers, raising the possibility that they are not as effective as OGB-1. There is at least one in vivo paper that used Fura-2 [23].

Many new users wish to use dyes other than OGB-1 because they are interested in combining calcium imaging with the use of green fluorescent protein (GFP) marker proteins. The single-photon excitation wavelengths of Alexa Fluor 488 and GFP are similar (peaking around 488 nm), so these molecules would be difficult to excite independently using a single-photon process (i.e., visible light) [52]. Many new users incorrectly assume that the wavelengths of infrared light that would be required to provide 2-photon excitation of these molecules would also be very similar. However, the exact wavelengths that provide strong 2-photon excitation of a molecule vary from molecule to molecule in a way that is not trivially predictable from its single-photon excitation spectrum [45, 53] and are best determined empirically. Fortunately, the wavelengths of infrared light that provide strong 2-photon excitation of OGB-1 and GFP differ considerably, such that 800 nm light can be used to excite OGB-1 and 960 nm light can be used to excite GFP. This difference allows OGB-1 and GFP fluorescence to be viewed independently in the same tissue [20].

2.6 Dye-Mixing Supplies

To mix the dye, one will need Oregon Green BAPTA-1AM packaged in 50 μg vials (Invitrogen #O-6807), 20 % F-127 Pluronic acid (Invitrogen #P-3000MP), a personal desktop centrifuge, water bath sonicator, 1.5 ml centrifuge tubes, 0.22 or 0.45 μm pore centrifuge filters (Ultrafree MC from Millipore UFC30GV25), dye pipette solution (*see* below), and Alexa Fluor 594 dye solution for counterstaining (*see* below).

2.7 Dye Pipette Solution Stock

Any ACSF may be used as a dye pipette solution, but the solution described here [11] doesn't need to be refrigerated. To a 1 l glass container with lid, add 1 l dH_2O, 8.77 g NaCl, 0.19 g KCl, and 2.38 g HEPES, and adjust to pH 7.3–7.5 with NaOH, store at room temperature for 6 months, and check pH every 2 months.

2.8 Alexa Fluor 594 Dye Solution Stock

Add 660 μl dH_2O to 1 mg Alexa Fluor 594 hydrazide power (Invitrogen #A-10438), store in a tightly capped centrifuge tube with parafilm at room temperature for months, and protect from light.

2.9 Pipettes, Puller, and Pressure Injector

Any patch pipette with a tip diameter of 2–6 μm should work well for AM dye delivery. We prefer pipettes pulled in a single heating step on a vertical puller (Kopf Instruments 700C; 5-turn coil, 14–16 A current, no solenoid/gravity only) because these pipettes exhibit a gradual taper and will enter the tissue with minimal damage. We buy glass pipettes with a small fiber that aids in bringing the fluid into the small pipette tip (FHC Catalog #30-30-1, 1.0 mm OD, 0.5 mm ID, with Omega dot fiber). Tips are manually broken (by touching them to a lint-free cloth) so they are between 3 and 6 μm. Tip size is verified under a 40× objective on a calibrated light microscope. Smaller tips tend to clog and are not recommended. Larger tips are also not recommended, as they allow cerebral spinal fluid or saline to wick up into the pipette when the pipette is lowered at or near the brain surface, which reduces the concentration of calcium dye in the pipette to ineffective levels.

Pipettes can be pulled days in advance of an experiment and stored in a micropipette storage jar (such as World Precision Instruments E210). During the experiment, pipettes are loaded with a pipettor capable of 2 μl volumes with a microloader tip.

A pressure injection system is needed to inject into the brain, such as an MPPI-3 (Applied Scientific Instrumentation, ASI).

The pressure injection system requires a compressed gas source (such as nitrogen) and a regulator capable of about 3.5 atmospheres (50 psi) pressure output.

2.10 Optical Chamber

The optical chamber is created during the experiment using agarose and a coverglass. We use either type III agarose (Sigma, A9793-100G), or type I-A agarose, low EEO (Sigma, A0169-100G). In advance of the experiment, we prepare 1–3 % agarose (in the pipette solution described below, any artificial cerebral spinal fluid, or Ringer's solution) and aliquot it in 3–5 ml portions into 20 ml glass vials (66022-004 from VWR). On the morning of the experiment, we heat a vial with any metal foil removed and with the top on tight until it just boils in a microwave oven (repeated heatings in 4 s steps until it just boils). It is important to be sure the agarose has fully melted in the jar. After initial heating, we keep the lid on the agarose and keep it warm on a hot block (Standard Heat Block, part 13259-030, from VWR Scientific), on the hotter setting, position 4, until it is ready for use.

We top the optical chamber with a coverglass. We buy small precut coverglass (#1 thickness, 0.15 mm) from Warner Instruments, which sells many precut diameters from 3 to 8 mm. For ferrets and tree shrews we generally use 5 mm diameter coverglass.

2.11 Surgical Supplies

Cotton-tipped applicators, gauze, scalpels (#10, #11, #15), sterile saline (0.9 %), or any common artificial cerebral spinal fluid (ACSF).

3 Methods

3.1 Surgery

Prepare the experimental animal according to typical in vivo procedures. For ferrets, we avoid using xylazine in 2-photon and intrinsic imaging experiments because we have the impression that it suppresses cortical responses; instead we use a sedating dose of ketamine (10–20 mg/kg intramuscular injection) followed by isoflurane anesthesia (1–3 %) with a mask. In both tree shrew and ferret, we perform a tracheostomy for delivery of gas anesthesia and mount the animal in the small and portable stereotaxic frame described in Sect. 2. In mouse, no tracheostomy is performed, but isoflurane anesthesia is delivered through a mask. After the animal is mounted in ear bars (ferrets and tree shrews) or ear cups (mice), an incision is made in the skin over the skull with a #10 scalpel blade, and the skin retracted. Detailed surgical methods are available for ferret [18, 19] and tree shrew [54]. In the mouse, it is important not to cut away too much skin, as it will help shield stray light from entering underneath the head plate. Further, in young mice, one can optionally paint the skull with Liquitex acrylic paint (Darice, Mars Black, BS1047-276) to aid in light blocking.

3.2 Head Plate Attachment

1. In order to mount the head plate, it is important to remove the skin and connective tissue from a relatively large region around where the head plate will be fixed. The connective tissue on the skull is scraped clean with the same #10 blade or a spatula. In the case of the tree shrew and ferret, muscle near the visual cortex should be bluntly resected by rubbing a spatula across the connective tissue until the muscle is disconnected and recedes. In all of these animals, when we are interested in attaching the head plate over visual cortex, we expose and clean most of the dorsal surface of the skull; in the ferret, we expose most of the temporal surface of the skull of the hemisphere we plan to use.
2. The dental caulk or dental cement that is used to attach the head plate does not adhere directly to the skull, so it is important to rigidly attach small pieces of metal to the skull. The dental caulk/cement will stick to the plate, which will in turn stick to the metal attached to the skull. This can be done by adding skull screws far away from the region to be recorded. We prefer to cut small pieces of a mesh screen with dedicated scissors. In the ferret, we can install a few square links of the mesh, but in the tree shrew and mouse, we cut a few loose pieces of metal about 3–5 mm long. We then glue these pieces onto the skull with cyanoacrylate adhesive (Loctite 401). The pieces of mesh are placed at some distance from the area to be imaged so that the mesh will not be in the way when the head plate is pressed tightly to the skull in the next step. In addition, it is important to use minimal cyanoacrylate adhesive so that the drying time is brief. The adhesive must be completely dry before proceeding to mounting the head plate or the head plate will not stick (wait several minutes, test with the wooden back end of a cotton-tipped applicator).
3. At this point, the head plate is ready to be attached and one must select a method of attachment. In ferrets and tree shrews, we typically use a dental caulk (Dycal, from Dentsply) because it dries much faster (1–5 min) than dental cement and does not have a strong, volatile odor. However, the dental caulk is not as strong as dental cement so it must be used with some finesse. Dental cement works well but requires some 30–60 min of drying time. We do use dental cement in mice because it seems to work better with the reduced skull surface area that is available in mice as compared to tree shrews and ferrets. In our most recent experiments in mice, we skip the dental cement in mice entirely and attach the head plate directly using cyanoacrylate glue.
4. To attach the head plate, use a lint-free towel, cotton-tipped applicator, or gauze to dry the skull. Hold the plate exactly where it will be placed and, optionally, draw a mark through the center hole with a pencil to help with the application. If using caulk, mix the two components well with a metal spatula for

only 10 s (it will begin setting after 30–60 s). Then, apply the caulk liberally in a circle around where the head plate center will go. Press the head plate firmly into the caulk, and hold the plate in place, tight against the skull, for 3–5 min. While the caulk is drying and the head plate is being held in place, it is helpful if a second experimenter can mix additional caulk and apply it underneath the head plate so that a strong seal is made. If using cement, mix the cement and apply a generous amount in the circle around where the head plate will go. Next, it is helpful to hold the head plate using one of the metal bars that will eventually attach the head plate to the ear bar slots and clamp it in place so that it remains tight against the skull for 30–40 min. During this time, additional dental cement can be integrated around the edges of the head plate.

5. When the head plate is firmly attached to the skull (i.e., the animal's skull moves with the head plate), remount the animal in the stereotaxic frame using the head plate. We carefully attach the metal bars that connect the head plate to the stereotaxic clamps that normally hold the ear bars. It is important not to put undue force on the head plate or it may become detached. If the head plate falls off, it can be reattached by cleaning the cement off the head plate and skull, removing the glued mesh (if it didn't come off with the head plate), and starting the procedure again. Caulk and cement can be removed from the skull and from metal tools with a dedicated surgical rongeur.
6. Finally, it is necessary to verify that the seal made with the skull is liquid-tight. Orient the head plate so the hole is facing up (i.e., opposite the direction of the force of gravity). Place a few drops of sterile saline or ACSF into the hole in the head plate and watch carefully for about 1–2 min to ensure the liquid does not leak, even slowly. If it leaks, remove the head plate, cement, or caulk, and mesh, and start again.
7. In the mouse, now is the time to glue the black weigh boat to the head plate. Light can leak in between the head plate and the black weigh boat, so we cover the underside of the hole of the black weigh boat with a thin amount of clay. This serves as a light-tight barrier in this small spot that is nearly impossible to seal with light-tight tape without blocking the animal's vision. Use cyanoacrylate glue such as Loctite to attach the weigh boat and clay.

   When the head plate is successfully attached with a good seal, one is ready for the craniotomy.

3.3 Craniotomy

1. In ferret and mouse, we drill a 2 mm × 2 mm craniotomy with a high-speed dental drill. If we expect the skull to be very thick (as in older ferrets), we thin the skull over a large area (5–7 mm × 5–7 mm) before drilling the hole so the pipette can

make an angled approach into the craniotomy. We believe a gradual craniotomy also has advantages over a "well" craniotomy because agarose can be removed more gently.

2. In tree shrew, we thin the skull above visual cortex with a #15 scalpel blade (back-and-forth movement for several minutes, using almost no downward pressure). We then put a drop of saline on the skull to observe the blood vessel pattern in the brain. (No further scraping is needed when the skull is just thin enough to see the blood vessels in the brain.) We choose an area that has relatively few blood vessels and then cut out a 2 mm × 1 mm or 2 mm × 2 mm window using a #11 scalpel blade. A small craniotomy is critical to avoid respiratory-locked pulsations of the brain, and therefore we try to keep them smaller than a few square millimeters in area.

3.4 Calcium Indicator Dye Preparation

It is helpful for a second experimenter to prepare the dye while the other experimenter is performing the craniotomy/durotomy.

1. Warm a 50 μg vial of Oregon Green BAPTA-1 (OGB-1) by hand for about 1 min (aids mixing). Remove the label.
2. Add 4 μl of 20 % Pluronic acid to the vial of OGB-1.
3. Centrifuge for 1–2 s to bring the Pluronic acid and OGB-1 into contact in the bottom of the vial. Vortex the vial for 2–3 min to mix well.
4. Add 44 μl of dye pipette solution (*see* above).
5. Add 1 μl Alexa Fluor 594 solution (*see* above).
6. Vortex for 2–3 min to mix well.
7. Centrifuge briefly and transfer the solution to a 1.5 ml centrifuge tube. Seal the tube with parafilm so that it is liquid-tight.
8. Sonicate on a float in a water bath for 20–30 min. To prevent the solution from overheating, add several ice chips to the water bath. There should always be some ice floating in the bath; typically one has to replace the initial ice after 10 min.
9. Vortex briefly (10 s) and then filter with a centrifuge filter (0.22 or 0.45 μm pore).
10. For best results, use immediately or store on ice and protect from light for 2–4 h.

3.5 Calcium Indicator Dye Injection Visualized with Epifluorescence Optics

An excellent alternate description is provided in reference [39].

1. In ferrets, remove the dura immediately prior to attempting the injection. In our experience, it is important to minimize the time between removing the dura and performing the injection (ideally less than 10 min to dye injection completion). The condition of the cortex underlying the dura is very important.

There should be very little blood on the cortical surface, both to avoid damaging the tissue and because one can't perform 2-photon imaging through blood or blood vessels. It is important to keep some sterile saline or ACSF on the exposed brain to keep it from drying, but it is also important to minimize the number of times that one puts new fluids on the brain or changes agarose before the dye loading is complete. In tree shrew, leave the thin dura intact or make a small nick in the dura to allow easier pipette access. In mice, leave the thin dura intact.

2. Move the animal so that it is underneath the 2-photon microscope and lock the stereotaxic frame in place. View the cortical surface in bright field mode under low power (4–5×) with a long working distance objective.
3. Study the craniotomy carefully to identify a region that is suitable for injection. The pipette will be inserted at a particular location (let's call it the insertion location) and will be driven at a shallow angle into the cortex until the pipette tip reaches a depth of about 300–500 μm. Good candidate regions for injection have the following characteristics: (a) The pipette insertion location must be free of blood vessels. (b) There must be no surface blood vessels above the location where the dye injection will actually take place; this "injection location" is usually 300–600 μm horizontally offset from the insertion location, depending upon the approach angle of the pipette. (c) There must be no vertically running blood vessels along the trajectory of the pipette from the insertion location to the dye injection location. Look carefully for vertically running blood vessels; when vessels disappear on the horizontal surface, it is typically because they are diving down vertically in the brain, and it is important not to break these vessels with the pipette.
4. Using a microloader and a pipettor, fill the injection pipette with 2 μl of calcium indicator dye. The dye solution is quite viscous, so wait several seconds for the dye to flow into the microloader. After the dye is in the microloader, place the microloader tip near the bottom of the pipette, and eject the dye into the pipette while removing the microloader from the glass pipette. Hold the glass pipette with the sharp end toward the ground, and tap the pipette gently for 1–2 min until the dye is loaded in the tip and there are no bubbles in the pipette. There can be a streak near the meniscus, but there should be no observable bubbles near the tip. We wear fresh gloves while loading the glass pipette so the oils from our hands do not clog the pipette.
5. Mount the pipette on the manipulator, and position it so that the tip is directly above the pipette insertion location

(7.5–10 mm vertically above, clear of any CSF or saline). The pipette can be visualized with bright field illumination and the filter cube appropriate for viewing Alexa Fluor 594. When viewing the Alexa Fluor 594, we take care not to shine the intense light of the fluorescence light source on the brain for very long; we open the shutter only long enough to take a quick look.

6. Examine the pipette tip under the microscope. If there are fine bubbles near the tip, remove the pipette and tap more until those bubbles are gone.
7. Briefly, dip the pipette into the saline/CSF above the brain, and give a couple of pressure pulses at 10 psi to verify that Alexa Fluor 594 can be seen flowing from the tip. The flow should be absolutely obvious; if the flow is not obvious, the pipette is at least partially clogged and the pipette should be changed. Immediately after establishing whether the pipette is patent or clogged, raise the pipette so it is no longer in the saline. The pipettes will wick up saline into the tip, which will dilute the concentration of dye in the pipette.
8. The next steps should be done in rapid succession in order to avoid clogging the pipette or drying out the brain. Confirm that the pipette is vertically above the insertion location. Dip a tightly rolled lint-free towel such as KimWipe into the edge of the craniotomy to remove the CSF or saline that is pooled on top. The craniotomy should not be completely dry but should not have a pool of liquid on top.
9. Lower the pipette vertically so that it drops down onto the cortical surface at the insertion location. One should be able to observe a tiny dimple (50–100 μm) on the brain surface in bright field illumination. If using a digital manipulator, zero the micrometer.
10. In the ferret, the brain pulses in time with the respiration, and the brain must be stabilized so the pipette does not damage the cortex on insertion. Draw 0.5 cc of warm agarose into a 1 ml syringe with the needle removed. Measure the temperature of the agarose in the tip of the syringe using a temperature probe, and begin adding agarose when the temperature near the tip drops to about 41 °C. Usually the temperature probe creates a bubble, so discard the first drops, and then put 5–8 drops (~0.2 cc) of agarose on the brain. Wait 10–15 s for the agarose to gel, at which point the brain will be gently held in place and will stop pulsing. In tree shrew and mice, the dura is thinner and brain pulsations are not as large as in the ferret, so adding a layer of agarose is not necessary or desired.
11. Drive the pipette diagonally until the vertical depth of the tip is about 300–500 μm. Give 40–100 brief (0.5–1 s) pressure

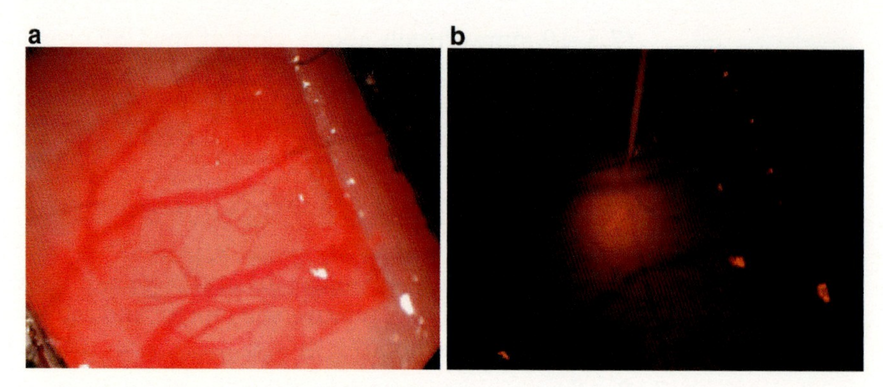

Fig. 5 Indicator dye injection visualized under low-power epifluorescence illumination (4× objective). (**a**) Craniotomy in the tree shrew. The approximate pipette insertion location is indicated by the faint cross hairs in the ocular reticle. The dura is still on at this stage in tree shrew. (**b**) Alexa Fluor 594 counterstain dye visualized with epifluorescence. The pipette and injection site at a depth of about 300 μm are *red*. The injection cloud built up slowly over the course of 40–60 short pulses and is about 500 μm in diameter

pulses at about 10 psi. After 10 or 20 pulses, a large sphere of dye about 500 μm in diameter should be evident. If the ball gets much larger than this, use fewer and shorter pulses. If the dye flows very quickly so that the field is bright after 2–3 pulses or if the dye flows backwards up around the shaft of the pipette, then the tip is likely broken, and one can either abort the injection or try very low pressure and very short pulses to try to rescue the injection. Ideally, the dye should accumulate slowly but be quite bright after the injection is complete (Fig. 5).

12. After injection, remove the pipette.
13. In ferrets, do not change the agarose for 45–60 min. Then, remove it with forceps prior to creating the optical chamber.
14. In tree shrew, place gauze saturated in saline on the craniotomy to prevent the brain from drying out and leave the brain alone for 45–60 min; after which time one can move the animal out of the microscope, make a small opening in the dura, and create the optical chamber.
15. In mice, one can move right ahead to creating the optical chamber.

3.6 Creating the Optical Chamber

We create an optical chamber using agarose and a coverslip as follows. The chamber serves to provide a flat and clear optical interface to the tissue.

1. Remove a coverslip from the case, and with forceps hold it up to the room lights to verify that it is clear and not dirty. If the coverslip is dirty, either choose another coverslip or clean the coverslip by adding a drop of water and rubbing it on clean lens paper. Place the coverslip on lens paper near the animal so it can be grabbed quickly with the forceps.

2. Dip a tightly rolled lint-free towel such as KimWipe into the edge of the craniotomy to remove the CSF or saline that is pooled on top. The craniotomy should not be completely dry but should not have a pool of liquid on top.
3. Draw 0.5 cc of warm agarose into a 1 ml syringe with the needle removed. Measure the temperature of the agarose in the tip of the syringe using a temperature probe, and begin adding agarose when the temperature near the tip drops to about 41 °C. The agarose temperature will drop rapidly when the agarose is out of the syringe, so this target syringe temperature of 41 °C is not too hot for the cortex. Usually the temperature probe creates a bubble, so discard the first drops in the syringe, and then put 5–8 drops (~0.2 cc) of agarose on the brain. Immediately, before the agarose begins to gel, place the coverslip on the agarose and use the forceps to gently press the coverslip into the solidifying agarose so that it touches the head plate and/or skull. Hold until the agarose becomes a gel (10–20 s).
4. Quickly examine the craniotomy using bright field illumination. If the agarose had begun to gel before the coverslip was applied, it is possible that the coverslip pressed solidified agarose onto the cortex, and blood vessels on the brain might be compressed and blocked. This situation is readily identifiable with bright field illumination through the oculars, as the blood vessels will appear empty in places and blood flow will not be evident. If blood flow appears to be blocked, this should be remedied quickly by removing the agarose with forceps and beginning the optical chamber process again.
5. Confirm that the craniotomy is clear and easy to view under the microscope. Place a few drops of saline or ACSF on top of the coverglass for liquid-immersion of the objective. The optical chamber is finished.

3.7 Applying and Testing the Light Block

Typically, we do a quick check of the cells using the procedures in Sect. 3.8 before setting up the light block. After confirming the presence of labeled cells, we proceed to the light-blocking steps.

1. Remove the objective, and dress it in the objective cloak (*see* Sect. 2).
2. Replace the objective, and pull down the cloak around the head plate. The elastic around the cloak should create a seal around the head plate. Be careful not to allow any part of the cloak to touch the saline/ACSF in the optical chamber, as it will wick the liquid away and leave none for immersion of the objective.
3. While looking through the oculars and using the epifluorescence cube for Alexa Fluor 488, lower the objective into the liquid and until the surface of the brain is in focus.

4. Apply a piece of light-tight tape between the front of the animal's skull and the cloak, and apply another piece of tape under each ear and up to the head plate.
5. Using light-tight tape, cover the openings of the microscope that were previously identified as needing to be covered (*see* Sect. 2).
6. If the microscope is inside an enclosure, lower the door of the enclosure as much as possible without blocking the visual field of the animal.
7. Test the quality of the light block. To do this, begin with the stimulus monitor turned off. Shutter the infrared laser so that no laser excitation is supplied to the fluorescent material. Turn up the PMT gain to maximum, and begin 2-photon scanning. The incoming images should exhibit weak shot noise; that is, there should be a few random bright pixels on a mostly black background. Now, turn on the stimulus monitor. There should be no increase or almost no increase in the weak shot noise on the incoming images. The clearest test is to turn off the monitor when a new image has scanned halfway; then the top and bottom of the image can be easily compared. If there is a dramatic increase in light when the monitor is turned on or a dramatic decrease when the monitor is turned off, then it is necessary to find the opening and patch it with light-tight tape. A small piece of black card stock (ThorLabs) taped on the wooden end of a cotton-tipped applicator can be used to quickly identify the location of the opening(s). Repeat the test until a high-quality light block has been achieved.

3.8 Imaging the Loaded Cells

1. Using epifluorescence through the oculars, focus on the surface of the cortex right at the injection location. Zero the *Z* axis on the microscope.
2. Switch to 2-photon mode, and start scanning with a relatively low power of 10–20 mW at the surface (blood vessels are usually evident) and PMT gain of about 600–800 units. Note: if you have an "offset" setting on your PMT, make sure it is at 0. Changing the PMT offset away from 0 will impact any attempt to measure the percentage change in fluorescence.
3. Focus down 150–300 μm and look around in *X* and *Y* for a field of cells. At this depth, one can use a laser power of 30–40 mW (the lower the better, but one still has to be able to see) and PMT gain of 750–950. Generally we begin searching at 150 μm and drop in 10–20 μm steps down to 350 μm. If one moved in *X* and *Y* to find the field of cells, it is a good idea to go back up to the surface and re-zero to identify the proper depth of the field.
4. The field of cells should look glorious, with dozens of cells present at a single *Z* position, like is seen in the papers [11, 14–16,

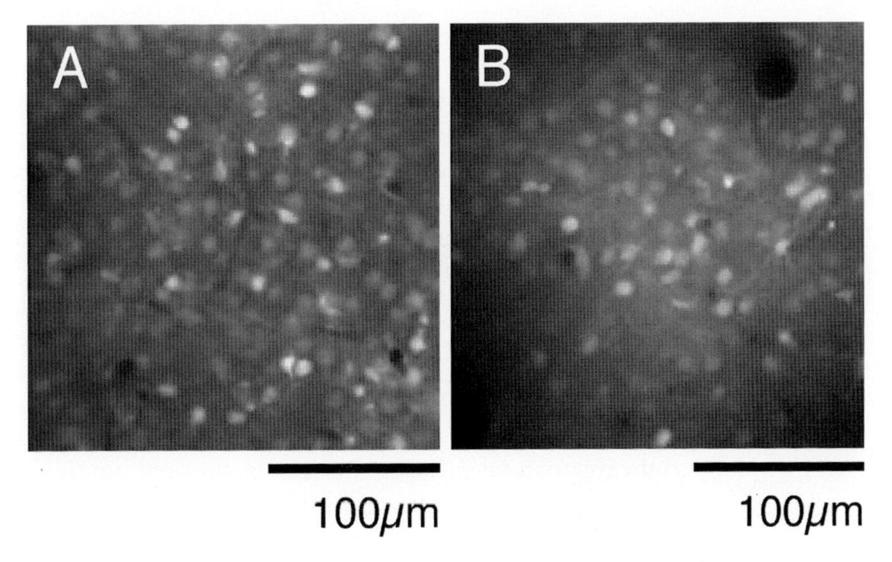

Fig. 6 Example image fields of bulk-labeled neurons from (**a**) ferret and (**b**) tree shrew

18, 19, 34, 40, 50, 54–58] (*see* Fig. 6). If there are only a handful of cells, then either the loading didn't work or there are optical issues. If there is green fuzziness, but the faint outlines of cells can be observed by averaging many scans together, then there may be an optical problem or the cells are too deep to be imaged. Change the agarose, remove the dura if it is present, make sure the objective is clean, and try again. If the problem cannot be resolved with these steps, then it is likely that the loading did not work.

4 Notes

4.1 Dye Injection Pipette Clog Rescue

If the pipette is in the brain and dye doesn't flow, try backing up 100 μm and give a few pulses. If dye flows, then take the pipette back down to the depth of interest and try again. Otherwise, back up another 100 μm and try to break the clog again. If that doesn't work, pull clear out, and try to break the clog with higher pressure (up to 20–30 psi) above the brain, and if successful, then go back in (remember to reduce pressure) and try again. If that doesn't work, we recommend not struggling further and beginning again with a new pipette. If a ball of dye does not form during the injection, then dye is not getting out. If it is at all unclear as to whether dye is coming out or not, then dye is probably not coming out.

4.2 Light Blocking when Pipette Access Is Required

In some experiments, it is necessary to have pipette access to the brain during visual stimulation. One can create a modified light-blocking solution for ferrets and tree shrew using a medium funnel.

Cut the bottom off the funnel so that the opening is big enough for the animal's head. In order to allow access for the tracheostomy tube, it is necessary to cut an opening in the funnel wall. If one imagines the funnel wall as a complete circle of 360° around the entry of the funnel, cut the funnel wall so that it only extends about 300° around the center (i.e., cut out a notch of about 60°). Note that this means the notch will be physically larger at the top of the funnel than at the bottom, since the opening at the top of the funnel is larger than at the bottom. Spray paint the funnel black, and add light-tight fabric (ThorLabs) to both sides of the notch so the opening can be completely covered. Attach the funnel to a laboratory clamp so it can be positioned close to the animal during the experiment. Add light-tight curtains (made with light-tight fabric) to the microscope enclosure that can be brought tightly around the edges of the funnel. During the experiment, wait until the animal is positioned under the 2-photon microscope, and then place the funnel around the animal's head, being careful not to disturb the trachea tube. Fill in any gaps around the animals' head with clay. When it is time to image, bring the curtains tightly around the funnel, securing with small binder clips or alligator clips. It is critical not to disturb the funnel, as moving the funnel can cause the animal's head to move and the head plate to become disconnected. For this reason, we prefer to use the funnel only when it is absolutely necessary to have pipette access during the experiment.

4.3 Troubleshooting

If the loading didn't work, one is left with the frustrating question of "why??!" It could be a failure to deliver enough dye, but if dye was observed coming out of the pipette, then this is not likely to be the problem. It could be bad dye, but we've never really been convinced that this has ever been the case. It could be that the dye was injected at the wrong depth or that the loaded cells are obscured under a blood vessel. The dye loading is very finicky, so if the cortex is damaged in any way, loading will often be prevented. Even when the cortex is not apparently damaged and one can still record physiological signals, the loading still sometimes doesn't work. We really have no idea why this is the case. In our mythology, we imagine that there may be some response of the tissue to exposure or damage that interferes with the loading process. To avoid this, we try to minimize the amount of time between the craniotomy and durotomy and bulk loading. In addition to the information provided here, there is an excellent troubleshooting section in Ohki and Reid [39]. We estimate that about 50 % of experiments yield very nice dye loading and responses when all the preparations have apparently gone smoothly.

If a particular injection didn't work and the dye was delivered successfully, then it is usually the case that subsequent injections into the same craniotomy will be unsuccessful. We don't recall seeing a successful loading in the same craniotomy when the first attempt was not successful. If we make multiple injections of dye, then when one works they all tend to work. This makes us think there is some sort of damage issue when the injections do not work. If one's injection didn't work and one is confident the dye was actually injected, there's no need to waste any more time with that craniotomy—go right to the other hemisphere or make a new craniotomy some distance away. The second hemisphere is often just fine even if the first hemisphere is not loading.

The beauty of observing the activity of hundreds of individual cells in the cortex is the best source of inspiration during difficult troubleshooting.

Acknowledgments

We thank Prakash Kara, Tom Mrsic-Flogel, Aaron Kerlin, and Clay Reid for their valuable advice as we were learning to perform 2-photon imaging. We also thank David Fitzpatrick and Leonard E. White for support and mentoring. We thank Frank Mello, machinist at Brandeis University, Don Pearce of the Medical Instrument Shop at Duke University Medical Center, and Janet Patterson of the Physics Machine Shop at Duke University for their creative input and expertise in building the stereotaxic devices we describe here. This work was supported by the National Institutes of Health, National Science Foundation, and the John Merck Foundation.

References

1. Denk W, Strickler JH, Webb WW (1990) Two-photon laser scanning fluorescence microscopy. Science 248(4951):73–76
2. Göppert-Mayer M (1931) Über Elementarakte mit zwei Quantensprüngen. Ann Phys 401(3): 273–294. doi:10.1002/andp.19314010303
3. Denk W, Svoboda K (1997) Photon upmanship: why multiphoton imaging is more than a gimmick. Neuron 18(3):351–357
4. Piston DW (1999) Imaging living cells and tissues by two-photon excitation microscopy. Trends Cell Biol 9(2):66–69
5. Takahashi A, Camacho P, Lechleiter JD, Herman B (1999) Measurement of intracellular calcium. Physiol Rev 79(4):1089–1125
6. Euler T, Detwiler PB, Denk W (2002) Directionally selective calcium signals in dendrites of starburst amacrine cells. Nature 418(6900):845–852, 10.1038/nature00931
7. Svoboda K, Denk W, Kleinfeld D, Tank DW (1997) In vivo dendritic calcium dynamics in neocortical pyramidal neurons. Nature 385(6612):161–165. doi:10.1038/385161a0
8. Yuste R, Denk W (1995) Dendritic spines as basic functional units of neuronal integration. Nature 375(6533):682–684. doi:10.1038/375682a0
9. Yuste R, Tank DW (1996) Dendritic integration in mammalian neurons, a century after Cajal. Neuron 16:701–716
10. Grienberger C, Konnerth A (2012) Imaging calcium in neurons. Neuron 73(5):862–885. doi:10.1016/j.neuron.2012.02.011
11. Stosiek C, Garaschuk O, Holthoff K, Konnerth A (2003) In vivo two-photon calcium imaging of neuronal networks. Proc Natl Acad Sci U S A 100(12):7319–7324
12. Tsien RY (1981) A non-disruptive technique for loading calcium buffers and indicators into cells. Nature 290(5806):527–528
13. Probes M (2010) Acetoxymethyl (AM) and acetate esters. http://tools.invitrogen.com/content/sfs/manuals/g002.pdf
14. Kara P, Boyd JD (2009) A micro-architecture for binocular disparity and ocular dominance in visual cortex. Nature 458(7238):627–631. doi:10.1038/nature07721

15. Ohki K, Chung S, Ch'ng YH, Kara P, Reid RC (2005) Functional imaging with cellular resolution reveals precise micro-architecture in visual cortex. Nature 433(7026):597–603
16. Ohki K, Chung S, Kara P, Hubener M, Bonhoeffer T, Reid RC (2006) Highly ordered arrangement of single neurons in orientation pinwheels. Nature 442(7105):925–928
17. Huber D, Gutnisky DA, Peron S, O'Connor DH, Wiegert JS, Tian L, Oertner TG, Looger LL, Svoboda K (2012) Multiple dynamic representations in the motor cortex during sensorimotor learning. Nature 484(7395):473–478. doi:10.1038/nature11039
18. Li Y, Vanhooser SD, Mazurek M, White LE, Fitzpatrick D (2008) Experience with moving visual stimuli drives the early development of cortical direction selectivity. Nature 456(7224): 952–956, doi:nature07417 [pii]10.1038/nature07417
19. Van Hooser SD, Li Y, Christensson M, Smith GB, White LE, Fitzpatrick D (2012) Initial neighborhood biases and the quality of motion stimulation jointly influence the rapid emergence of direction preference in visual cortex. J Neurosci 32(21):7258–7266. doi:10.1523/JNEUROSCI.0230-12.2012
20. Kerlin AM, Andermann ML, Berezovskii VK, Reid RC (2010) Broadly tuned response properties of diverse inhibitory neuron subtypes in mouse visual cortex. Neuron 67(5):858–871. doi:10.1016/j.neuron.2010.08.002
21. Kuhlman SJ, Tring E, Trachtenberg JT (2011) Fast-spiking interneurons have an initial orientation bias that is lost with vision. Nat Neurosci 14(9):1121–1123. doi:10.1038/nn.2890
22. Runyan CA, Schummers J, Van Wart A, Kuhlman SJ, Wilson NR, Huang ZJ, Sur M (2010) Response features of parvalbumin-expressing interneurons suggest precise roles for subtypes of inhibition in visual cortex. Neuron 67(5):847–857. doi:10.1016/j.neuron.2010.08.006
23. Sohya K, Kameyama K, Yanagawa Y, Obata K, Tsumoto T (2007) GABAergic neurons are less selective to stimulus orientation than excitatory neurons in layer II/III of visual cortex, as revealed by in vivo functional Ca2+ imaging in transgenic mice. J Neurosci 27:2145–2149
24. Li Y, Lu H, Cheng PL, Ge S, Xu H, Shi SH, Dan Y (2012) Clonally related visual cortical neurons show similar stimulus feature selectivity. Nature 486(7401):118–121. doi:10.1038/nature11110
25. Jarosiewicz B, Schummers J, Malik WQ, Brown EN, Sur M (2012) Functional biases in visual cortex neurons with identified projections to higher cortical targets. Curr Biol 22(4):269–277. doi:10.1016/j.cub.2012.01.011
26. Ko H, Hofer SB, Pichler B, Buchanan KA, Sjostrom PJ, Mrsic-Flogel TD (2011) Functional specificity of local synaptic connections in neocortical networks. Nature 473(7345):87–91. doi:10.1038/nature09880
27. Bock DD, Lee WC, Kerlin AM, Andermann ML, Hood G, Wetzel AW, Yurgenson S, Soucy ER, Kim HS, Reid RC (2011) Network anatomy and in vivo physiology of visual cortical neurons. Nature 471(7337):177–182. doi:10.1038/nature09802
28. Briggman KL, Helmstaedter M, Denk W (2011) Wiring specificity in the direction-selectivity circuit of the retina. Nature 471(7337):183–188. doi:10.1038/nature09818
29. Chen X, Leischner U, Varga Z, Jia H, Deca D, Rochefort NL, Konnerth A (2012) LOTOS-based two-photon calcium imaging of dendritic spines in vivo. Nat Protoc 7(10):1818–1829. doi:10.1038/nprot.2012.106
30. Grienberger C, Adelsberger H, Stroh A, Milos RI, Garaschuk O, Schierloh A, Nelken I, Konnerth A (2012) Sound-evoked network calcium transients in mouse auditory cortex in vivo. J Physiol 590(Pt 4):899–918. doi:10.1113/jphysiol.2011.222513
31. Jia H, Rochefort NL, Chen X, Konnerth A (2010) Dendritic organization of sensory input to cortical neurons in vivo. Nature 464(7293): 1307–1312, doi:nature08947 [pii]10.1038/nature08947
32. Jia H, Rochefort NL, Chen X, Konnerth A (2011) In vivo two-photon imaging of sensory-evoked dendritic calcium signals in cortical neurons. Nat Protoc 6(1):28–35. doi:10.1038/nprot.2010.169
33. Garaschuk O, Milos RI, Grienberger C, Marandi N, Adelsberger H, Konnerth A (2006) Optical monitoring of brain function in vivo: from neurons to networks. Pflugers Arch 453(3):385–396. doi:10.1007/s00424-006-0150-x
34. Garaschuk O, Milos RI, Konnerth A (2006) Targeted bulk-loading of fluorescent indicators for two-photon brain imaging in vivo. Nat Protoc 1(1):380–386, doi:nprot.2006.58 [pii]10.1038/nprot.2006.58
35. Golshani P, Portera-Cailliau C (2008) In vivo 2-photon calcium imaging in layer 2/3 of mice. J Vis Exp 13. doi:10.3791/681
36. Helmchen F, Waters J (2002) Ca2+ imaging in the mammalian brain in vivo. Eur J Pharmacol 447(2–3):119–129
37. Kleinfeld D, Denk W (2000) Two-photon imaging of neocortical microcirculation. In: Yuste R, Konnerth A, Lanni F (eds) Imaging

neurons: a laboratory manual. Cold Spring Harbor Press, Cold Spring Harbor, NY

38. Mostany R, Portera-Cailliau C (2008) A method for 2-photon imaging of blood flow in the neocortex through a cranial window. J Vis Exp 12. doi:10.3791/678
39. Ohki K, Reid RC (2011) In Vivo Two-Photon Calcium Imaging in the Visual System. In: Helmchen F, Konnerth A (eds) Imaging in neuroscience: a laboratory manual. Cold Spring Harbor Press, Cold Spring Harbor, NY, pp 511–528
40. Rochefort NL, Grienberger C, Konnerth A (2011) In Vivo Two-Photon Calcium Imaging Using Multicell Bolus Loading of Fluorescent Indicators. In: Helmchen F, Konnerth A (eds) Imaging in neuroscience: a laboratory manual. Cold Spring Harbor Press, Cold Spring Harbor, NY, pp 491–500
41. Shih AY, Mateo C, Drew PJ, Tsai PS, Kleinfeld D (2012) A polished and reinforced thinned-skull window for long-term imaging of the mouse brain. J Vis Exp 61. doi:10.3791/3742
42. Svoboda K, Tank DW, Stepnoski RA, Denk W (2000) Two-photon Imaging of Neuronal Function in the Neocortex In Vivo. In: Yuste R, Konnerth A, Lanni F (eds) Imaging neurons: a laboratory manual. Cold Spring Harbor Press, Cold Spring Harbor, NY
43. Helmchen F, Denk W (2005) Deep tissue two-photon microscopy. Nat Methods 2(12):932–940. doi:10.1038/nmeth818
44. Helmchen F, Denk W (2002) New developments in multiphoton microscopy. Curr Opin Neurobiol 12(5):593–601
45. Svoboda K, Yasuda R (2006) Principles of two-photon excitation microscopy and its applications to neuroscience. Neuron 50(6): 823–839
46. Margrie TW, Brecht M, Sakmann B (2002) In vivo, low-resistance, whole cell recordings from neurons in the anaesthetized and awake mammalian brain. Pflugers Arch 444:491–498
47. Judkewitz B, Rizzi M, Kitamura K, Hausser M (2009) Targeted single-cell electroporation of mammalian neurons in vivo. Nat Protoc 4(6):862–869. doi:10.1038/nprot.2009.56
48. Nimmerjahn A (2011) Two-Photon Imaging of Neuronal Structural Plasticity in Mice during and after Ischemia. In: Helmchen F, Konnerth A (eds) Imaging in neuroscience: a laboratory manual. Cold Spring Harbor Press, Cold Spring Harbor, NY, pp 961–980
49. Nauhaus I, Nielsen KJ, Callaway EM (2012) Nonlinearity of two-photon Ca2+ imaging yields distorted measurements of tuning for V1 neuronal populations. J Neurophysiol 107(3):923–936. doi:10.1152/jn.00725.2011
50. Schummers J, Yu H, Sur M (2008) Tuned responses of astrocytes and their influence on hemodynamic signals in the visual cortex. Science 320(5883):1638–1643
51. Nauhaus I, Nielsen KJ, Disney AA, Callaway EM (2012) Orthogonal micro-organization of orientation and spatial frequency in primate primary visual cortex. Nat Neurosci 15(12):1683–1690. doi:10.1038/nn.3255
52. Probes M (2013) Fluorescence SpectraViewer. http://www.invitrogen.com/site/us/en/home/support/Research-Tools/Fluorescence-SpectraViewer.html
53. Xu C (2000) Two-photon Cross Sections of Indicators. In: Yuste R, Konnerth A, Lanni F (eds) Imaging neurons: a laboratory manual. Cold Spring Harbor Press, Cold Spring Harbor, NY
54. Johnson EN, Van Hooser SD, Fitzpatrick D (2010) The representation of S-cone signals in primary visual cortex. J Neurosci 30(31): 10337–10350, doi:30/31/10337 [pii]10.1523/JNEUROSCI.1428-10.2010
55. Histed MH, Bonin V, Reid RC (2009) Direct activation of sparse, distributed populations of cortical neurons by electrical microstimulation. Neuron 63(4):508–522, doi:S0896-6273(09)00545-5 [pii]10.1016/j.neuron.2009.07.016
56. Kerr JN, Greenberg D, Helmchen F (2005) Imaging input and output of neocortical networks in vivo. Proc Natl Acad Sci U S A 102(39):14063–14068
57. Rochefort NL, Narushima M, Grienberger C, Marandi N, Hill DN, Konnerth A (2011) Development of direction selectivity in mouse cortical neurons. Neuron 71(3):425–432. doi:10.1016/j.neuron.2011.06.013
58. Sato TR, Gray NW, Mainen ZF, Svoboda K (2007) The functional microarchitecture of the mouse barrel cortex. PLoS Biol 5(7):e189. doi:10.1371/journal.pbio.0050189

Chapter 7

Optogenetic Dissection of Neural Circuit Function in Behaving Animals

Carolina Gutierrez Herrera, Antoine Adamantidis, Feng Zhang, Karl Deisseroth, and Luis de Lecea

Abstract

One of the major challenges of modern neuroscience is to identify the anatomical and functional wiring of specific brain circuits to understand their respective role in higher brain functions. Gain- or loss-of-function assays using electrical, lesion, genetic, and pharmacological manipulation of molecular and cellular targets as well as correlative analysis of neural activities during selective behavior have substantially contributed to our knowledge on the function of neural network interactions. Technologies for imaging and controlling neural activities have progressively open new perspectives in the investigation of the neural substrates of brain functions. Optogenetics combines optical stimulation of genetically defined cell types through activation of microbial opsin-mediated insertion of light-sensitive ion channels to provide robust gain- or loss-of-function modulation of specific cell types at high temporal resolution. This powerful technology has emerged as a revolutionary force within modern neuroscience and has provided significant new insights into brain functions. Here, we will describe the successive steps for in vitro and in vivo optogenetics dissection of arousal circuits in mice.

Key words Viral targeting, Genetic engineering, Neural circuits, Behavior, Optogenetics

1 Introduction

The mammalian brain consists of heterogeneous cell types, including excitatory and inhibitory neurons and glial cells. Those cells are organized in complex circuits whose wiring diagram often remains unclear. Although the anatomy of some brain structure show a clear laminar structure (hippocampus, cortex, cerebellum), most of them lack evident structure or restricted cell clusters. Specific manipulation of neural circuits is accessible through the use of slow-acting, nonselective pharmacological or genetic interventions, whereas temporal precision was inevitably associated with poor spatial and genetic specificity, typically through electrical stimulation of target pathways. Although these approaches remain

Benjamin R. Arenkiel (ed.), *Neural Tracing Methods: Tracing Neurons and Their Connections*, Neuromethods, vol. 92, DOI 10.1007/978-1-4939-1963-5_7,

very useful in many experimental strategies, their inherent limitations have often hampered our understanding of brain functions.

Thus, understanding the organization and dynamics of neural networks in the brain represents one of the major goals of neuroscience which is to dissect the molecular and cellular components of neural circuits underlying selective brain functions. The identification of temporally precise control of specific neuronal subtypes may lead to insights into clinically relevant neuromodulation with fewer side effects and more robust therapeutic efficacy, as well as insights into the cellular basis of systems-level neural circuit function.

Recent developments in microbial opsin tools now allow to reverse-engineer intact neural circuits by directly testing their necessity and sufficiency to control specific behaviors. The high-speed kinetics of light-sensitive neuromodulatory opsins and their selective expression in genetically identified neurons have been successfully used to either activate or inhibit neuronal circuits in a variety of rodent models. Perturbations of targeted circuits with such unprecedented spatial and temporal resolutions have identified neuronal circuits that are causally controlling or modulating mammalian behaviors including arousal and sleep [1–3], reward [4–13], locomotion, memory [14–18], anxiety [19, 20], depression [21], aggression [22], and social interaction [23].

Here we detail the technical considerations for employing microbial opsins in both in vitro and in vivo systems and summarize protocols for integrating opsin-based neuromodulation with electrophysiological and behavioral readout methods in free-moving rodents.

2 Materials

This chapter provides a list of materials for applying in vitro and in vivo optogenetics. These typical reagents, equipments, and the accompanying procedures (commercial distributors, animal care committee, etc.) might slightly differ between laboratories.

3 Stereotactic Injection of Recombinant Viruses and Optical Fiber Implantation

Reagents

- Anesthetics (Mice: ketamine 80 mg/kg and xylazine 12 mg/kg; rats: 80 mg/kg and xylazine 6 mg/kg). Alternatively, isoflurane may also be used
- Analgesics (Buprenorphine: mice: 0.05–0.1 mg/kg; rats: 0.01–0.05 mg/kg; subcutaneous injection)
- Eye ointment

- Sterile PBS
- Ethanol
- Hydrogen peroxide 1 %
- C&B Metabond (Parkell)
- Dental cement
- Tissue adhesive
- Paraffin oil
- (Iso-)betadine
- High-titer lentivirus or AAV virus preparation (vector core facilities)

Equipments

- Small animal stereotactic frame with holder (Kopf Instruments)
- Surgical tools (scissors, tweezer, hemostats, scalpel blades, etc.)
- Programmable microsyringe pump
- 10 μl Hamilton microsyringe and corresponding PET tubing
- Optical fiber implants or cannula guide
- Stereoscope
- High-speed micro-drill with 0.3–0.9 mm stainless steel burrs
- 1 ml syringes with subcutaneous needles
- Surgical suture or skin glue (e.g., Vetbond 3 M)
- Cotton swabs
- Heating blanket

3.1 Simultaneous In Vivo Optical Stimulation and Electrophysiological Recording

Reagents

- Anesthetics (xylazine/ketamine, isoflurane)
- Sterile PBS
- Ethanol
- Hydrogen peroxide 1 %

Equipments

- High-power light sources (lasers, LED, other sources with appropriate filters)
- 200 μm multimode fiber patch (ThorLabs, Doric) with proper connectors (FC/PC, SMA, etc.)
- Fiber stripping tool
- Tungsten electrodes or glass pipette (1–10 m Ω or Optrode [24])
- Analog-to-digital board and amplifier
- Data recording system
- TTL/analog waveform generator

4 In Vivo Optical Stimulation for Behavioral Studies

Equipments

- High-power light sources (lasers, LED, other sources with appropriate filters)—200 μm multimode fiber patch with proper connectors (FC/PC, SMA, etc.)
- Optical swivel (e.g., rotary joint from Doric Inc.)
- Arbitrary waveform generator
- Light power meter
- Animal transduced with appropriate virus showing selective expression of opsin

5 Methods

This chapter provides a comprehensive method for applying optogenetics technology to circuit of interest in vitro and in vivo, including the choice of the opsin, the procedures for stereotactic virus injection and implantation of optical fibers into the rodent brain, and the experimental requirements for combining optogenetics with behavioral studies.

5.1 Choice of the Opsins

Phototactic orientation and photophobic behavioral responses observed in alga are dependent on the retinal-binding proteins, called rhodopsins, that mediate light-sensitive changes in the intracellular ionic constituency [25, 26]. Optogenetics results from the identification of some of those light-sensitive membrane proteins from lower organism and their use to control ionic influx in mammalian tissue with light [27–29]. A large number of synthetic opsins with distinct properties are now available to enable optical modulation of electrical activity in neurons. Most of them require retinal as a cofactor; however, its concentration in the mammalian is sufficient for functional expression of the opsins [27, 30, 31]. The discovery of light-sensitive channel and pumps and the progressive improvement of their kinetics through site-directed mutagenesis provided a large repertoire of distinct opsins for fast activation or silencing of individual cells [31]. Detailed comparison of the properties and experimental application of opsins are provided below.

5.1.1 Fast Activatory Opsins

In 2002, two independent groups isolated the genetic sequence that coded for two single-component light-sensitive channels, termed channelrhodopsins (ChR): channelrhodopsin-1 (ChR1), which becomes permeable to protons upon illumination with green light (510 nm), and channelrhodopsin-2 (ChR2), which allows cation transport upon blue light (470 nm) illumination [25, 26]. In contrast to ChR1, which showed poor temporal activation and deactivation kinetics, ChR2 provided millisecond time-scale activation of cells and led to significant depolarization of cellular membrane. In 2005, Boyden and collaborators [27] used viral technology to express ChR2 in cultured hippocampal neurons and showed that brief illumination with blue light (470 nm) was sufficient to reliably depolarize individual brain cells. Codon humanization and site-directed mutagenesis of ChR2 original sequence have provided a number of variants that address different limitations of the original form (hChR2) [1, 30, 32] but also highlighted some interesting properties in newly generated ChR2 mutants [33].

In 2010, Gunaydin and colleagues [34] identified a ChR2 variant with faster on- and off-kinetics. This mutant, called ChETA (ChR2 E123T accelerated), which substitutes Glu123 for threonine at E123, showed faster onset and decay kinetics compared to wild-type ChR2 with comparable photocurrent amplitudes. Wavelength for peak excitation was slightly red-shifted (~500 nm). Compared to wild-type ChR2, the use of ChETA opsin allowed a better temporal control with improved fidelity (i.e., less spike doublets, better light sensitivity) in response to higher stimulation frequencies (up to 200 Hz). Thus, it indicates that transfection of naturally fast-spiking neurons with ChETA can allow for precisely timed and sustained light-driven spikes at physiological frequencies.

Small conductances observed with ChR2 require high levels of channel expression at the cell membrane and stimulation with high-intensity blue light. Additionally, prolonged stimulation is known to lead to depolarization-dependent deactivation of ChR2 [35] therefore limiting long-term activation with sustained illumination. To overcome this, Berndt and colleagues have engineered high-efficiency channelrhodopsins that show larger photocurrents with very low light intensity (as low as 1.9 mW/mm^2) through the substitution of threonine 159 to cysteine, called ChR2(T159C). To overcome the slower decay kinetic of this new mutant at high light intensities and the occurrence of doublets or triplets, the authors combined the fast photocycle kinetic mutant E123T with T159C to develop a ChR2(ET/TC) mutant that exhibited increased sensitivity to low level of illumination while improving off-kinetics. Importantly, this ET/TC variant showed an increased sensitivity to red-shifted spectral excitation (up to 530 nm), a feature that is particularly useful for combinatorial use of 470- and 530 nm-sensitive opsins.

Besides these major ChR2 mutants, genomic screens and site-directed mutagenesis identified a large repertoire of opsins with slightly different properties. Those include the red-shifted C1V1 (also called VChR1 from the colonial alga *Volvox carteri*) [36], a cationic channelrhodopsin that shows peak responsiveness to 545 nm light (and up to 589 nm), as well as other opsins such as ChIEF and ChARGR. Detailed description of those opsins has been recently described [33].

5.1.2 Slow Activatory Opsins

Additional H134R ChR2 variant (and likewise the T159C and L132C variants) shows a twofold increase in photocurrent amplitude but also has a twofold slowing of the decay kinetics of the channel, resulting in poor temporal precision [23]. Such slowing of decay kinetic has resulted in the engineering of step-function opsins (SFO), which show a prolonged photocycle due to mutations in the amino acid sequence.

Under control conditions, ChR2 shows a decay kinetic ~10 ms, which is sufficient to elicit a spiking activity. In contrast, mutations of cysteine-128 and aspartate-156 in ChR2 result in a significantly slower closing, with decay kinetics up to 100 s in the ChR2 C128T mutant and 20 min for stabilized step-function opsin (SSFO). Therefore, illumination with blue light can induce a persistent state of depolarization to threshold and near-threshold potentials. These bistable opsin channels offer a useful tool in optogenetic research to increase cellular excitability of target cells. While blue light triggers depolarization of the cell, this persistent excitability can be terminated with a brief yellow light pulse (590 nm), enabling a "steplike" activation and deactivation of SFO-expressing cell to allow for precise control of "up" states.

Importantly, both electrical and optogenetic activation of neurons with ChR2 may induce synchronization of a large assemble of cells which may not mimic natural firing states. In contrast, SFO-mediated excitability provides an alternative tool to overcome such hypersynchrony and allow neural network to fire at more physiologically relevant firing patterns.

5.1.3 Fast Silencing Opsins

In contrast to gain of function (causality) approaches, loss-of-function protocols provide more robust insight into the significance (necessity) of a specific cell type contribution to the studied behavior. Pharmacological, lesion-based intervention or genetic ablation of genes, cells, or circuits lacks physiologically relevant temporal resolution. In addition those are often nonreversible and rarely spatially limited (i.e., passive diffusion of infused molecules can reach distant targets). For these purposes, fast optogenetic inhibition can provide reversible, specific, and fast optical silencing of genetically targeted cell types, allowing for accurate microdissection of neural circuit function in the brain.

Previous biochemical work had demonstrated that *halobacteria* within the *Archaea* contained a number of light-sensitive proteins that, when activated by specific wavelengths of light, can hyperpolarize cells through a number of divergent and efficient processes [37]. These rhodopsins, which include bacteriorhodopsins (BR) and halorhodopsins (HR), are single component-based proteins that act as proton (expulse protons) or chloride pump (pump in chloride ions), respectively, to hyperpolarize the target cell.

Genomic screening of halobacteria identified a light-driven protein from the archaebacterium *Natronomonas pharaonis*, halorhodopsin (NpHR), that pumps in chloride influx when illuminated by yellow light [38–40]. Five hundred and ninety nanometer light illumination of virally transduced cultured hippocampal neurons expressing a mammalian codon-optimized variant of NpHR induced a strong and sustained outward currents (and concomitant hyperpolarization).

In contrast to ChR2, NpHR requires constant light illumination to activate its photocycle, and hyperpolarization is terminated upon cessation of light delivery. Although the original NpHR opsin worked efficiently to hyperpolarize neurons, enhanced NpHR2.0 (eNpHR2.0) was introduced to improve expression levels and trafficking of the protein to the membrane through addition of endoplasmic reticulum export motifs cloned from Kir 2.1 ion channels [41]. Further development in eNpHR3.0 added a neurite trafficking sequence, resulting in the widespread membrane expression through the soma and dendrites of the transfected neuron. Interestingly, the addition of the neurite trafficking sequence in eNpHR3.0 has increased the sensitivity of the opsin to further red-shifted light spectrums (up to 680 nm, which is close to the infrared border). This enhanced expression resulted in significantly improved light-induced hyperpolarizations averaging near 40–50 mV and up to 5 nA photocurrents.

Finally, alternative powerful methods to remotely control neural activity to varying degree exist based on single or dual components. First and second generation of orthologous GPCR-ligand pairs include receptors activated solely by synthetic ligand (*RASSLs*) and designer receptors exclusively activated by designer drug (*DREADDs*—review [42]), respectively, that responds to small (non-endogenous) molecule. Upon activation, they enable manipulation of G proteins (Gq, Gi, and Gs) signaling pathways. For instance, nonnative or peptide receptors—e.g., TRPV1 ligand-gated ion channel [43] and the *Drosophila melanogaster* allatostatin receptor (AlstR), FMRF-amide channel [44], as well as light-gated glutamate receptor, ChARGe [45], ivermectin-gated chloride channel [46], and optoXR receptor chimeras [47]—have successfully provided control over neuronal activity with various degree of temporal, spatial, and directional control (see review in [42]).

5.2 Stereotactic Injection of Lentivirus (LV)/Adeno-Associated Viruses (AAV) and Optical Fiber Implants

Opsins can be selectively expressed in specific subsets of neurons in the rodent brain using a variety of modern genetic targeting strategies [48]. Those include stereotactic injection of recombinant viruses or the generation of transgenic mice models. Recombinant virus (LV or AAV) represents a versatile tool to target the local brain area or a subset of cells in a given area. Once injected into the brain target, the expression of viral construct is controlled by the promoter. Thus, the identification of a minimal promoter–recombinant virus vector is limited in size, and the use of small construct often results in better expression and is strongly advised for efficient viral transduction. Several excellent reviews have recently summarized the use of selective promoters [23, 30].

Alternatively, we and others have developed an alternative strategy using cre-inducible AAV vector with a double-floxed inverted open reading frame, wherein the gene encoding the opsin fused to an eYFP fluorescent tag was inserted in the antisense orientation. After infusion of cre-dependent AAV in the targeted brain area of cre-driver animals (transgenic animal expressing the cre-recombinase in targeted neurons only), the cassette flipped upon cre-mediated recombination, and the opsin expression in targeted neurons reached a maximum within ~20 days after AAV infusion.

In some case, transgenic animals expressing opsins in subsets of neurons (neurons expressing parvalbumin, somatostatin, vesicular GABA transporter, dopamine transporter, etc.) are already available and can be used directly for in vitro or in vivo optogenetic experiments. A more exhaustive list of available cre-driver is available from different databases and institutes (http://www.alleninstitute.org, http://www.informatics.jax.org/recombinase.shtml, http://www.credrivermice.org, http://www.creportal.org, http://creline.org). If a transgenic model is not available, then new transgenic model must be generated. In that case, it is strongly recommended to use a knock-in approach since it is often more selective than regular bacterial artificial chromosome (BAC) transgenic approach (i.e., with less ectopic expression of the transgene).

Recently, cre- or flp-dependent constitutive opsin transgenic animals have been generated [6, 49, 50]. Those animals do not require virus injection since opsins are expressed by breeding cre-driver animals with cre-dependent constitutive animals. Note that this approach will target every subpopulation expressing cre, making it difficult to target a single circuit. Finally, in both viral and transgenic targeting, long-term expression (up to 6 months) of opsins has been reported.

- Surgical procedures and animal handling must follow institutional and national ethic guidelines.
- Sterilize field (ethanol 70 %) and all surgical tools (autoclaving or immersion in a disinfectant).

- Anesthetize animals using isoflurane (1.5 %, for surgeries longer than 1 h) or xylazine/ketamine cocktail (IP). Check for the absence of toe-pinch reflex and apply ophthalmic ointment to prevent eye drying. Keep the animal warm by using a heating pad during the surgical procedure and the recovery step.
- Place the animal in a stereotactic frame. Correct head position in the stereotactic frame allows vertical, but no lateral movement of the head. Use appropriate ear bars for mice and rats.
- Inject saline solution subcutaneously to prevent dehydration during surgery (30 ml/kg).
- Shave and clean the head, wipe with 70 % ethanol followed by application of (iso)-betadine. Make a midline scalp incision of ~1 cm using scalpel and expose the skull. Use hydrogen peroxide to clean the skull with cotton swabs. Stop the reaction by applying sterile PBS or saline.
- Align bregma and lambda to the same dorsal–ventral coordinates by adjusting the height of the nose clamp on the stereotactic frame (clamp a metallic needle on the holder to mark skull suture). Place the needle above the target area (use brain atlas stereotactic coordinates) and use a micro-drill to make a small craniotomy (slightly larger diameter than the optical fiber). Drill can be mounted on the stereotactic frame for ease of use. Use fine forceps to remove the dura while avoiding contact with the surface of the cortex.
- Thaw the virus on ice (virus aliquots are stored a –80 °C for long-term storage). Lentivirus can be kept on ice for 6 h, whereas AAV can be stored at +4 °C for at least 1 month without the loss of significant titer. Fill the syringe (10 μl microsyringe), the tubing, and the glass needle or injection needle (e.g., 34 G metal needle) with paraffin oil. Load the virus solution in the metal or glass needle using the programmable microsyringe pump (withdrawal function).
- Dry the skull with cotton swabs. Perform virus injection. Inject 0.2–0.5 μl of virus into the brain. Injection volume should not exceed 1 μl and flow rate should be low to avoid tissue damage (0.05–0.1 μl/min) (see Note 1). After injection leave the needle in place for 10 additional minutes to allow the virus to diffuse in the brain. Withdraw the needle slowly afterwards. After completely withdrawing the needle, check that it is not clogged by pumping out a small droplet (0.1 μl).

5.3 Optical Fiber Implantation for Optical Control

Selective expression of opsins in genetically defined neurons makes it possible to control a subset of neurons without affecting nearby cells and processes in the intact brain. There are several important factors for achieving effective optical stimulation in vivo (Fig. 1).

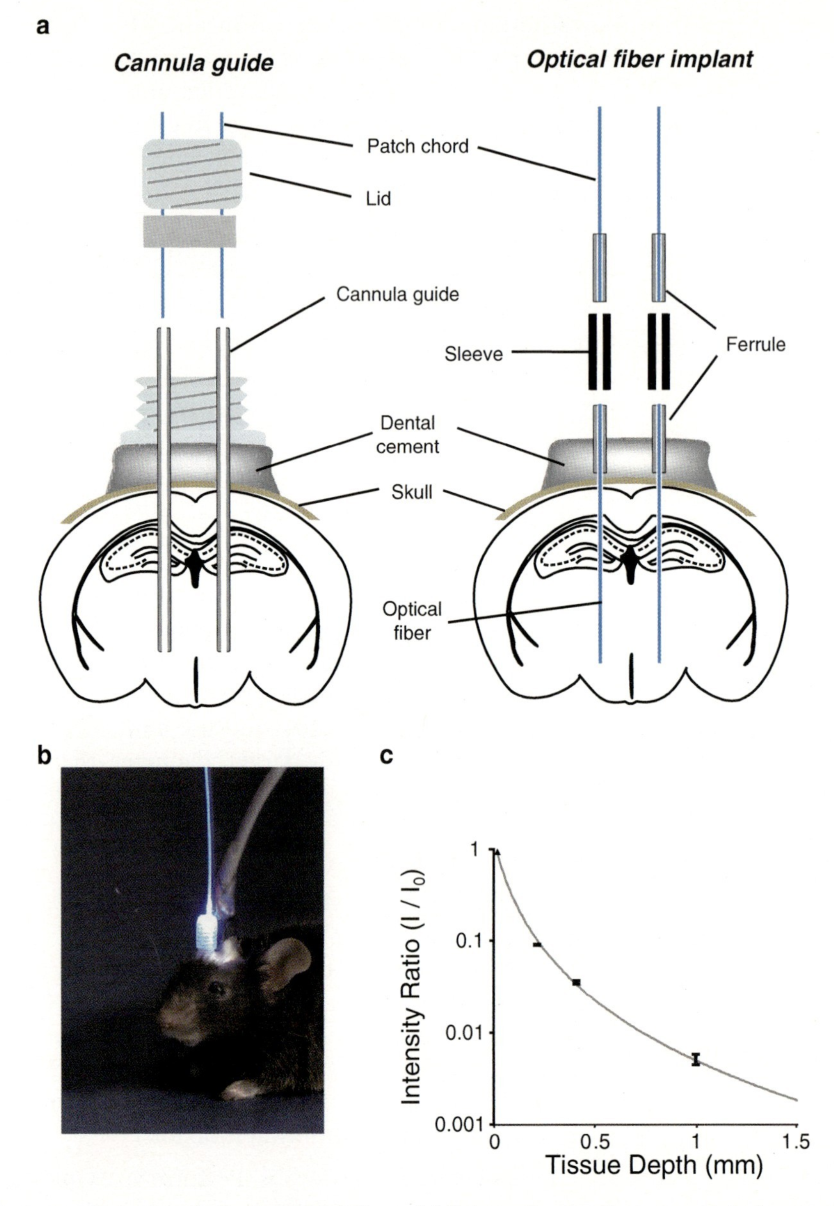

Fig. 1 Optical neural interface for in vivo light delivery. (**a**) Schematic of optical neural interfaces using cannula guide (*left*) or optical fiber implant (*right*) for optogenetic control of neurons located in the lateral hypothalamus. Note that the tip of the optical fiber is positioned ~200 μm above the target area. (**b**) Behavioral setup used for in vivo deep brain optical stimulation and electroencephalogram/electromyogram (EEG/EMG) recording in free-behaving mice. Picture shows a unilateral optical fiber inserted through a cannula guide chronically implanted on the skull of the animal. For illustration purpose, the furcation tubing has been removed from the optical fiber that is no longer opaque. EEG/EMG tether for polysomnographic recording is positioned caudal to the optical fiber (*gray cable*). (**c**) Normalized light intensity (mW/mm^2) as a function of lateral hypothalamus tissue depth. Values were experimentally determined by measuring the light intensity after transmission through a given tissue thickness and dividing by the intensity of the light emanating from the optical fiber tip. Tissue of different thickness was prepared in the form of acute brain slices from adult C57BL6 mice. Error bars indicate 1 SD from the mean. Curve fits were produced using the Kubelka–Munk model of light transmission through diffuse scattering media. Modified from Adamantidis et al. [1]

First, the light source must be of sufficient power and stable (<5 % variation) for reliable optical control. It is estimated that ChR2 and eNpHR require 1 and 10 mW/mm^2, respectively, of light for proper modulation of expressing cells [31]. For in vitro applications, light-emitting diodes (LEDs) or solid-state diode lasers can be used. For most in vivo applications, solid-state diode lasers are preferred since they provide more light power than LEDs.

Second, multimode optical fibers (62–400 μm diameter) are used to deliver light into the target brain regions. Virus injection and optical fibers can be guided via the same stereotactically implanted cannula guide to ensure co-registration of light illumination and opsin-expressing cells. An alternative strategy consists to implant optical fiber with a ferrule/sleeve connectors on the skull of the animals. Those will reduce the invasiveness of the implant and the risk of breaking optical fibers during their insertion in the cannula guide. However, they won't allow the combination of optogenetics with local drug delivery concomitant with optical stimulations.

In the following sections, we provide a detailed protocol for implanting chronic optical fiber for in vivo optogenetic control [1, 32, 51]. See Note 2 for fabrication and details regarding optical fiber implants. Detailed protocols for stereotactic surgery have been described previously [52].

- To ensure proper light power delivery in the brain tissue during in vivo experiment, measure the light transmission of every optical fiber implants before starting the surgery.
- Hold the optical fiber implant with the holder from the stereotactic apparatus (some holder are commercially available). Place the optical fiber implant above the targeted structure (use bregma for reference of stereotactic coordinates [53]) and drill a hole as described above.
- Remove the dura with a needle and slowly lower the optical fiber implant in the brain tissue. Apply a thin layer of C&B Metabond on the exposed skull and around the optical fiber implant.
- Once the C&B Metabond hardens, release the implant from the holder and secure the implant and other implants (e.g., cannula guide for drug delivery, probe connectors for electrophysiology) with dental cement. Glue the skin back with Vetbond surgical adhesive. Suture is preferable if skin opening is large.
- Administrate painkiller (e.g., buprenorphine 0.03 mg/kg) subcutaneously following the surgical procedure to minimize pain/discomfort and let the animal fully recover in a clean cage placed over a heating blanket. Monitor the recovery of the animal every 10 min.

5.4 Simultaneous In Vivo Optical Stimulation and Electrophysiological Recording

Optogenetics allows high temporal control of neural activities in the brain. In combination with in vivo electrophysiology, it can provide useful information on the activity of targeted cells or downstream circuits using equipment to record single- or multi-unit activity, local field potentials, or electroencephalogram.

Here, we detail procedure for recording single-unit activity in an anesthetized animal preparation. There are several probes that can be coupled to an optical fiber including a tungsten wire, a glass pipette, or an optical probe. Although they are different, their use to record optically evoked action potentials requires identical preparation.

- Design the tip of the electrode (tungsten or glass pipette) with a >0.3 mm projection and place the electrode on the holder from the stereotactic frame.
- Perform craniotomy around the stereotactic coordinates in stereotactic frame as described above.
- Remove the dura with a needle and lower the optrode to the desired depth without damaging the electrode tip. Avoid touching the skull or the cannula directly with the electrode tip. Attach a ground wire (silver chloride wire) to the skull, classically above the cerebellum, as a reference signal.
- Connect the optical fiber to a laser diode of desired wavelength and the electrode to an amplifier. For single-unit recordings, the recorded signal is band pass filtered between 300 Hz and 5 kHz. For field potentials and EEG, low-pass filter the signal at 300 or 100 Hz, respectively. Use a Faraday cage to shield out electrical noise.
- Anesthetized procedures are usually considered as terminal procedure and animal is euthanized immediately after recording. Note that the brain can be extracted anatomical studies.

5.5 In Vivo Optical Activation/Silencing for Behavioral Studies

Gain- or loss-of-function optogenetic experiments have provided significant insights into the neural substrate of several brain functions, including arousal, learning and memory, goal-oriented behaviors, and emotional perception. Here, we summarize important parameters to consider when conducting in vivo optogenetic experiments. Those parameters vary with the choice of species, strain, gender, age, housing condition, handling, and sample size.

- When conducting behavioral experiment for the first time, we recommend conducting a pilot behavioral test (see Note 4). Although those animals usually do not undergo surgical procedure (virus infusion or optical fiber implantation), result from such pilot experiments will serve as reference value for further comparison of control as well as gain- and loss-of-function optogenetic experiments.

- Because rodents are nocturnal, in general, tests that require the subjects to be spontaneously active and explore the apparatus for prolonged period of time are preferentially done during the dark phase of the light/dark cycle. This holds for most of the behavioral tasks, in particular cognitive task and instrumental conditioning.
- Once the animal have been injected with optogenetic virus and implanted with optical fiber implant or cannula guide, there is a mandatory recovery period. During this recovery period (3–10 days) and virus incubation (15–20 days), mice are usually single (or pair) housed to prevent fighting and damage of the implanted fiber (see Note 4). This period should be used for habituation to the testing conditions (behavioral setup and handling) before any behavioral assay to reduce the stress level during the actual experiment. It is also important to handle animals for habituation to the testing conditions.
- Before the behavioral experiment, insert the optical fiber in to the cannula guide or connect optical fiber implant to the patch cord using a zirconia sleeve. Connect the FC end of the patch cord to the PC connector on the output port of the laser. Use a waveform generator to generate the desired stimulation protocol (i.e., pulse duration, frequency, intensity). Optical stimulation can be triggered manually or time lock to specific cue through a feedback connection [4]. Note that the use of SSFO opsins avoids long-term connection and habituation of patch cord [23].
- In addition to the recovery period, most behavioral experiment will require the animal to habituate to the fiber-patch cord connection (the experimenter should practice fiber insertion/removal several times before the actual experiment) and the testing room (on general animals are placed in the testing room few hours before).
- Behavioral instruments used for in vivo optogenetic experiments can be easily adapted to the optical fiber tether. Use optical swivels to relieve torsion during long-term connection of animals to light source.
- The parameters of the behavioral paradigm (number of trials, sessions, their duration, number of animals to be used) largely depend on the task to be used. For each experiment, we recommend to use parameters based on pilot experiment, statistical significance, and the literature.
- At the end of behavioral assay, all experimental subjects need to be analyzed to verify the location of the optical fiber implant or the cannula guide, viral expression, and any unusual brain damage or infection.

6 Notes

Optogenetics and related technologies offer a large repertoire of powerful tools for interrogating the wiring and function of neural circuitries. Applying optogenetic to specific neuronal circuit requires careful verification steps. Each of them involves technical skills in molecular, cellular, and behavioral techniques. We have summarized an extensive list of troubleshooting notes that shall support the successful use of optogenetics.

6.1 Note 1: Efficiency of Virus Transduction In Vivo

The efficiency of viral transduction depends on many factors that require injection of important steps of optimization. The first step is to choose a targeting strategy and the virus to be used (e.g., lentivirus, AAV, cre-dependent-AAV). Then, for successful targeting, various amounts of virus should be stereotactically injected in the brain area of interest, and parameters such as the spread of virus, the efficiency and selectivity of the virus transduction, and the expression delay should be defined. Classically, the virus transduction efficiency is established empirically by injecting various amounts of virus in the target area (0.2 μl increment, up to 1 μl) and very slow injection speed (max. 0.1 μl per min). Although packaging and purification methods for virus preparation are highly standardized, each virus preparation has different titer and various degrees of purity (e.g., autofluorescent cellular debris). Once the transduction method has been established, it is recommended to test every new preparation of virus. Note that the efficiency of the transduction depends on the cell type to be targeted, the density of the brain area, and the serotype of the AAV virus used for the genetic targeting. Many different AAV serotypes have been used to transfect neurons (e.g., AAV2, AAV2/5, AAV5, etc.) with different efficiency. It is recommended to avoid the injection of virus solution close to bundle of fibers, since it is likely that the virus solution will spread along the fibers.

6.2 Note 2: Light Transmission: Optical Fiber Implant

There are two types of optical neural interfaces that can be used to deliver light deep into the brain tissue in free-moving mice or rats, cannula guide, and optical fiber implants; both can be uni- or bilateral. Note that several optical fiber implants can be implanted to optimize light delivery to large brain structure (e.g., hippocampus) or multiple site optogenetic experiments [3]. If the brain area is large, cannula guide is used when intracerebral or intracerebroventricular injection is combined to optogenetic control of target cells. Optical fiber implant does not allow concomitant pharmacology experiment but minimizes invasiveness. In addition, their use reduces the risk of breaking optical fiber during insertion into the cannula guide during behavioral experiments. Their chronic implantation should avoid sinuses by using mediolateral or rostrocaudal angle (5–30 C).

Optical fibers can be easily engineered before their insertion into the cannula guide. This has been described in Zhang et al. [54]. Briefly, it consists of a screwing cap that secures the fiber to the cannula guide implanted on the animal's skull, a fiber guard, and a bare fiber whose length is customized based on the depth of the target region.

Optical fiber implants are available from several optic companies; however, their fabrication is easy and cost effective [51]. Briefly, bare or coated optical fiber, ferrule, and sleeves can be cut with diamond knife and assembled with epoxy glue. It is important to check the cut of the optical fiber under a microscope. The edges of the optical fiber must be perpendicular for proper light dispersion (avoid beveled cut). It is recommended to polish optical fiber for optimized light transmission. Finally, it is important to measure the light transmission of each implant using a power meter before surgically implanting the optical fiber implant. To calculate light scattering through the brain, use the formula provided on optogenetics.org.

6.3 Note 3: Behavioral Control Conditions

When conducting behavioral experiment, it is important to first validate the experimental strategy by running a sham experiment on wild-type animal from the same genetic background of the ones to be used in the optogenetic experiments. We recommend using mice ~8–10 weeks of age with a cohort size of 10–12 for each testing condition. At this age, their skulls have matured to adult size, ensuring precise stereotactic targeting. We also recommend conducting a 3-min open field test on the test and control cohorts using the same stimulation paradigm to determine any potential stimulation-induced side effects on stereotypic behavior, exploration, locomotor activity, memory, and anxiety.

Several control experiments can be conducted:

- A control cohort that has received a sham surgery with control virus (i.e., YFP expressing virus).
- Another useful control is to optically stimulate a control cohort that has received a sham surgery with no virus injection.
- Test optical stimulation in the experimental cohort using a different wavelength that does not excite the opsin (e.g., 630 nm red light for ChR2).

Such control experiments validate the absence of side effect from the surgical procedure, viral vector expression, or light-induced perturbations (light-responsive cells, temperature, etc.), as well as the reliability of the behavioral task to be used. This is particularly important when the experimenter has no previous behavioral experience or data to compare with.

In control experiments, light stimulation should not produce any behavioral effect. When applicable, positive control experiments can also be conducted using electrical stimulation, pharmacological intervention, or natural reward. In addition, it is important to visually observe the test subject after surgical procedures to check for any abnormal behavior (e.g., circling behavior, asymmetrical posture, normal feeding/drinking behaviors, circadian locomotor activity, etc.).

6.4 Note 4: Long-Term Optical Fiber Tethering

Longitudinal studies require acute or chronic tethering of the optical fiber to the optical neural interface (e.g., cannula guide or optical fiber implants). For several reasons, it is often advised to minimize animal handling during behavioral experiments to reduce stress and confounding factors as much as possible. Thus, to reduce animal handling and risk of optical fiber breakage during connection and deconnection of optical fiber during behavioral experiment, one can use optical swivel (e.g., from Doric Inc.) and keep animal chronically tethered during the entire time of the experiment. Note that optical fiber should be opaque to avoid visual stimulus to the animal. This can be achieved by using black (or colored but opaque) furcation tubing, which also makes the optical fiber more resistant to animals' movement. Several companies are currently developing optical–electrical swivel for concomitant recording of electrical activity and optogenetic modulation of neural activities in free-moving rodents.

References

1. Adamantidis AR, Zhang F, Aravanis AM, Deisseroth K, de Lecea L (2007) Neural substrates of awakening probed with optogenetic control of hypocretin neurons. Nature 450:420–424
2. Carter ME, Adamantidis A, Ohtsu H, Deisseroth K, de Lecea L (2009) Sleep homeostasis modulates hypocretin-mediated sleep-to-wake transitions. J Neurosci 29:10939–10949
3. Carter ME et al (2012) Mechanism for hypocretin-mediated sleep-to-wake transitions. Proc Natl Acad Sci U S A 109:E2635–E2644
4. Adamantidis AR et al (2011) Optogenetic interrogation of dopaminergic modulation of the multiple phases of reward-seeking behavior. J Neurosci 31:10829–10835
5. Tsai H-C et al (2009) Phasic firing in dopaminergic neurons is sufficient for behavioral conditioning. Science 324:1080–1084
6. Witten IB et al (2011) Recombinase-driver rat lines: tools, techniques, and optogenetic application to dopamine-mediated reinforcement. Neuron 72:721–733
7. Brown MTC et al (2012) Ventral tegmental area GABA projections pause accumbal cholinergic interneurons to enhance associative learning. Nature 492:452–456
8. Stamatakis AM, Stuber GD (2012) Activation of lateral habenula inputs to the ventral midbrain promotes behavioral avoidance. Nat Neurosci 15:1105–1107
9. Stuber GD et al (2011) Excitatory transmission from the amygdala to nucleus accumbens facilitates reward seeking. Nature 475:377–380
10. Stamatakis AM, Stuber GD (2012) Optogenetic strategies to dissect the neural circuits that underlie reward and addiction. Cold Spring Harb Perspect Med 2. doi:10.1101/cshperspect.a011924
11. van Zessen R, Phillips JL, Budygin EA, Stuber GD (2012) Activation of VTA GABA neurons disrupts reward consumption. Neuron 73: 1184–1194
12. Jennings JH et al (2013) Distinct extended amygdala circuits for divergent motivational states. Nature 496:224–228
13. Vaziri A, Emiliani V (2012) Reshaping the optical dimension in optogenetics. Curr Opin Neurobiol 22:128–137

14. Ramirez S et al (2013) Creating a false memory in the hippocampus. Science 341:387–391
15. Liu X et al (2012) Optogenetic stimulation of a hippocampal engram activates fear memory recall. Nature 484:381–385
16. Nakashiba T et al (2012) Young dentate granule cells mediate pattern separation, whereas old granule cells facilitate pattern completion. Cell 149:188–201
17. Goshen I et al (2011) Dynamics of retrieval strategies for remote memories. Cell 147: 678–689
18. Warden MR et al (2012) A prefrontal cortex-brainstem neuronal projection that controls response to behavioural challenge. Nature 492:428–432
19. Tye KM et al (2011) Amygdala circuitry mediating reversible and bidirectional control of anxiety. Nature 471:358–362
20. Kim S-Y et al (2013) Diverging neural pathways assemble a behavioural state from separable features in anxiety. Nature 496:219–223
21. Tye KM et al (2013) Dopamine neurons modulate neural encoding and expression of depression-related behaviour. Nature 493: 537–541
22. Lin D et al (2011) Functional identification of an aggression locus in the mouse hypothalamus. Nature 470:221–226
23. Yizhar O et al (2011) Neocortical excitation/inhibition balance in information processing and social dysfunction. Nature 477:171–178
24. LeChasseur Y et al (2011) A microprobe for parallel optical and electrical recordings from single neurons in vivo. Nat Methods 8:319–325
25. Nagel G et al (2002) Channelrhodopsin-1: a light-gated proton channel in green algae. Science 296:2395–2398
26. Nagel G et al (2003) Channelrhodopsin-2, a directly light-gated cation-selective membrane channel. Proc Natl Acad Sci U S A 100: 13940–13945
27. Boyden ES, Zhang F, Bamberg E, Nagel G, Deisseroth K (2005) Millisecond-timescale, genetically targeted optical control of neural activity. Nat Neurosci 8:1263–1268
28. Miesenbock G (2009) The optogenetic catechism. Science 326:395–399
29. Deisseroth K (2012) Optogenetics and psychiatry: applications, challenges, and opportunities. Biol Psychiatry 71:1030–1032
30. Zhang F, Wang L-P, Boyden ES, Deisseroth K (2006) Channelrhodopsin-2 and optical control of excitable cells. Nat Methods 3:785–792
31. Yizhar O, Fenno LE, Davidson TJ, Mogri M, Deisseroth K (2011) Optogenetics in neural systems. Neuron 71:9–34
32. Aravanis AM et al (2007) An optical neural interface: in vivo control of rodent motor cortex with integrated fiberoptic and optogenetic technology. J Neural Eng 4:S143–S156
33. Mattis J et al (2012) Principles for applying optogenetic tools derived from direct comparative analysis of microbial opsins. Nat Methods 9:159–172
34. Gunaydin LA et al (2010) Ultrafast optogenetic control. Nat Neurosci 13:387–392
35. Berndt A, Yizhar O, Gunaydin LA, Hegemann P, Deisseroth K (2009) Bi-stable neural state switches. Nat Neurosci 12:229–234
36. Zhang F et al (2008) Red-shifted optogenetic excitation: a tool for fast neural control derived from volvox carteri. Nat Neurosci 11: 631–633
37. Mukohata Y (1994) Comparative studies on ion pumps of the bacterial rhodopsin family. Biophys Chem 50:191–201
38. Han X, Boyden ES (2007) Multiple-color optical activation, silencing, and desynchronization of neural activity, with single-spike temporal resolution. PLoS One 2:e299
39. Han X et al (2009) Millisecond-timescale optical control of neural dynamics in the nonhuman primate brain. Neuron 62:191–198
40. Zhang F et al (2007) Multimodal fast optical interrogation of neural circuitry. Nature 446: 633–639
41. Gradinaru V et al (2010) Molecular and cellular approaches for diversifying and extending optogenetics. Cell 141:154–165
42. Rogan SC, Roth BL (2011) Remote control of neuronal signaling. Pharmacol Rev 63: 291–315
43. Arenkiel BR, Klein ME, Davison IG, Katz LC, Ehlers MD (2008) Genetic control of neuronal activity in mice conditionally expressing TRPV1. Nat Methods 5:299–302
44. Schanuel SM, Bell KA, Henderson SC, McQuiston AR (2008) Heterologous expression of the invertebrate FMRFamide-gated sodium channel as a mechanism to selectively activate mammalian neurons. Neuroscience 155:374–386
45. Zemelman BV, Lee GA, Ng M, Miesenböck G (2002) Selective photostimulation of genetically chARGed neurons. Neuron 33: 15–22
46. Li P, Slimko EM, Lester HA (2002) Selective elimination of glutamate activation and introduction of fluorescent proteins into a Caenorhabditis elegans chloride channel. FEBS Lett 528:77–82
47. Airan RD, Thompson KR, Fenno LE, Bernstein H, Deisseroth K (2009) Temporally precise

in vivo control of intracellular signalling. Nature 458:1025–1029
48. Luo L, Callaway EM, Svoboda K (2008) Genetic dissection of neural circuits. Neuron 57:634–660
49. Taniguchi H et al (2011) A resource of Cre driver lines for genetic targeting of GABAergic neurons in cerebral cortex. Neuron 71: 995–1013
50. Tanaka KF et al (2012) Expanding the repertoire of optogenetically targeted cells with an enhanced gene expression system. Cell Rep 2:397–406
51. Sparta DR et al (2012) Construction of implantable optical fibers for long-term optogenetic manipulation of neural circuits. Nat Protoc 7:12–23
52. Cetin A, Komai S, Eliava M, Seeburg PH, Osten P (2006) Stereotaxic gene delivery in the rodent brain. Nat Protoc 1:3166–3173
53. Paxinos G, Franklin K (2014) The mouse brain in stereotaxic coordinates.
54. Zhang F et al (2010) Optogenetic interrogation of neural circuits: technology for probing mammalian brain structures. Nat Protoc 5:439–456

Chapter 8

Remote Control of Neural Activity Using Chemical Genetics

Andrew J. Murray and Peer Wulff

Abstract

Understanding how the nervous system functions requires a methodological toolbox for the manipulation of neuronal circuits. Over the last decade, there has been an explosion in the availability of methods to map and manipulate genetically defined populations of neurons. The control of neural signaling via pharmacological receptor-ligand interactions, or chemical genetics, allows for the modulation of neural signaling at an intermediate time scale and provides a complimentary approach to other technologies such as optogenetics. Here, we review the variety of chemical genetic techniques that are currently available and discuss the considerations that must be undertaken when choosing a technique for a particular experimental system.

Key words Chemical genetics, pharmacogenetics, Ligand-receptor pairs, Circuit, TRPV1, GluCl, Zolpidem, GABA receptors, PSEMs, GPCRs, Allatostatin, DREADDs, hM3Dq, hM4Di

1 Introduction

Understanding the cellular and molecular basis of behavior is the major challenge in neuroscience. To this end, the last decade has seen an exponential increase in the availability of methods to map and manipulate specific genetically defined populations of neurons, making a comprehensive understanding of how specific subcircuits contribute to behavior a reasonable possibility. A major advantage that modern-day neuroscientists have over their historical counterparts is the breadth and variety of technologies that can be used to manipulate neuronal output (see Fig. 1). These approaches range from more classical lesion studies [1] and the cell-type-specific block of transmitter release [2, 3] to the temporal control of neural activity by pharmacology—termed chemical genetics—or light, termed optogenetics [4–6]. Individually, each of these techniques can provide a degree of insight into circuit activity; however, each also comes with their own unique set of advantages and disadvantages.

As an example, optogenetics has gained recent preeminence for its unprecedented temporal control of neural activity but there

Benjamin R. Arenkiel (ed.), *Neural Tracing Methods: Tracing Neurons and Their Connections*, Neuromethods, vol. 92, DOI 10.1007/978-1-4939-1963-5_8,

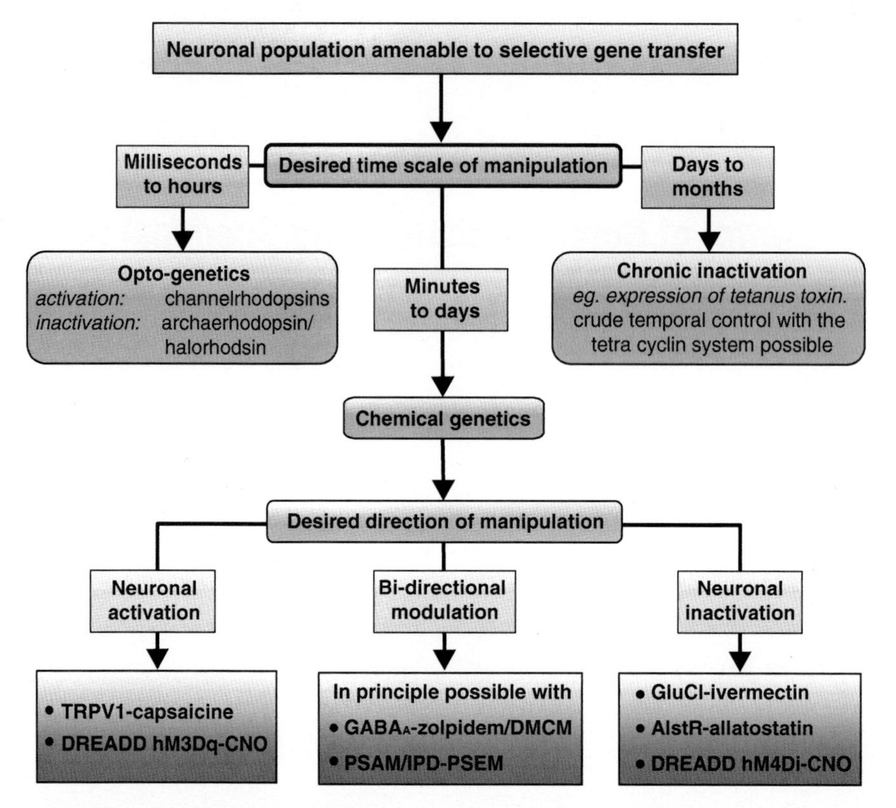

Fig. 1 A rough guide to neuronal manipulations. The question which system is best suited will depend on the scientific question to be addressed. Chemical genetics may be of particular interest for manipulations on an intermediate time scale (minutes or days). To choose from the different receptor-ligand pairs, not only directionality of interference but also the unique properties of the individual systems have to be taken into account (see text for details)

remain some inherent limitations with the technology. These include the requirement for an invasive optic fiber, the challenges of manipulating diffuse signaling networks in opaque brain tissue, the potential toxicity of certain opsin proteins, or the relatively limited ability to control neural activity over the long term (weeks or months). Lesions or blockade of transmitter release on the other hand are generally irreversible and can be susceptible to unforeseen compensation in neural circuits, potentially clouding experimental results. Here, we review how chemical genetics, the manipulation of neuronal activity via receptor-ligand interactions (sometimes referred to as pharmacogenetics), can be utilized to understand neural circuits and consequently animal behavior.

2 General Considerations for Chemical Genetic Manipulation

The general principle of chemical genetics is based on the transgenic expression of an ionotropic or metabotropic receptor, which is subsequently activated by an exogenous ligand to modulate selectively those cells that express the receptor. Whereas some techniques permit bidirectional modulation, that is, neuronal excitation or inhibition (depending on ligand or receptor), other methods only allow unidirectional control. Besides directionality several other important factors must be considered in the design and use of pharmacological ligand-receptor pairs for neural circuit analysis. These include background activity, cell type specificity, spatial restriction, as well as route of ligand application and activation and decay kinetics of the system. Firstly, the ligand-receptor pairs should be orthogonal. This means neither the receptor nor the ligand should have any baseline activity in the organism that is to be studied. For example, the receptor itself when expressed should not be active in the absence of the exogenous ligand. Similarly the ligand should have no effect on any endogenous receptors. To achieve this, the strategy has often been to either transplant receptor-ligand pairs from other tissues and species or to genetically reengineer endogenous ionotropic or metabotropic receptors. Expression of the receptor can be accomplished by various transfection and transduction methods in vitro or for in vivo use by generation of transgenic animals or injection of viral vectors. To target the receptor to a genetically identified type of neuron, it can either be directly expressed under the control of a cell-type-selective promoter or its expression can be made conditional to the expression of a second factor, whose expression in turn is controlled by a cell-type-selective promoter. These conditional methods mainly comprise the Cre-loxP system and tetracycline controlled gene regulation with cell-type-selective expression of Cre-recombinase or tetracycline transactivator (tTA), respectively. If the manipulation of neuronal activity is to be restricted to a particular brain region (e.g., particular hypothalamic nuclei) but the activity of the promoter is not confined to a single region, spatial control can be achieved to a reasonable degree by targeted injections of viral vectors for circumscribed receptor expression or by localized application of the ligand. Finally, appropriate consideration has to be given to temporal aspects of the manipulation as the various techniques that we discuss below may differ in their on-off kinetics, the duration of activity after single-dose ligand application, and the effects of chronic or repetitive ligand application. The kinetics of the system, however, will also depend on dose and application route of the ligand. While local application of the ligand, e.g., via intracerebral injection, allows the most rapid control of neuronal activity, not all experiments may be compatible with intracerebral injections and instead systemic applications may be preferable.

Following from the above, there will be no single chemical genetic system that fits all applications. Instead, the most suitable ligand-receptor pair has to be chosen according to the specific requirements of the experiment (see Fig. 1).

3 Manipulating Neural Activity via Ionotropic Receptors

3.1 Transient Receptor Potential (TRP) Channels

One of the first chemical genetic approaches for neural manipulation was based on the transient receptor potential cation channel subfamily V member 1 (Fig. 2; TRPV1). TRPV1 is a homotetrameric, nonselective cation channel gated by heat, low pH, and pungent substances such as the vanilloid-like ligand capsaicin from hot chili peppers [7]. Accordingly, activation of TRPV1 by capsaicin in neuronal cultures induces inward currents and membrane depolarization [7, 8]. It is mainly expressed in nociceptive neurons of the peripheral nervous system but can also be found in other tissues including the brain [7, 9]. The first use of TRPV1 and its ligand capsaicin as artificial neuronal activation device was a by-product of a genetic rescue experiment in the nematode *C. elegans*. In this elegant study, the mammalian TRPV1 receptor channel was transgenically transplanted into ASH nociceptive neurons of *C. elegans*, which normally neither expresses TRPV1 nor responds to capsaicin. Subsequent exposure of the transgenic worms to capsaicin induced a robust avoidance behavior through artificial activation of TRPV1-expressing ASH nociceptive neurons [10]. One year later, Zemelman and colleagues demonstrated that the capsaicin-gated TRPV1 as well as another member of the TRP subfamily, the menthol gated TRPM8, could be used to "remote control" action potential firing in rat hippocampal primary neurons in culture. Using photo release of caged ligands, they showed that neuronal firing could be controlled in the range of a few seconds [11]. This proof of principle study was later extended to allow cell-type-specific neuronal activation in vivo using knock-in mice with

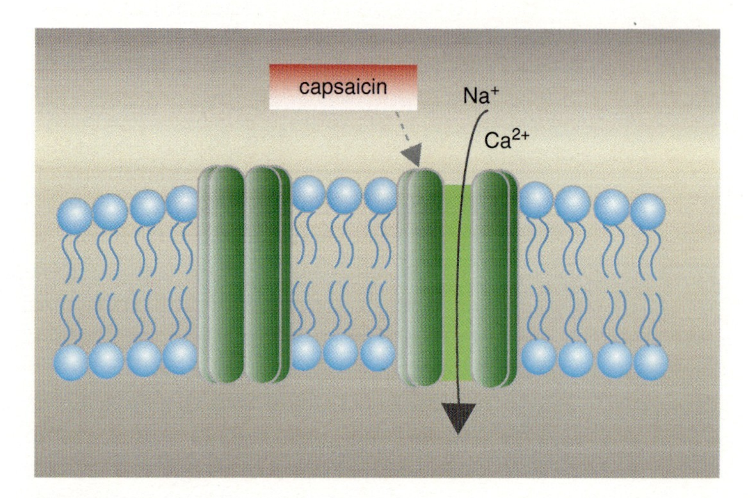

Fig. 2 TRPV1-mediated neuronal activation. Binding of the endogenous ligand capsaicin to TRPV1 opens the nonspecific cation channel, resulting in an inward flow of cations and consequent depolarization. Channels can also be activated directly by heat

Cre-loxP-dependent conditional expression of TRPV1 [12]. In this knock-in mouse line, expression of TRPV1 can be induced either by mating with a Cre-recombinase expressing transgenic mouse line (e.g., to achieve cell-type-specific TRPV1 expression) or by local expression of Cre-recombinase, for example, through a Cre-expressing virus. The authors showed that intracerebral application of capsaicin caused dose-dependent firing in TRPV1-expressing neurons. Moreover, within 5 min unilateral striatal capsaicin infusion was capable of evoking a stereotyped contralateral turning behavior lasting for minutes [12].

However, there are several challenges regarding the use of TRP channels for circuit manipulations in vivo. (1) TRPV1 channels are endogenously expressed in the mammalian CNS. Excitation by capsaicin may thus not be limited to neurons that have been targeted to transgenically express TRPV1. (2) TRPV1 exhibits light baseline effects. Specifically, the resting membrane potential of hippocampal neurons transfected with TRPV1 has been shown to be more positive than in non-transfected neurons, suggesting a mild depolarizing current in the absence of capsaicin [11]. Along these lines, it is worth mentioning that TRPV1 can also be activated by endogenous ligands such as endocannabinoids [13]. (3) High concentrations of the ligand cause excitotoxicity, presumably due to excessive calcium influx [12]. (4) Capsaicin can only be applied via direct intracranial infusion as it does not cross the blood-brain barrier. In addition, systemic application will activate nociceptive neurons due to the peripheral expression of TRP channels [14]. An interesting methodological variant that may circumvent the localized application of capsaicin exploits the temperature sensitivity of these channels and uses superparamagnetic nanoparticles to generate heat in response to an alternating magnetic field. The localized thermal effects of membrane-targeted nanoparticles are sufficient to open and close nearby TRPV1 channels [15]. Although promising, this system awaits validation in mouse and so far has only been used in a proof of concept study in *C. elegans*.

3.2 Ivermectin-Gated Chloride Channels

Nematodes express a glutamate-gated chloride channel (GluCl) which assembles as heteropentamer from α and β subunits. This receptor channel is a target of the anti-worm drug ivermectin which acts as a receptor agonist (Fig. 3; [16]). Expression of GluCl α and β subunits from C. elegans in mammalian neurons renders these cells ivermectin sensitive. In vitro ivermectin-mediated activation of GluCl indeed caused hyperpolarization and action potential shunting in transfected hippocampal neurons within seconds with protracted ligand unbinding and recovery over several hours [17]. Subsequent modifications of GluCl α and β subunit cDNAs reduced GluCl glutamate sensitivity, improved protein expression in mammalian cells, and allowed detection of the individual subunits by insertion of fluorescent tags [18, 19]. Unilateral viral

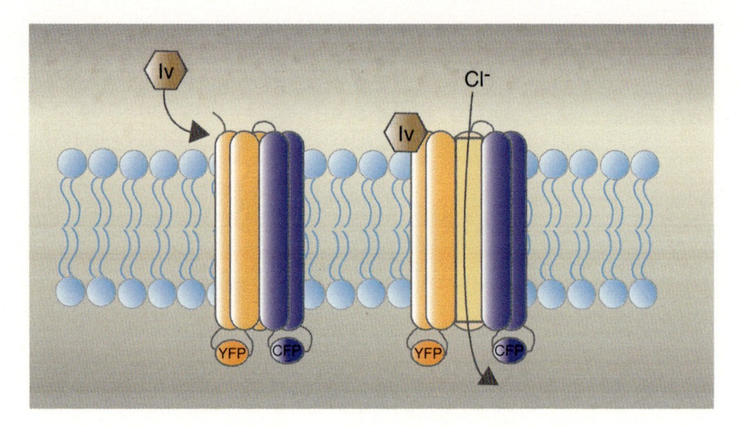

Fig. 3 Ivermectin-gated chloride channels. Combined expression of the α and β subunits (tagged with the fluorescent proteins YFP and CFP) of the *C. elegans* GluCl renders neurons sensitive to ivermectin-mediated neuronal inhibition. Binding of ivermectin causes channel opening and consequent influx of chloride ions

expression of these improved cDNAs in mouse striatum allowed a detailed characterization of the time course of behavioral effects by measuring turning behavior in response to intraperitoneal ivermectin injections: behavioral effects started after 12 h, peaked after 24 h, and lasted for several days [20]. The ivermectin-based technique has been successfully applied in studies of fear conditioning in the amygdala [21], studies of hypothalamic circuits involved in aggression and mating behavior [22], and the role of the prefrontal cortex in emotionally induced cataplexy in a mouse model of narcolepsy [23]. Potential limitations of the technique include the need for expression of two individual subunits, the variability in receptor expression levels, the protracted on-off kinetics (on the order of days with intraperitoneal application), and the high concentrations of ivermectin required for consistent silencing after systemic drug application, the latter giving rise to concerns about interference with endogenous ligand-gated receptor channels such as $GABA_A$ and glycine receptors. Recently, rational protein engineering strategies have yielded modified GluCl channels that show improved expression at the plasma membrane and require much lower ivermectin concentrations for activation, thus reducing the risk of unwanted side effects [24].

3.3 Allosteric Modulation of $GABA_A$ Receptor Conductance

The "zolpidem" method takes advantage of the rich pharmacology of the hetero pentameric $GABA_A$ receptors and in particular of the fact that several drugs can bind to a site on the $GABA_A$ receptor (the benzodiazepine binding site) that is distinct from the GABA-binding site. These drugs act as allosteric modulators: they alter the shape of the receptor oligomer so that the efficiency of GABA for opening the chloride channel is changed. The efficiency of GABA

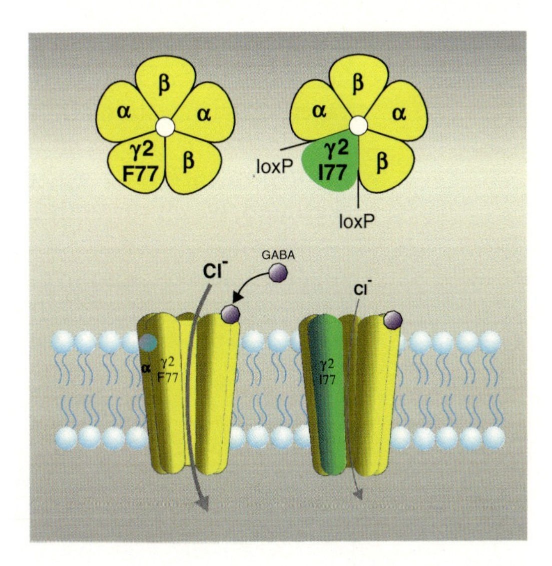

Fig. 4 The zolpidem method for modulation of GABAergic transmission. Binding of zolpidem to the benzodiazepine site on the $GABA_A$ receptor results in increased channel conductance when GABA is bound. Substitution of the phenylalanine (F) with isoleucine (I) at position 77 of the γ2 subunit abolishes binding of zolpidem, but does not affect GABA binding. Mice with the I77 substitution are thus insensitive to the actions of zolpidem. Reintroduction of the F77 subunit via genetic means allows cell-type-selective enhancement of GABAergic transmission and thus inhibition of neuronal subtypes

can be increased (positive allosteric modulation) by the actions of diazepam or zolpidem, thus increasing chloride influx and decreasing neuronal activity (Fig. 4). Conversely, chloride influx can be decreased and neuronal activity enhanced through the actions of negative allosteric modulators such as the β-carboline DMCM (methyl-6,7-dimethoxy-4-ethyl-beta-carboline-3-carboxylate) [25]. Zolpidem and DMCM binding at the $GABA_A$ receptor occurs in a pocket created between an α subunit (α1, α2, or α3) and the γ2 subunit, and substitution of phenylalanine (F) with isoleucine (I) at position 77 of the γ2 subunit abolishes binding of zolpidem and DMCM, without altering the actions of GABA itself (Fig. 4) [26, 27]. Mice engineered to carry this mutation (γ2I77) are thus insensitive to zolpidem and DMCM [28, 29]. However, transgenic Cre-mediated substitution of γ2F77 for γ2I77 in these mice permits cell-type-selective restoration of drug sensitivity as first shown for cerebellar Purkinje cells, where zolpidem specifically enhanced inhibitory postsynaptic currents causing motor deficits within minutes after intraperitoneal application [30]. Later, this approach was adapted to allow for virus-mediated restoration of drug sensitivity to cortical pyramidal cells [31]. The special feature of this allosteric method is that it modulates physiological neurotransmission rather than controlling neuronal activity independent of context. An advantage is that potentially bidirectional modulation of neuronal activity can be achieved in the same animal with the application of either zolpidem or DMCM, both of which act within minutes after intraperitoneal injection. In addition, both drugs can be acutely antagonized with flumazenil. The main disadvantage of the system is the dependence on a genetically modified zolpidem-insensitive background, which currently limits its application to mice.

3.4 Modular Combination of Modified Ligand-Binding and Ion Pore Domains

This approach takes advantage of the fact that within the family of Cys-loop receptors, the ligand-binding domain of the α7 nicotinic acetylcholine receptor (nAChR) is capable of functioning as an independent actuator module, which can be combined with various ion pore domains of other Cys-loop receptor family members [32–34]. In a first step, the ligand-binding domain of the α7 nAChR was mutated to be unresponsive to the endogenous ligand acetylcholine but responsive to different synthetic ligands, termed pharmacologically selective effector molecules (PSEMs; Fig. 5) [32]. In a second step, these modified α7 nAChR ligand-binding domains—termed pharmacologically selective actuator modules (PSAM)—were transplanted onto the functionally distinct ion pore domains (IPDs) of other Cys-loop receptors to produce chimeric receptor channels with selectivities for calcium, cations, or chloride (Fig. 5). Combination of the modified nAChR ligand-binding domain with the serotonin 5HT3 ion pore domain allowed activation of cation influx (depolarization and neuronal activation),

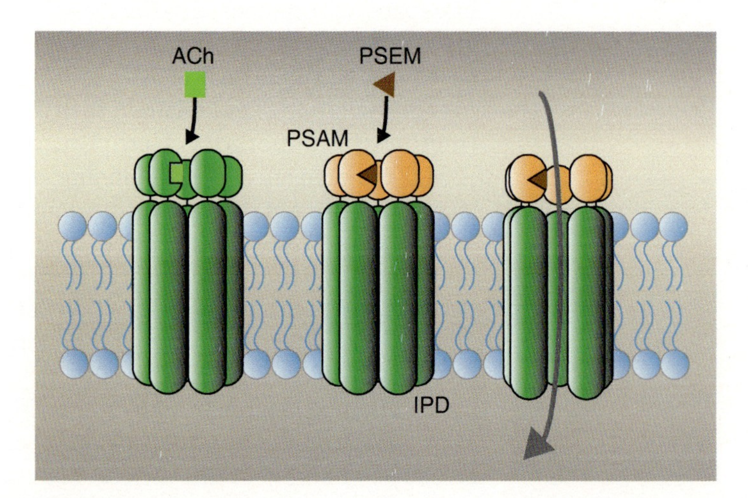

Fig. 5 Neuronal modulation with pharmacologically selective actuator modules (PSAMs) and pharmacologically selective effector molecules (PSEMs). Mutation of the ligand-binding domain of the α7 nAChR makes the Cys-loop receptor unresponsive to its endogenous ligand, but responsive to the synthetic ligands PSEMs. These modified α7 nAChR ligand-binding domains (PSAMs) can then be transplanted onto the ion pore domains (IPDs) of other Cys-loop receptors with different ion selectivities to allow for neuronal activation or inhibition

and combination with the glycine receptor ion pore domain allowed activation of chloride influx (hyperpolarization and neuronal silencing) in brain slices within seconds to minutes after addition of synthetic ligands. In vivo experiments showed that a behavioral effect mediated by the chimeric glycine receptor occurred about 30 min after intraperitoneal injection of the synthetic ligand [32, 35]. The kinetics of this system still have to be characterized in more detail. A clear advantage of the system is that in principle different ionic conductances can be modulated.

4 Targeting G Protein-Coupled Receptors

G-Protein-coupled receptors (GPCRs) are membrane-bound proteins that convert extracellular cues to intracellular signals, thus modulating the intracellular state. GPCRs consist of seven transmembrane domains, connected via intracellular and extracellular protein domains. Depending on the G protein involved, three major GPCR pathways have been distinguished: via Gs, which increases levels of cyclic adenosine monophosphate (cAMP); via Gi, which decreases levels of cAMP; and via Gq, which activates phospholipase C (PLC) and thus raises levels of diacylglycerol (DAG) and inositol trisphosphate (IP3) [36]. In neurons, GPCR activity can activate or inhibit neuronal firing depending on the intracellular signaling cascades that are activated by the receptor and the downstream effector proteins that are expressed by the cell [37]. In general the strength and temporal profile of GPCR-based neuronal manipulation depends on the availability of associated signaling pathways and effector molecules, which will differ between cell types. In addition, the activation of second messenger pathways will have multiple cellular effects besides changing the cells excitability.

4.1 Allatostatin Receptor

Evolutionary conservation of GPCRs and associated intracellular signaling cascades provides the opportunity to transplant receptors from simpler species to mammals, maintaining the ability to modulate endogenous signaling cascades. The *Drosophila* allatostatin receptor when expressed in mammalian neurons induces neuronal silencing in the presence of the insect peptide allatostatin via Gi-mediated activation of G protein-coupled inwardly rectifying potassium (GIRK) channels, thereby hyperpolarizing the cell membrane and silencing neural activity (Fig. 6) [38]. Efficient silencing within minutes of allatostatin application has been reported for slice preparations from a variety of mammalian CNS tissues expressing the allatostatin receptor, including the spinal cord, hippocampus, cortex, amygdala, and brainstem [38–42]. In vivo local application of allatostatin results in neuronal inactivation ranging from minutes to hours [43]. Allatostatin-based manipulations

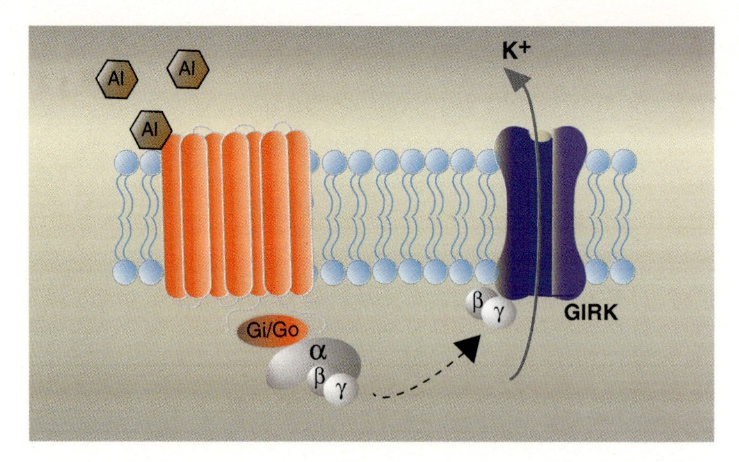

Fig. 6 Allatostatin-mediated neuronal silencing. Expression of the *Drosophila* allatostatin receptor in mammalian neurons causes Gi-mediated activation of G protein-coupled inwardly rectifying potassium (GIRK) channels in the presence of allatostatin (AI). This hyperpolarizes the cell and silences neuronal activity

have been successfully used to elucidate the function of several neuronal classes in locomotion, respiration, and fear in mice [39, 40, 42–44]. Importantly, expression of the allatostatin receptor is one of the few manipulations that has been validated for reversible neuronal manipulation in nonhuman primates [45]. However, allatostatin is a peptide and can thus not cross the blood-brain barrier, requiring local application, which can potentially complicate the experimental design. In addition, tissue diffusion is limited, which can lead to variability in silencing efficiency and recovery times as a function of distance from the injection site.

4.2 Designer Receptors Exclusively Activated by Designer Drugs (DREADDs)

Perhaps the most commonly used group of chemical genetic-based neural manipulators is the so-called DREADDs (Fig. 7; designer receptors exclusively activated by designer drugs). Starting with muscarinic acetylcholine receptors (mAChRs), DREADDs were engineered to have no baseline activity, to be unresponsive to the natural ligand acetylcholine, and, as their name suggests, to only be activated by a biologically inert designer ligand. Armbruster et al. [46] generated several DREADD variants coupled to Gq, Gi, or Gs that could be activated by the pharmacologically inert ligand clozapine *N*-oxide (CNO) [47]. Of these DREADDs, the Gi-coupled hM4Di and the Gq-coupled hM3Dq have become established tools for inhibition and activation of neuronal activity, respectively. Since all currently used DREADDs share the same ligand, CNO, bidirectional manipulations in the same cell are not possible.

DREADDs have been shown to be affective when expressed either via transgenic mouse lines (i.e., ref. [48]) or when expressed through the application of viral vectors [49]. In brain

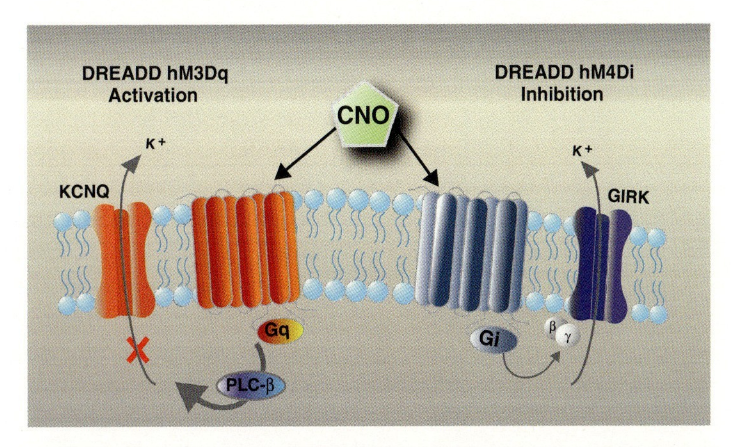

Fig. 7 Modulation of neuronal activity with the DREADD system. Neuronal activation can be achieved with the Gq-coupled hM3Dq (*left*), whereas inhibition is mediated via the Gi-coupled hM4Di DREADD (*right*). Activation is achieved via the closing of KCNQ channels by a phospholipase C-β (PLC-β)-dependent mechanism. Neuronal inhibition via hM4Di involves the Gi-mediated opening of GIRK channels. In both cases the ligand is CNO

slices, Gq-coupled hM3Dq causes depolarization and increased firing minutes after CNO application—presumably through phospholipase C-dependent closure of KCNQ channels (Fig. 7) [48]. In vivo experiments have reported that behavioral effects begin around 10–30 min after intraperitoneal injection of CNO and last for about 8–10 h [48, 50]. Recently, in a clever study, hM3Dq receptor expression under the control of the activity-dependent c-Fos promoter has been used to reactivate acquired fear memories [51].

Neural inhibition can be achieved with the expression of hM4Di, which probably activates GIRK channels and thus leads to neuronal hyperpolarization and silencing minutes after CNO application in slices (Fig. 7) [46, 49, 50]. Intraperitoneal CNO application has revealed similar in vivo on-off kinetics for hM4Di as for hM3Dq [49, 52]. The long decay kinetics of hM4Di have allowed the coupling of this technology with whole-brain μPET imaging to analyze whole-brain effects of neuron-specific silencing [53]. Importantly, DREADD systems have also been combined with optogenetic technology to analyze neural circuits involved in feeding behavior [35, 54], demonstrating compatibility and synergy of these two systems. However, repeated application of CNO or longterm exposure may cause receptor desensitization.

Becnel et al. [55] again took advantage of the evolutionary conservation of GPCRs and translated the DREADD technology for use in *Drosophila*, altering activity of defined neuronal classes. Intriguingly, in these studies, altering the dose of CNO also altered neural activity in a graded manner. However, high doses of CNO may cause DREADD-independent side effects.

5 Conclusions

Chemical genetic strategies have yielded numerous insights into the function of specific circuits throughout the nervous system. However, as discussed above, each of these techniques comes with its own set of advantages and disadvantages, and the choice which of them is best suited depends on the specific scientific context. Some important points that need to be considered before choosing a particular chemical genetic approach include the following: Does the ligand have to cross the blood-brain barrier or is local application sufficient? Is neuronal inhibition or activation desired or should the system allow bidirectional modulation? How rapid should the effect start and how long should the manipulation last (minutes, hours, days)? In general the major strength of chemical genetic techniques compared with optogenetics is the ability to target distributed neuronal networks over long periods without the requirement to send an optic fiber into the brain. Optogenetic techniques on the other hand provide temporal precision on the millisecond time scale and allow better physical restriction of the manipulated area (e.g., if only a specific axonal projection is to be manipulated). The choice between optogenetics and chemical genetics will thus also depend on the specific experimental question, and many studies may profit from a combination of these complementary approaches.

References

1. Figee M, Wielaard I, Mazaheri A, Denys D (2013) Neurosurgical targets for compulsivity: what can we learn from acquired brain lesions? Neurosci Biobehav Rev 37(3):328–339. doi:10.1016/j.neubiorev.2013.01.005
2. Yamamoto M, Wada N, Kitabatake Y, Watanabe D, Anzai M, Yokoyama M, Teranishi Y, Nakanishi S (2003) Reversible suppression of glutamatergic neurotransmission of cerebellar granule cells in vivo by genetically manipulated expression of tetanus neurotoxin light chain. J Neurosci 23(17):6759–6767
3. Murray AJ, Sauer JF, Riedel G, McClure C, Ansel L, Cheyne L, Bartos M, Wisden W, Wulff P (2011) Parvalbumin-positive CA1 interneurons are required for spatial working but not for reference memory. Nat Neurosci 14(3):297–299. doi:10.1038/nn.2751
4. Wulff P, Arenkiel BR (2012) Chemical genetics: receptor-ligand pairs for rapid manipulation of neuronal activity. Curr Opin Neurobiol 22(1):54–60. doi:10.1016/j.conb.2011.10.008
5. Boyden ES, Zhang F, Bamberg E, Nagel G, Deisseroth K (2005) Millisecond-timescale, genetically targeted optical control of neural activity. Nat Neurosci 8(9):1263–1268. doi:10.1038/nn1525
6. Fenno L, Yizhar O, Deisseroth K (2011) The development and application of optogenetics. Annu Rev Neurosci 34:389–412. doi:10.1146/annurev-neuro-061010-113817
7. Caterina MJ, Schumacher MA, Tominaga M, Rosen TA, Levine JD, Julius D (1997) The capsaicin receptor: a heat-activated ion channel in the pain pathway. Nature 389(6653):816–824. doi:10.1038/39807
8. Tominaga M, Caterina MJ, Malmberg AB, Rosen TA, Gilbert H, Skinner K, Raumann BE, Basbaum AI, Julius D (1998) The cloned capsaicin receptor integrates multiple pain-producing stimuli. Neuron 21(3):531–543
9. Gibson HE, Edwards JG, Page RS, Van Hook MJ, Kauer JA (2008) TRPV1 channels mediate long-term depression at synapses on hippocampal interneurons. Neuron 57(5):746–759. doi:10.1016/j.neuron.2007.12.027
10. Tobin D, Madsen D, Kahn-Kirby A, Peckol E, Moulder G, Barstead R, Maricq A, Bargmann C (2002) Combinatorial expression of TRPV

channel proteins defines their sensory functions and subcellular localization in C. elegans neurons. Neuron 35(2):307–318

11. Zemelman BV, Nesnas N, Lee GA, Miesenbock G (2003) Photochemical gating of heterologous ion channels: remote control over genetically designated populations of neurons. Proc Natl Acad Sci U S A 100(3):1352–1357. doi:10.1073/pnas.242738899
12. Arenkiel BR, Klein ME, Davison IG, Katz LC, Ehlers MD (2008) Genetic control of neuronal activity in mice conditionally expressing TRPV1. Nat Methods 5(4):299–302. doi:10.1038/nmeth.1190
13. Ross RA (2003) Anandamide and vanilloid TRPV1 receptors. Br J Pharmacol 140(5):790–801. doi:10.1038/sj.bjp.0705467
14. Fernandes ES, Fernandes MA, Keeble JE (2012) The functions of TRPA1 and TRPV1: moving away from sensory nerves. Br J Pharmacol 166(2):510–521. doi:10.1111/j.1476-5381.2012.01851.x
15. Huang H, Delikanli S, Zeng H, Ferkey DM, Pralle A (2010) Remote control of ion channels and neurons through magnetic-field heating of nanoparticles. Nat Nanotechnol 5(8):602–606. doi:10.1038/nnano.2010.125
16. Raymond V, Sattelle DB (2002) Novel animal-health drug targets from ligand-gated chloride channels. Nat Rev Drug Discov 1(6):427–436. doi:10.1038/nrd821
17. Slimko EM, McKinney S, Anderson DJ, Davidson N, Lester HA (2002) Selective electrical silencing of mammalian neurons in vitro by the use of invertebrate ligand-gated chloride channels. J Neurosci 22(17):7373–7379
18. Li P, Slimko EM, Lester HA (2002) Selective elimination of glutamate activation and introduction of fluorescent proteins into a Caenorhabditis elegans chloride channel. FEBS Lett 528(1–3):77–82
19. Slimko EM, Lester HA (2003) Codon optimization of Caenorhabditis elegans GluCl ion channel genes for mammalian cells dramatically improves expression levels. J Neurosci Methods 124(1):75–81
20. Lerchner W, Xiao C, Nashmi R, Slimko EM, van Trigt L, Lester HA, Anderson DJ (2007) Reversible silencing of neuronal excitability in behaving mice by a genetically targeted, ivermectin-gated Cl- channel. Neuron 54(1):35–49. doi:10.1016/j.neuron.2007.02.030
21. Haubensak W, Kunwar PS, Cai H, Ciocchi S, Wall NR, Ponnusamy R, Biag J, Dong HW, Deisseroth K, Callaway EM, Fanselow MS, Luthi A, Anderson DJ (2010) Genetic dissection of an amygdala microcircuit that gates conditioned fear. Nature 468(7321):270–276. doi:10.1038/nature09553
22. Lin D, Boyle MP, Dollar P, Lee H, Lein ES, Perona P, Anderson DJ (2011) Functional identification of an aggression locus in the mouse hypothalamus. Nature 470(7333):221–226. doi:10.1038/nature09736
23. Oishi Y, Williams RH, Agostinelli L, Arrigoni E, Fuller PM, Mochizuki T, Saper CB, Scammell TE (2013) Role of the medial prefrontal cortex in cataplexy. J Neurosci 33(23):9743–9751. doi:10.1523/JNEUROSCI.0499-13.2013
24. Frazier SJ, Cohen BN, Lester HA (2013) An engineered glutamate-gated chloride (GluCl) channel for sensitive, consistent neuronal silencing by ivermectin. J Biol Chem 288(29):21029–21042. doi:10.1074/jbc.M112.423921
25. Campo-Soria C, Chang Y, Weiss DS (2006) Mechanism of action of benzodiazepines on GABAA receptors. Br J Pharmacol 148(7):984–990. doi:10.1038/sj.bjp.0706796
26. Sancar F, Ericksen SS, Kucken AM, Teissere JA, Czajkowski C (2007) Structural determinants for high-affinity zolpidem binding to GABA-A receptors. Mol Pharmacol 71(1):38–46. doi:10.1124/mol.106.029595
27. Buhr A, Baur R, Sigel E (1997) Subtle changes in residue 77 of the gamma subunit of alpha1beta2gamma2 GABAA receptors drastically alter the affinity for ligands of the benzodiazepine binding site. J Biol Chem 272(18):11799–11804
28. Cope DW, Wulff P, Oberto A, Aller MI, Capogna M, Ferraguti F, Halbsguth C, Hoeger H, Jolin HE, Jones A, McKenzie AN, Ogris W, Poeltl A, Sinkkonen ST, Vekovischeva OY, Korpi ER, Sieghart W, Sigel E, Somogyi P, Wisden W (2004) Abolition of zolpidem sensitivity in mice with a point mutation in the GABAA receptor gamma2 subunit. Neuropharmacology 47(1):17–34. doi:10.1016/j.neuropharm.2004.03.007
29. Ogris W, Poltl A, Hauer B, Ernst M, Oberto A, Wulff P, Hoger H, Wisden W, Sieghart W (2004) Affinity of various benzodiazepine site ligands in mice with a point mutation in the GABA(A) receptor gamma2 subunit. Biochem Pharmacol 68(8):1621–1629. doi:10.1016/j.bcp.2004.07.020
30. Wulff P, Goetz T, Leppa E, Linden AM, Renzi M, Swinny JD, Vekovischeva OY, Sieghart W, Somogyi P, Korpi ER, Farrant M, Wisden W (2007) From synapse to behavior: rapid modulation of defined neuronal types with engineered GABAA receptors. Nat Neurosci 10(7):923–929. doi:10.1038/nn1927

31. Sumegi M, Fukazawa Y, Matsui K, Lorincz A, Eyre MD, Nusser Z, Shigemoto R (2012) Virus-mediated swapping of zolpidem-insensitive with zolpidem-sensitive GABA(A) receptors in cortical pyramidal cells. J Physiol 590(Pt 7):1517–1534. doi:10.1113/jphysiol.2012.227538
32. Magnus CJ, Lee PH, Atasoy D, Su HH, Looger LL, Sternson SM (2011) Chemical and genetic engineering of selective ion channel-ligand interactions. Science 333(6047):1292–1296. doi:10.1126/science.1206606
33. Eisele JL, Bertrand S, Galzi JL, Devillers-Thiery A, Changeux JP, Bertrand D (1993) Chimaeric nicotinic-serotonergic receptor combines distinct ligand binding and channel specificities. Nature 366(6454):479–483. doi:10.1038/366479a0
34. Grutter T, de Carvalho LP, Dufresne V, Taly A, Edelstein SJ, Changeux JP (2005) Molecular tuning of fast gating in pentameric ligand-gated ion channels. Proc Natl Acad Sci U S A 102(50):18207–18212. doi:10.1073/pnas.0509024102
35. Atasoy D, Betley JN, Su HH, Sternson SM (2012) Deconstruction of a neural circuit for hunger. Nature 488(7410):172–177. doi:10.1038/nature11270
36. Masseck OA, Rubelowski JM, Spoida K, Herlitze S (2011) Light- and drug-activated G-protein-coupled receptors to control intracellular signalling. Exp Physiol 96(1):51–56. doi:10.1113/expphysiol.2010.055517
37. Gainetdinov RR, Premont RT, Bohn LM, Lefkowitz RJ, Caron MG (2004) Desensitization of G protein-coupled receptors and neuronal functions. Annu Rev Neurosci 27:107–144. doi:10.1146/annurev.neuro.27.070203.144206
38. Lechner HA, Lein ES, Callaway EM (2002) A genetic method for selective and quickly reversible silencing of mammalian neurons. J Neurosci 22(13):5287–5290, doi:20026527
39. Marina N, Abdala AP, Trapp S, Li A, Nattie EE, Hewinson J, Smith JC, Paton JF, Gourine AV (2010) Essential role of Phox2b-expressing ventrolateral brainstem neurons in the chemosensory control of inspiration and expiration. J Neurosci 30(37):12466–12473. doi:10.1523/JNEUROSCI.3141-10.2010
40. Zhou Y, Won J, Karlsson MG, Zhou M, Rogerson T, Balaji J, Neve R, Poirazi P, Silva AJ (2009) CREB regulates excitability and the allocation of memory to subsets of neurons in the amygdala. Nat Neurosci 12(11):1438–1443. doi:10.1038/nn.2405
41. Wehr M, Hostick U, Kyweriga M, Tan A, Weible AP, Wu H, Wu W, Callaway EM, Kentros C (2009) Transgenic silencing of neurons in the mammalian brain by expression of the allatostatin receptor (AlstR). J Neurophysiol 102(4):2554–2562. doi:10.1152/jn.00480.2009
42. Gosgnach S, Lanuza GM, Butt SJ, Saueressig H, Zhang Y, Velasquez T, Riethmacher D, Callaway EM, Kiehn O, Goulding M (2006) V1 spinal neurons regulate the speed of vertebrate locomotor outputs. Nature 440(7081):215–219. doi:10.1038/nature04545
43. Tan EM, Yamaguchi Y, Horwitz GD, Gosgnach S, Lein ES, Goulding M, Albright TD, Callaway EM (2006) Selective and quickly reversible inactivation of mammalian neurons in vivo using the Drosophila allatostatin receptor. Neuron 51(2):157–170. doi:10.1016/j.neuron.2006.06.018
44. Zhang Y, Narayan S, Geiman E, Lanuza GM, Velasquez T, Shanks B, Akay T, Dyck J, Pearson K, Gosgnach S, Fan CM, Goulding M (2008) V3 spinal neurons establish a robust and balanced locomotor rhythm during walking. Neuron 60(1):84–96. doi:10.1016/j.neuron.2008.09.027
45. Nielsen KJ, Callaway EM, Krauzlis RJ (2012) Viral vector-based reversible neuronal inactivation and behavioral manipulation in the macaque monkey. Front Syst Neurosci 6:48. doi:10.3389/fnsys.2012.00048
46. Armbruster BN, Li X, Pausch MH, Herlitze S, Roth BL (2007) Evolving the lock to fit the key to create a family of G protein-coupled receptors potently activated by an inert ligand. Proc Natl Acad Sci U S A 104(12):5163–5168. doi:10.1073/pnas.0700293104
47. Bender D, Holschbach M, Stocklin G (1994) Synthesis of n.c.a. carbon-11 labelled clozapine and its major metabolite clozapine-N-oxide and comparison of their biodistribution in mice. Nucl Med Biol 21(7):921–925
48. Alexander GM, Rogan SC, Abbas AI, Armbruster BN, Pei Y, Allen JA, Nonneman RJ, Hartmann J, Moy SS, Nicolelis MA, McNamara JO, Roth BL (2009) Remote control of neuronal activity in transgenic mice expressing evolved G protein-coupled receptors. Neuron 63(1):27–39. doi:10.1016/j.neuron.2009.06.014
49. Ferguson SM, Eskenazi D, Ishikawa M, Wanat MJ, Phillips PE, Dong Y, Roth BL, Neumaier JF (2011) Transient neuronal inhibition reveals opposing roles of indirect and direct pathways in sensitization. Nat Neurosci 14(1):22–24. doi:10.1038/nn.2703
50. Krashes MJ, Koda S, Ye C, Rogan SC, Adams AC, Cusher DS, Maratos-Flier E, Roth BL, Lowell BB (2011) Rapid, reversible activation

of AgRP neurons drives feeding behavior in mice. J Clin Invest 121(4):1424–1428. doi:10.1172/JCI46229

51. Garner AR, Rowland DC, Hwang SY, Baumgaertel K, Roth BL, Kentros C, Mayford M (2012) Generation of a synthetic memory trace. Science 335(6075):1513–1516. doi:10.1126/science.1214985
52. Ray RS, Corcoran AE, Brust RD, Kim JC, Richerson GB, Nattie E, Dymecki SM (2011) Impaired respiratory and body temperature control upon acute serotonergic neuron inhibition. Science 333(6042):637–642. doi:10.1126/science.1205295
53. Michaelides M, Anderson SA, Ananth M, Smirnov D, Thanos PK, Neumaier JF, Wang GJ, Volkow ND, Hurd YL (2013) Whole-brain circuit dissection in free-moving animals reveals cell-specific mesocorticolimbic networks. J Clin Invest 123(12):5342–5350. doi:10.1172/JCI72117
54. Carter ME, Soden ME, Zweifel LS, Palmiter RD (2013) Genetic identification of a neural circuit that suppresses appetite. Nature 503(7474): 111–114. doi:10.1038/nature12596
55. Becnel J, Johnson O, Majeed ZR, Tran V, Yu B, Roth BL, Cooper RL, Kerut EK, Nichols CD (2013) DREADDs in Drosophila: a pharmacogenetic approach for controlling behavior, neuronal signaling, and physiology in the fly. Cell Rep 4(5):1049–1059. doi:10.1016/j.celrep.2013.08.003

Chapter 9

Generation of BAC Transgenic Mice for Functional Analysis of Neural Circuits

Jonathan T. Ting and Guoping Feng

Abstract

Functional analysis of neural circuits in the living brain is an exceptional challenge that has been greatly advanced in the modern molecular genetics era by the development of elegant techniques for marking and manipulating genetically defined neuronal subsets. Technologies of this nature are enabling neuroscientists to probe the causal role of specific neuronal populations in sensory information processing, complex animal behaviors, and brain dysfunction in a myriad of neurological disorders. Gains in these areas have been especially catalyzed by the development and expansion of optogenetics, specifically, the use of channelrhodopsin-2 (ChR2) for inducible and reversible control of neuronal firing with blue light. Here we provide detailed methods based on our recent success in developing BAC transgenic mouse lines with functional ChR2 expression in diverse genetically-defined neuronal subsets. In principle this BAC transgenic strategy can be implemented to express other transgenes of interest or can be further applied to achieve stable transgenic expression of ChR2 and other microbial opsin variants in additional diverse cell types of the mammalian nervous system.

Key words Optogenetics, Bacterial artificial chromosome, Pronuclear injection, Channelrhodopsin-2, Pulsed-field gel electrophoresis

1 Introduction

Transgenic approaches based on the modification of bacterial artificial chromosomes (BACs) are well suited for achieving transgene expression in highly restricted, genetically defined cellular populations [1, 2]. This fact is attributable to the large size of genomic DNA fragments contained in BAC clones (generally hundreds of kilobases), which provides a high likelihood that a single BAC clone can be identified which spans an entire gene of interest (GOI) plus large flanking regions of DNA that may be crucial for instructing the precise pattern of gene expression. Thus, the insertion of a target transgene in place of the coding region of a GOI contained within a BAC can enable the transgene to be expressed under the control of the same regulatory elements and, consequently, in

Benjamin R. Arenkiel (ed.), *Neural Tracing Methods: Tracing Neurons and Their Connections*, Neuromethods, vol. 92, DOI 10.1007/978-1-4939-1963-5_9,

the same cellular populations and pattern as for the GOI. The immense value of this approach is exemplified by the Gene Expression Nervous System Atlas (GENSAT) project at Rockefeller University, a large-scale effort that successfully created and characterized over a thousand BAC transgenic mouse lines with expression of green fluorescent protein (GFP) or Cre recombinase in genetically defined cellular subsets throughout the nervous system [1, 3, 4]. These BAC transgenic lines have provided ample new experimental inroads for investigating the structure and function of the mammalian nervous system in both healthy and diseased states [5–20]. Furthermore, analogous BAC transgenic strategies have been utilized to develop rodent lines for specialized neuroscience applications such as labeling of newborn neurons in the adult brain [21], modeling features of human brain disorders [22, 23], cell-type-specific biochemical tagging of signaling molecules [24], and large-scale transcriptional profiling in a cell-type and behavioral state-dependent manner [25–27].

In our own laboratory we recently developed a collection of BAC transgenic mice for cell-type-specific expression of ChR2-EYFP in the nervous system [28] (Fig. 1). These BAC transgenic lines enable precise optical control of neuronal firing with blue light for several major classes of neurons within the brain, including cholinergic, serotonergic, GABAergic, and Pvalb-expressing populations. These tools can be implemented to interrogate synaptic function and dysfunction with temporal and spatial precision and to probe the circuitry basis of complex animal behavior in an inducible and reversible manner. Already these BAC transgenic lines have been utilized to investigate diverse facets of neuronal circuitry function, such as: (1) defining a causal role of cortical spindle activity in relation to sleep states [29], (2) mapping of synaptic connectivity in diverse brain regions [30, 31] and (3) demonstrating functional co-release of the neurotransmitters glutamate and acetylcholine [32]. In addition, a Vglut2-ChR2 BAC transgenic mouse line was successfully developed and utilized to demonstrate that photostimulation of glutamatergic neurons in the spinal cord was sufficient to drive rhythmic locomotor behavior [33]. Collectively, these findings further confirm that diverse endogenous cellular promoters which are active in distinct neuronal subsets can be efficiently co-opted to drive transgene expression using BAC transgenic strategies—even for transgenes that require exceptionally high-level expression for full functionality such as ChR2.

The BAC transgenic strategy we have employed can be applied to express virtually any transgene of interest or can be further applied to achieve stable transgenic expression of ChR2 (and other microbial opsin variants for optogenetics research) in additional

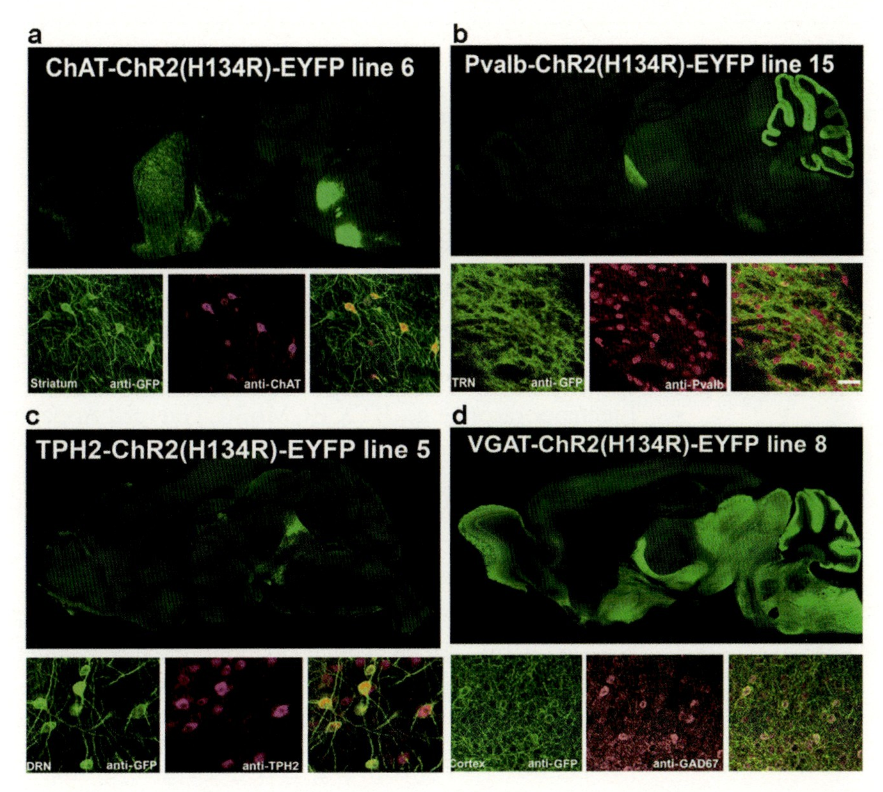

Fig. 1 BAC transgenic mouse lines exhibit functional expression of ChR2-EYFP in genetically defined neuronal subsets. (**a**) Sagittal whole-brain montage (*top*) from a ChAT-ChR2(H134R)-EYFP line 6 mouse (*top*). High-magnification confocal images demonstrating co-localization of ChR2-EYFP (anti-GFP) and ChAT (anti-ChAT) in the dorsal striatum region (*bottom*). (**b**) Sagittal whole-brain montage (*top*) from a Pvalb-ChR2(H134R)-EYFP line 15 mouse (*top*). High-magnification confocal images demonstrating co-localization of ChR2-EYFP (anti-GFP) and Pvalb (anti-Pvalb) in the thalamic reticular nucleus (TRN) region (*bottom*). (**c**) Sagittal whole-brain montage (*top*) from a TPH2-ChR2(H134R)-EYFP line 5 mouse (*top*). High-magnification confocal images demonstrating co-localization of ChR2-EYFP (anti-GFP) and TPH2 (anti-TPH2) in the dorsal raphe nucleus (DRN) region (*bottom*). (**d**) Sagittal whole-brain montage (*top*) from a VGAT-ChR2(H134R)-EYFP line 8 mouse (*top*). High-magnification confocal images demonstrating co-localization of ChR2-EYFP (anti-GFP) and GAD67 (anti-GAD67) in the cortex (*bottom*). Scale bar: 100 μm (applies to all high-magnification confocal image panels)

diverse cell types of the mammalian nervous system. Here we describe several important aspects pertaining to the development of BAC transgenic mice, including the selection and verification of appropriate BAC clones from BAC libraries, the design and construction of BAC targeting vectors, the manipulation of BAC DNA in *E. coli* (BAC recombineering), and the purification of high-integrity BAC DNA for pronuclear injections. Furthermore, we describe effective strategies for avoiding undesirable overexpression of "extra" genes contained within large BAC clones.

2 Materials

2.1 General Equipment and Reagents

- Thermal cycler and general PCR reagents
- MilliQ water purification system
- Shaking incubator for bacterial culture
- Shaking water bath (orbital shaker preferred over linear shaker)
- Standard incubator for bacterial plates
- Tabletop microcentrifuge with refrigeration capability
- Ultracentrifuge with appropriate rotors
- Electroporator (e.g., BTX ECM630, Harvard Apparatus)
- Orbital platform shaker
- Standard DNA agarose gel electrophoresis system and related reagents
- UV transilluminator system
- 50× Tris/acetate/EDTA (TAE) buffer (per liter): 242 g Tris base, 57.1 mL glacial acetic acid, 100 mL 0.5 M EDTA
- LB-Miller (per liter): 10 g NaCl, 10 g tryptone, 5 g yeast extract
- LB-Lennox (per liter): 5 g NaCl, 10 g tryptone, 5 g yeast extract
- Antibiotics (carbenicillin 50 mg/mL in Milli-Q water, kanamycin 30 mg/mL in Milli-Q water, chloramphenicol 12.5 mg/mL in absolute ethanol)
- Molecular cloning reagents (restriction enzymes, T4 DNA ligase, alkaline phosphatase)
- Competent *E. coli* for standard cloning (DH5α or DH10B strains are adequate)

2.2 DNA Plasmids and Plasmid Purification Kits

- iTV1: Targeting vector backbone containing multiple cloning sites for BoxA and BoxB homology arms, BGHpA, and FRT-NEO-FRT cassette)
- cDNA for transgene of interest, e.g., ChR2(H134R)-EYFP
- Qiagen Plasmid Miniprep Kit
- Zymo Research Gel DNA Recovery Kit
- Zymo Research DNA Clean & Concentrator-25 Kit

2.3 DNA Sequence Analysis and Editing Software

- Vector NTI, SnapGene, or other suitable program for sequence analysis

2.4 BAC Recombineering and BAC DNA Purification

- BAC clones (BACPAC resource center CHORI; http://bacpac.chori.org/)
- *E. coli* strain for recombineering (EL250)
- 50 % glycerol
- 10 % L-arabinose
- NucleoBond BAC 100 or NucleoBond Xtra MAXI Kit (Clontech)
- Oak Ridge high-speed centrifuge tubes (clear, screw cap, round bottom, 50 mL)
- Reagents for BAC DNA miniprep: Qiagen buffers P1/P2/P3, ethanol, isopropanol, 10 mM Tris/1 mM EDTA
- 0.1 cm gap cuvettes

2.5 Pulsed-Field Gel Electrophoresis (PFGE)

- Bio-Rad CHEF-DR II or -DR III system: controller module, driver module, chiller module, pump, gel chamber, gel cast and comb, tubing)
- 10× Tris/Borate/EDTA (TBE) buffer (per liter): 108 g Tris base, 55 g boric acid, 40 mL 0.5 M EDTA
- Pulsed-field gel certified agarose
- Ethidium bromide
- NEB MidRange I PFG Marker (15–300 kb range)
- Bio-Rad 2.5 kB Molecular Ruler (2.5–30 kb range)
- 6× DNA loading buffer

2.6 Electroelution and Spot Dialysis

- Poly-Prep chromatography columns, 0.8 cm × 4 cm.
- Dialysis tubing (Spectra/Por membrane, 6–8 K cutoff).
- G50 Sephadex beads solution: layer ~1 cm of dry G50 Sephadex beads at the bottom of a clean 250 mL bottle. Add 200 mL of 1× TE and autoclave. Beads should swell up to about half the volume of the bottle. Store at room temperature.
- Spot dialysis discs, 0.025 μm VSWP.
- 100 mM 6-aminocaprioc acid solution.
- 10 mM Tris/1 mM EDTA.
- Concentrated phenol red stock.
- Razor blades.
- Ring stand and clamps.
- Column end cap.
- Custom positive and negative lead wires.

- Power supply (must be capable of sustaining continuous low current of 1–2 mA without error. Older units may work better for this purpose).
- Microinjection buffer: 100 mM NaCl, 10 mM Tris–HCl, 0.1 mM EDTA, pH 7.4 ± 0.05.
- Ultrapure water for embryo transfer.
- 1,000× polyamine stock: (30 mM spermine and 70 mM spermidine in MilliQ water).

3 Methods

3.1 Identification of Suitable BAC Clones

The first step in the design process is to perform a database search to identify suitable BAC clones for making a BAC transgenic mouse line. Ensembl (http://ensemble.org) is an excellent online resource for this purpose. The database can display BAC clones from various sources aligned to the mouse genomic DNA sequence, which facilitates analysis and selection of suitable clones.

1. Go to http://ensemble.org and enter "mouse" as the species of interest and the name of your gene of interest in the search field. For example, for designing a transgenic line with expression of ChR2-EYFP under the control of the choline acetyltransferase (ChAT) gene regulatory elements (transgene expression in cholinergic neurons) enter the gene name "ChAT."
2. Click on the location link to view the chromosomal region harboring the ChAT gene and flanking DNA. In the case of the ChAT gene, the location link reads "14:32408203-32465909:-1," which indicates a position on chromosome 14 of the mouse genome in the antisense (-1) orientation.
3. To show BAC clone alignments, open the tab labeled "configure this page" and select "Clones" under the "Sequence and Assembly" menu option. Enable "BAC map m38" to display BAC clones from the mouse C57Bl/6 strain BAC library (RP23 and RP24) and/or enable "129S7/AB2.2 clones m38" to display BAC clones from the mouse 129S strain BAC library. The BAC clones should now appear in alignment with the mouse genomic DNA region of interest.
4. Identify BAC clones that cover the coding region of the gene of interest plus ~50 kb of 5′ sequence and ~50 kb of 3′ sequence. In some cases it may be necessary to select a BAC clone that is not ideal, in which case it is preferable to retain as much 5′ sequence as possible and a minimum of 15 kb 3′ sequence (*see* **Note 1**). Particular attention should be paid to the orientation of the gene (e.g., sense versus antisense) since the 5′ and 3′ regions are defined relative to the orientation of the gene of interest. Figure 2 shows an example of BAC clone selection for BAC clones spanning the ChAT gene in the

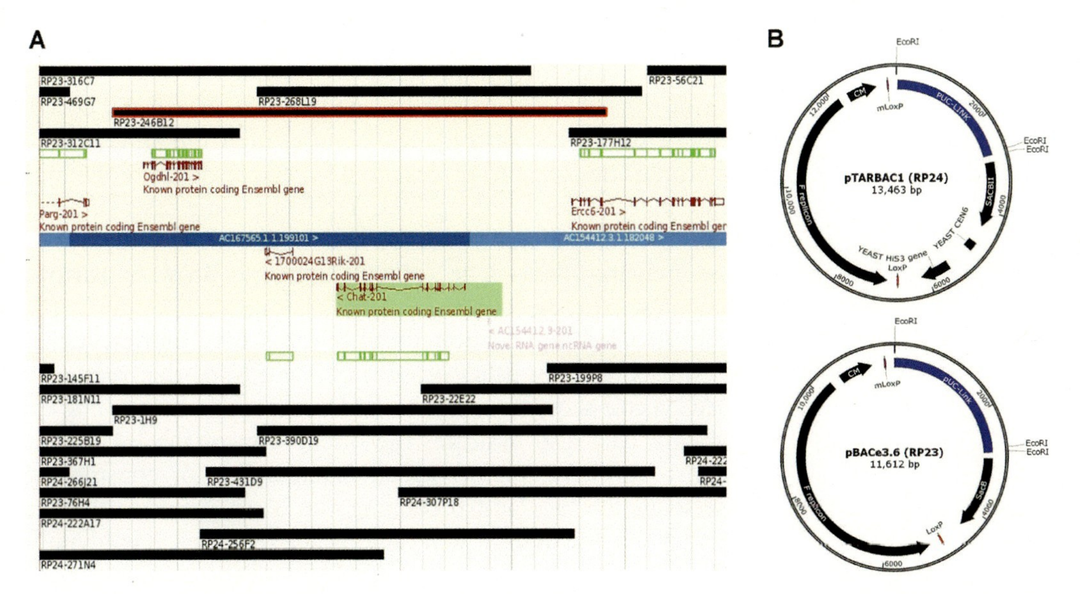

Fig. 2 Selection of BAC clones from the RP23/RP24 C57Bl/6 mouse BAC library. (**a**) Genomic alignment of BAC clones covering the mouse ChAT locus and neighboring regions on chromosome 14. The view shown is a screen capture during a BAC clone search in the Ensembl genome browser. Note the small arrow heads indicating the directionality of each transcript. Each division in the image spans 10 kb for a total of 300 kb represented in this view. The selected BAC clone (RP23-246B12) containing at least 50 kb of 5′ and 3′ sequence flanking the ChAT gene is shown outlined in *red*. (**b**) Vectors used to construct the RP23/RP24 C57Bl/6 mouse BAC library. The RP24 (female) library utilizes the pTARBAC1 vector while the RP23 library (male) utilizes the pBACe.3.6 vector. Critical features of the vectors are nearly identical, including the F replicon, chloramphenicol resistance cassette (CM), loxP, and mloxP sequence. Genomic fragments are cloned into the BAC vectors using EcoRI sites to replace the pUC link

mouse genome. We selected BAC clone RP23-246B12 due to the fact that the same clone was successfully employed to make multiple BAC transgenic mouse lines in the GENSAT project; however, there are other potentially suitable clones available such as RP23-431D9 and RP23-390D19. Please note that for Fig. 2 an archived version of the Ensembl program was intentionally used in order to show the BAC clone identification tags as a default option for display purposes only.

3.2 Verifying BAC Clone Ends and Cloning Orientation

Once one or more suitable BAC clones have been identified and obtained, it is desirable to ensure that the selected BAC clones received from the repository correctly match the digital clone data from Ensembl. In our experience with the RP23 and RP24 BAC libraries, we estimate that as many as 1 in 10 clones may be incorrectly archived. Others have estimated the error rate at 5 % for mammalian BAC clones [34]. These errors may occur at the level of digital data handling and entry or in the physical archiving of BAC clone stocks. In many cases an incorrect clone will cover a similar region of DNA to the desired BAC clone, in which case very specific measures for error detection will be required. For example, assaying for the presence of a sequence in the middle region of the BAC would not be adequate to ensure the correct identity of the clone. Instead, BAC end sequencing and/or PCR using primers designed to recognize BAC end regions are the most accurate method for verifying BAC DNA. In some cases it is possible to verify BAC clone ends by searching online databases for end sequencing data. The UCSC Genome Browser (http://genome.ucsc.edu/cgi-bin/hgGateway) is one such resource that provides detailed information about selected BAC clones, such as BAC end sequencing using primers (T7 and SP6) that read from the BAC vector sequence into the BAC insert at each end. The end sequencing data also indicates the orientation of the BAC insert relative to the BAC vector, which is important for constructing digital sequence files and predicting the band pattern following restriction enzyme digestion of the BAC DNA. Thus, it is advisable to select a BAC clone that can be verified with the UCSC Genome Browser data when possible, as this will save considerable time and effort (*see* **Note 2**).

1. Create a digital sequence file for selected BAC clones:
   (a) The BAC clones from the RP23 and RP24 library are all constructed by ligation of BAC inserts with the vectors pBACe3.6 and pTARBAC1 (EcoRI digest), respectively. Information about these vectors including maps and Genbank format digital sequence files are available online (http://bacpac.chori.org/vectorsdet.htm).
   (b) The entire BAC clone DNA sequence can be obtained in Ensembl by clicking on the desired BAC clone in the genomic alignment window. A pop-up box appears with the BAC start and end locations. Enter the start and end locations into the location box on the main page. For example, for the ChAT containing BAC RP23-246B12, enter 14:32307888-32528398 in the location box and click "Go." Now open the "export data" tab on the left-hand side and select the desired format for the sequence. Copy the sequence information into a separate file.
   (c) Determine the orientation of the BAC insert relative to the BAC vector using the UCSC Genome Browser (*see* **Note 3**). Insert the digital BAC sequence into the BAC vector sequence at the EcoRI sites to replace the pUC link element. Annotate any important features of the complete BAC clone (exons, genes, start and end of the BAC insert, etc.).
2. Order BAC clones from the BAC PAC Resource Center at CHORI (http://bacpac.chori.org/order_clones.php). Enter the BAC clone identification tags (e.g., RP23-246B12) in the box and select "verify clones."

3. BAC clones are shipped as stab cultures. All BAC clones in the RP23 and RP24 library contain the chloramphenicol resistance gene. Streak the culture on a plate of LB/Agar plus + 12.5 μg/mL chloramphenicol and grow for 20+h at 32 °C (*see* **Note 4**).
4. Isolate small-scale BAC miniprep DNA for analysis:
   (a) Inoculate 5 mL LB + 12.5 μg/mL chloramphenicol with a single bacterial colony. Grow overnight (~16–20 h) at 32 °C in a shaking incubator set to ~250 RPM. Make sure the cap is slightly loose, but securely in place.
   (b) From the saturated overnight culture, add 300 μL of the overnight culture to 200 μL 50 % glycerol to make a 20 % glycerol stock for future use. Store at −80 °C.
   (c) Pellet cells in the remaining cull culture by centrifugation for 4 min at 4,000 RPM.
   (d) Resuspend the bacterial pellet in 250 μL of cold Qiagen P1 resuspension buffer. (If the media was divided into separate tubes for centrifugation, the tubes should now be consolidated into only one tube with 250 μL of resuspension buffer.)
   (e) Add 250 μL of Qiagen P2 lysis buffer and mix by gently inverting 5–10 times.
   (f) Add 350 μL of cold Qiagen P3 neutralization buffer and mix by inverting 5–10 times. (Do not let the lysis reaction proceed for more than 5 min.)
   (g) Pellet cellular debris by centrifugation for 10 min at 13,200 RPM.
   (h) Transfer the supernatant to a new tube with a P1000 tip. Do not disturb or transfer the pellet of white precipitate.
   (i) Add 700 μL of room temperature isopropanol to precipitate the BAC DNA. Incubate for 5 min.
   (j) Pellet the BAC DNA by centrifugation at top speed (15,000 RPM) for 10 min.
   (k) Carefully remove the supernatant being careful to not disturb the BAC DNA pellet.
   (l) Rinse BAC DNA pellet with 500 μL of 70 % ethanol. Centrifuge for 5 min at 15,000 RPM. Carefully remove the supernatant without disturbing the BAC DNA pellet. Try to remove as much ethanol as possible with a P200 tip. Perform a quick spin if necessary to get residual ethanol out of the cap.
   (m) Briefly dry the BAC DNA pellet (approximately 10–15 min) with the tube cap open. Use caution not to overdry the

BAC DNA pellet, as this may make it significantly more difficult to redissolve the BAC DNA.

(n) Gently resuspend the BAC DNA pellet in 50 μL of 1× TE by lightly tapping the tube several times. The presence of 1 mM EDTA stabilizes the BAC DNA and helps to prevent degradation. Do not vortex or re-pipette BAC DNA at any time. Harsh handling of BAC DNA will result in shearing and degradation. Store BAC DNA at 4 °C. Freezing and subsequent thawing of supercoiled BAC DNA can result in degradation and thus should be avoided.

5. Design primers to check the BAC ends. The best strategy is to design a reverse primer that binds within the first 1 kb of the predicted BAC insert and a forward primer that binds within the last 1 kb of the predicted BAC insert. These primers can be used in PCR reactions in combination with T7 and SP6 primers that recognize sequences at opposite ends of the BAC vector. It is important to try each primer with both T7 and SP6 to verify the orientation of cloning as referenced against the digital sequence file (*see* **Note 5**).

3.3 Verification of BAC Clones by Pulsed-Field Gel Electrophoresis

Due to the large size of BAC DNA, it is not feasible to perform analysis with standard agarose gel electrophoresis that utilizes a constant electric field. Under such conditions, large DNA species exceeding ~20 kb will co-migrate regardless of size. The efficient electrophoretic separation of large DNA fragments requires a specialized technique called pulsed-field gel electrophoresis (PFGE) in which the electric field periodically alternates to different angles, thereby enabling far greater separation of very large DNA species up to the Mb range [35]. Commercial PFGE systems such as the contour-clamped homogeneous electric field (CHEF) devices from Bio-Rad (CHEF-DR II and CHEF-DR III) are of immense utility for analysis of BAC DNA throughout the process of manipulating the DNA to develop BAC transgenic mouse lines (Fig. 3a, b). To facilitate analysis of BAC DNA using PFGE, first the BAC DNA is digested with restriction enzymes that are carefully selected to yield fragments over a desirable size range and that can be readily separated.

1. Verification of BAC clone insert size by NotI digestion. The pBACe3.6 and pTARBAC1 vectors contain NotI restriction sites adjacent to the start and end of the BAC inserts. Thus, NotI digestion can be used to liberate the BAC vector (NotI digest: pBACe3.6 vector band is 8.7 kb, pTARBAC1 vector band is 10.6 kb) and evaluate the size of the BAC insert by PFGE. If NotI sites are present within the BAC insert, AscI and PvuI also cut within the BAC vector and can be used to determine the BAC insert size (although a portion of the vector will migrate with the BAC insert DNA.

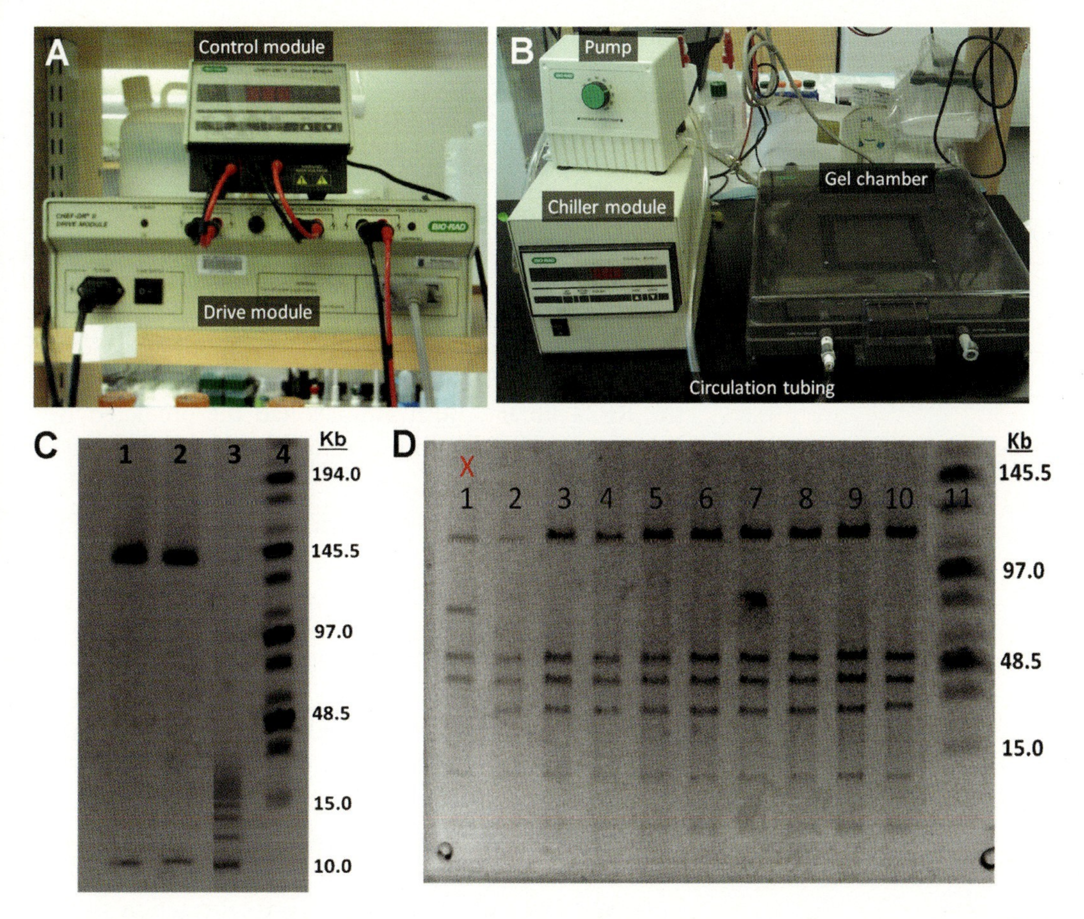

Fig. 3 BAC DNA analysis by pulsed-field gel electrophoresis (PFGE). (**a**, **b**) Components of the Bio-Rad CHEF-DR II PFGE system. (**c**) Determination of BAC insert size by NotI digestion and PFGE. *Lanes 1* and *2*: analysis of two modified Drd2 BAC clones confirms a correct BAC insert size of 144 kb for both clones. Note the presence of the 10 kb BAC vector. *Lane 3*: 2.5 kb marker. *Lane 4*: MidRange I PFG Marker. (**d**) BAC fingerprint analysis of ten modified Drd1a BAC clones. *Lane 1*: BAC clone with unexpected fingerprint. *Lanes 2–10*: correct BAC clones with identical fingerprint. *Lane 11*: MidRange I PFG Marker

(a) Perform restriction digests with 15 μL of freshly prepared BAC miniprep DNA using 1–2 μL of NotI enzyme in appropriate buffer with a 30 μL total reaction volume. When possible, the use of high-fidelity restriction enzymes is desirable to ensure complete and rapid digestion of BAC DNA. The restriction digest should proceed for 2–4 h.

(b) Prepare a 1 % agarose gel in 0.5× TBE buffer and insert the gel comb in the gel cast. Fill the PFGE chamber with 2 L of 0.5× TBE buffer. Turn on the power module, drive module, controller module/pump (set to 70), and chiller module (set to 14 °C). Remove the comb and load the solidified gel into the chamber and pre-run the system for 30 min (*see* **Note 6** for run parameters). Ensure that the system is operating properly and the buffer is circulating with no obstructions or air bubbles.

(c) Pause the drive module, stop the pump, and turn off the chiller. Load the digested BAC DNA into the wells. Load the MidRange I PFG Marker and 2.5 kb Molecular Ruler. Restart the drive module and run the gel for 20 min before turning back on the pump and chiller (*see* **Note** 7).

(d) At the end of the run, remove the gel and stain with 0.5 mg/mL ethidium bromide solution for ~30 min. If necessary, destain the gel with MilliQ water. Image the gel on a UV transilluminator system to visualize DNA bands. Figure 3c shows the result of NotI digestion and PFGE for two independent samples of a BAC clone derived from the RP24 library. Both samples had the predicted BAC insert size of 140 kb with a 10.6 kb vector band from the pTARBAC1 vector.

2. BAC clone "fingerprinting" by restriction digestion and PFGE.

(a) Using the digital sequence file for the BAC clone, identify candidate restriction enzymes that can be used to cut the BAC DNA at well-spaced intervals. In general we find that restriction enzymes having 5–10 predicted restriction sites within the BAC DNA and that generate predicted DNA fragments that span from 10–80 kb work the best. Avoid enzymes that generate multiple fragments at a given size since this complicates analysis.

(b) Perform restriction digests with 15 μL of freshly prepared BAC miniprep DNA using 1–2 μL of restriction enzyme in appropriate buffer with a 30 μL total reaction volume. BAC DNA should be prepared from several colonies from each BAC clone in order to screen for potential rearrangements or deletions of the BAC DNA. The restriction digest should proceed for at least 2–4 h. Figure 3d demonstrates BAC DNA fingerprint analysis following restriction digestion and PFGE for ten BAC clones that were expected to be identical. Note that clones #2–10 have the identical pattern of DNA fragments while clone #1 appears to have undergone a spontaneous rearrangement.

(c) Subject samples to PFGE as outlined above in step 1(b–d).

It is not uncommon to find unexpected restriction digestion patterns for certain enzymes. Some enzymes are not well suited for digestion of BAC DNA and do not cut at all predicted sites. It is recommended to try multiple restriction enzymes to determine which enzyme provides an acceptable and reliable result, and then to use this same restriction enzyme digestion for analysis following the BAC recombineering steps. If the DNA bands are weak and smeared, this indicates that the BAC DNA has degraded. BAC DNA should be freshly prepared to repeat the analysis.

3.4 Design and Construction of BAC Targeting Vectors

A targeting vector must be constructed with the goal of inserting a transgene of interest at a specific location within the selected BAC DNA sequence via homologous recombination (Fig. 4). We employed a targeting vector backbone iTV1 (based on the high-copy-number plasmid pBluescript) that contains generic features including a multiple cloning site for inserting the "BoxA" homology arm and the cDNA of interest, the bovine growth hormone polyadenylation sequence (BGHpA), an FRT-NEO-FRT cassette for positive selection, and a second multiple cloning site for inserting the "BoxB" homology arm. The targeting vector is built sequentially by adding BoxA and BoxB homology arms and then the cDNA for the transgene of interest (e.g., ChR2-EYFP) between the two homology arms.

1. The BoxA homology arm is selected as the ~500 bp region immediately upstream of the translation initiation site (ATG) of the targeted gene within the BAC (*see* **Note 8**). Design forward and reverse primers that will bind to the start and end of the BoxA region, respectively. Incorporate restriction sites to the ends of each primer that are compatible with the first multiple cloning site in iTV1 (SacI, SacII, NotI, NheI) and that do not cut within the BoxA sequence. PCR amplify the BoxA homology arm using the BAC DNA as the template for the reaction. Purify the PCR product using the Zymo DNA Clean and Concentrator Kit, and then digest the ends of the PCR product with the selected restriction enzymes. Purify the digested PCR product from a 1 % agarose gel using the Zymo Gel Recovery Kit and verify the correct size of the BoxA homology arm.
2. Digest 1 μg of the iTV1 vector with the same restriction enzymes that were incorporated to the start and end of the BoxA homology arm. Treat with alkaline phosphatase to dephosphorylate the vector and purify the linear vector DNA from a 1 % agarose gel.
3. Ligate the purified vector and homology arm using T4 ligase. Perform a vector-only control ligation to determine the ligation efficiency. Transform the ligation mixtures into competent DH5α or DH10B cells, recover the cells for 1 h at 37 °C, and then plate the cells on LB/Agar plus 50 μg/mL carbenicillin. Incubate the plates at 37 °C for 12–16 h or until colonies are clearly visible.
4. If there are few or no colonies on the vector-only plate and many colonies on the vector plus insert plate (*see* **Note 9**), pick 2 colonies for growing overnight cultures (3 mL LB plus 50 μg/mL carbenicillin). Prepare DNA from the overnight cultures using the Qiagen Plasmid Miniprep Kit following the manufacturer provided protocol. Sequence the DNA samples using T7 forward primer to verify the sequence of the BoxA homology arm insert.

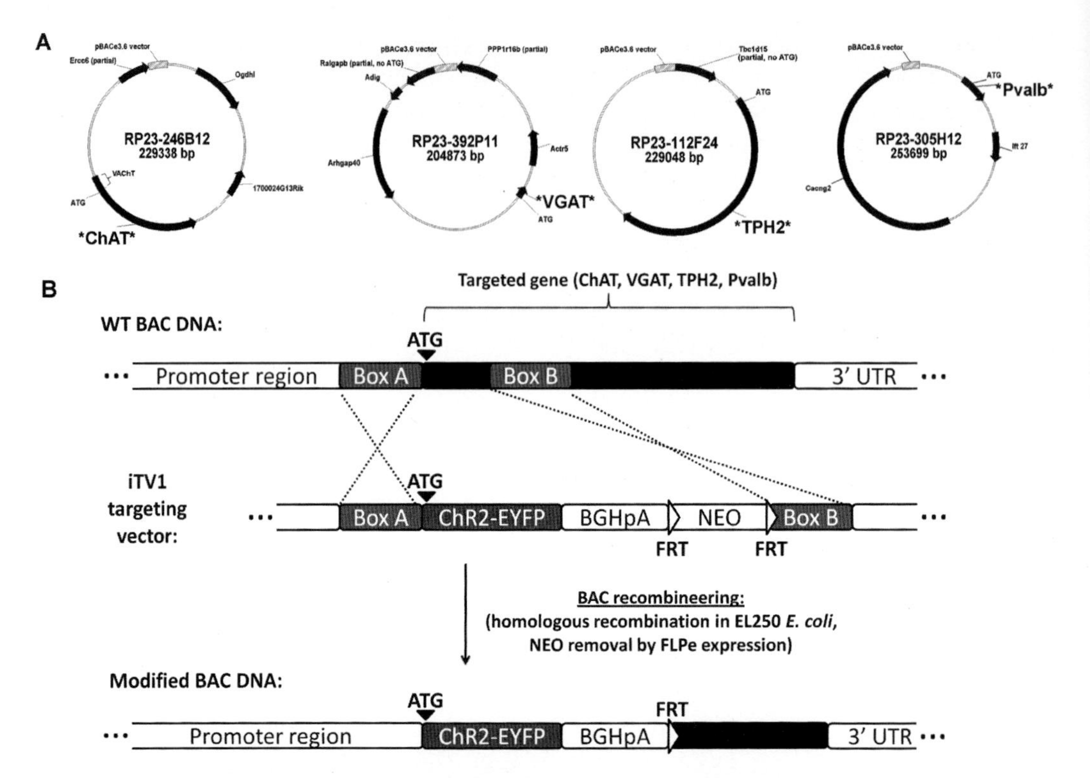

Fig. 4 Overview of BAC clones and targeting strategy for generation of ChR2-EYFP BAC transgenic mice. (**a**) Diagram of BAC clones selected for modification. (**b**) BAC recombineering strategy to introduce ChR2-EYFP under the control of the cell-type-specific promoter elements. The vector iTV1 was constructed and used to introduce the ChR2-EYFP-BGHpA-FRT-NEO-FRT cassette immediately upstream of the known translation initiation site (labeled ATG) for each targeted gene in the four different BAC clones. Each targeting vector contains BAC-specific BoxA and BoxB fragments for homologous recombination in the EL250 strain of *E. coli*. The presence of the FRT-NEO-FRT confers kanamycin resistance for selection of EL250 cells containing the properly targeted BACs. The NEO cassette is subsequently removed by arabinose-induced FLPe expression in the EL250 cells, thereby leaving only a single residual FRT site. Modified from [28]

5. The BoxB homology arm is selected as a ~500 bp region located downstream of the ATG site of the targeted gene within the BAC. Design primers for PCR amplification of the BoxB homology arm as outlined for BoxA above, but in this case incorporating restriction sites that are compatible with the second multiple cloning site in iTV1 (SmaI, XhoI, SpeI, KpnI). PCR amplifies the BoxB homology arm using the BAC DNA as the template for the reaction. Purify the PCR product using the Zymo DNA Clean and Concentrator Kit, and then digest the ends of the PCR product with the selected restriction enzymes. Purify the PCR product from a 1 % agarose gel using the Zymo Gel Recovery Kit and verify the correct size of the PCR product.
6. Digest 1 μg of the iTV1-BoxA vector with the same restriction enzymes that were incorporated to the start and end of the

BoxB homology arm. Treat with alkaline phosphatase to dephosphorylate the vector and purify the linear vector DNA from a 1 % agarose gel using the Zymo Gel Recovery Kit.

7. Repeat steps 3 and 4 as outlined above but noting that in this case, the putative iTV1-BoxA-BoxB DNA is sequenced using T3 reverse primer to verify the sequence of the BoxB homology arm insert.
8. Using the general procedural guidelines above, insert the cDNA for the transgene of interest using any of the remaining restriction site(s) between the BoxA homology arm and BGHpA of the iTV1-BoxA-BoxB vector. It is important to ensure that the transgene of interest cDNA begins with an optimal Kozak consensus sequence (GCCACCATGG) and ends with a stop codon (*see* **Note 10**). Prepare DNA using the Qiagen Plasmid Miniprep Kit and confirm the entire transgene insert of the completed targeting vector by DNA sequencing (*see* **Note 11**).
9. Linearize ≤ 1 μg of targeting vector DNA by restriction digestion with BssHII (or other suitable restriction enzymes that excise the vector backbone). Allow sufficient time and add sufficient enzyme to ensure complete DNA digestion. Run the linearized targeting vector DNA sample on a 1 % agarose gel. It is highly recommended to use a low-voltage setting for extended time (e.g., 40 V for overnight, or alternatively 70 V for several hours) in order to ensure the complete segregation of the vector backbone from the desired DNA fragment on the agarose gel. Excise the upper band (linearized targeting vector DNA) from the agarose gel and purify using the Zymo Gel Recovery Kit (*see* **Note 12**). Store at −20 °C until use.

3.5 Transfer of BAC DNA to the EL250 Strain in Preparation for Recombineering

Once the BAC clones have been verified to cover the correct region of the mouse genome, the BAC DNA must be introduced into the recombination proficient EL250 strain of *E. coli* [36]. This can be accomplished by making the EL250 cells competent for uptake of the BAC miniprep DNA by electroporation.

1. From a glycerol stock of EL250 cells, streak out the bacteria on an LB/Agar plate without antibiotics. Be very careful to avoid contamination since the EL250 cells will grow in the absence of antibiotic selection. Incubate ~16 h (overnight) at 32 °C.
2. Inoculate 5 mL LB (Lennox) with a single colony from the EL250 plate. Incubate ~16 h (overnight) at 32 °C (*see* **Note 13**).
3. Transfer 500 μL of the saturated culture to 25 mL of fresh LB (Lennox) in a 250 mL baffled Erlenmeyer flask. Incubate for

2 h at 32 °C with shaking at 180–200 RPM. Do not overgrow the cells.

4. Place the flask into wet ice on an orbital shaker for 5 min to cool the cells (*see* **Note 14**). Divide flask contents into (2) prechilled 50 mL conical tubes, 12.5 mL into each tube. Centrifuge at 4,000 RPM at 4 °C for 5 min to pellet the cells.
5. Gently resuspend pelleted cells from each tube with 1 mL ice-cold MilliQ water and transfer the cells into two prechilled 1.7 mL eppendorf tubes on ice. Do not vortex or re-pipette the cell suspension.
6. Centrifuge for 4 min at 4,000 RPM in a refrigerated tabletop microcentrifuge at 4 °C to pellet the cells. Carefully remove the supernatant and gently resuspend the pelleted cells from each tube with 1 mL ice-cold MilliQ water. Repeat wash step two more times. At this point the EL250 cells are competent and ready for electroporation.
7. Resuspend each EL250 cell pellet in a total volume of 100 μL of MilliQ water, accounting for the volume of the cell pellet itself. Dispense cells into 50 μL aliquots on ice.
8. Add 1–5 μL of BAC miniprep DNA to one 50 μL aliquot of electrocompetent EL250 cells and mix gently with the pipette tip. Transfer the cells plus DNA into a prechilled 0.1 cm gap cuvette on ice. It may be desirable to test variable amounts of BAC DNA (e.g., 1 μL and 3 μL), making use of the additional 50 μL aliquots of competent cells.
9. Electroporate using the following settings: 1.75 KV, 25 μF, 200 Ohms. The tau value should be between 4.0 and 5.0 ms. Place the cuvettes on ice for 2 min.
10. Add 1 mL LB (Lennox) to the cuvette to collect and transfer the EL250 cells into a 14 mL round-bottom Falcon tube. Incubate at 32 °C at 220 RPM for 1 hr to recover cells.
11. Pellet the recovered EL250 cells for 4 min at 4,000 RPM in 1.7 mL eppendorf tubes. Discard all but 100 μL of the supernatant, then gently resuspend the cells and spread onto an LB/Agar plate plus 12.5 μg/mL chloramphenicol. Incubate plates at 32 °C for ~20 h. It may take ≥20 h for good size colonies to form.
12. The successful introduction of the BAC clone into the EL250 strain can be verified by PCR screening using primers that amplify the end regions of the BAC (to ensure no deletions have occurred) or by isolating BAC miniprep DNA and performing fingerprint analysis by restriction digestion and PFGE as described in Sect. 3.3.
13. Prepare a glycerol stock of the EL250 cells containing the target BAC clone. Store at −80 °C.

3.6 BAC Recombineering in E. coli

The modification of BAC DNA is carried out by homologous recombination in *E. coli*, a process that has been termed BAC "recombineering," i.e., recombination-based genetic engineering [37–39]. In order to develop a homologous recombination proficient bacterial strain, the EL250 strain was engineered with a defective λ prophage inserted into the bacterial genome [36]. The relevant phage genes that are required for homologous recombination (*exo*, *bet*, and *gam*) are repressed when EL250 cells are grown at 32 °C but become transiently expressed when EL250 cells are subjected to a brief 15-min heat shock at 42 °C. Following heat shock the linear double-stranded targeting vector DNA is introduced into the EL250 cells by electroporation and the homologous recombination events can proceed with high efficiency. In addition, the EL250 strain exhibits arabinose-inducible Flp expression for excision of the FRT-flanked NEO cassette.

1. Perform the homologous recombination procedure in EL250 cells:
   (a) Briefly remove the glycerol stock of EL250 cells containing the BAC clone of interest from the −80 °C freezer and streak out the bacteria on an LB/Agar plate plus 12.5 μg/mL of chloramphenicol. Incubate ~16–20 h (overnight) at 32 °C.
   (b) Inoculate 5 mL LB (Lennox) plus 12.5 μg/mL of chloramphenicol with a single bacterial colony. Incubate ~16 h (overnight) at 32 °C.
   (c) Transfer 500 μL of the saturated culture to 25 mL of fresh LB (Lennox) in a 250 mL baffled Erlenmeyer flask (*see* **Note 13**). Incubate for 2 h at 32 °C with shaking at 180–200 RPM. Do not overgrow the cells.
   (d) Heat shock the cells at 42 °C for exactly 15 min in a shaking water bath (*see* **Note 15**). If desired, perform a control treatment without heat shock for comparison.
   (e) Place the flask into wet ice on an orbital shaker for 5 min to rapidly cool the cells. Divide flask contents into (2) prechilled 50 mL conical tubes, 12.5 mL into each tube. Centrifuge at 4,000 RPM at 4 °C for 5 min to pellet the cells.
   (f) Gently resuspend pelleted cells from each tube with 1 mL ice-cold MilliQ water and transfer the cells into two prechilled 1.7 mL eppendorf tubes on ice. Do not vortex or re-pipette the cell suspension.
   (g) Centrifuge for 4 min at 4,000 RPM in a refrigerated tabletop microcentrifuge at 4 °C to pellet the cells. Carefully remove the supernatant and gently resuspend the pelleted cells from each tube with 1 mL ice-cold MilliQ water. Repeat wash step two more times. At this point the EL250 cells are competent and ready for electroporation.

(h) Resuspend each EL250 cell pellet in a total volume of 100 μL of MilliQ water, accounting for the volume of the cell pellet itself. Dispense cells into 50 μL aliquots on ice.

(i) Add 1–5 μL of linearized targeting vector DNA into one 50 μL aliquot of electrocompetent EL250 cells and mix gently with the pipette tip. Transfer the cells plus DNA into a prechilled 0.1 cm gap cuvette on ice. It may be desirable to test variable amounts of linearized DNA (e.g., 1 μL and 3 μL), making use of the additional 50 μL aliquots of competent cells.

(j) Electroporate using the following settings: 1.75 kV, 25 μF, 200 Ω. The tau value should be between 4.0 and 5.0 ms. Place the cuvettes on ice for 2 min.

(k) Add 1 mL LB (Lennox) to the cuvette to collect and transfer the EL250 cells into a 14 mL round-bottom Falcon tube. Incubate at 32 °C at 220 RPM for 1 hr to recover cells.

(l) Pellet the recovered EL250 cells for 4 min at 4,000 RPM in 1.7 mL eppendorf tubes. Discard all but 100 μL of the supernatant, then gently resuspend the cells and spread onto an LB/Agar plate plus 30 μg/mL kanamycin. Incubate plates at 32 °C for ~20 h. It may take ≥20 h for good size colonies to form. Antibiotic selection with kanamycin provides positive selection for successful recombinants that now have the neomycin resistance gene (NEO) located within the BAC.

2. Identify clones with successful homologous recombination by colony PCR screening.

(a) Design two independent PCR primer pairs that amplify from upstream of the BoxA homology arm to the transgene cDNA region and from the terminal portion of the NEO cassette to downstream of the BoxB homology arm. Perform the PCR reaction and analyze the PCR products on a 1 % agarose gel. A positive PCR screen for both primer pairs indicates that the transgene cassette was correctly inserted at the intended location within the BAC DNA sequence.

(b) Select multiple positive colonies for growing overnight cultures. Inoculate 5 mL LB (Miller) plus 12.5 μg/mL chloramphenicol and 30 μg/mL kanamycin with a single positive colony. Incubate ~16–20 h (overnight) at 32 °C.

(c) Prepare glycerol stocks for the EL250 cells containing the correctly modified BAC clone. Store at −80 °C.

(d) For each positive clone, prepare BAC miniprep DNA as outlined previously under step 4 in Sect. 3.2.

3. Induce Flp expression in EL250 cells to excise the FRT-NEO-FRT cassette from the modified BAC.
   (a) Briefly remove the glycerol stock of EL250 cells containing the correctly modified BAC from the −80 °C freezer and streak out the bacteria onto a fresh LB/Agar plate plus 30 μg/mL kanamycin and 12.5 μg/mL chloramphenicol. Incubate the plate at 32 °C for ~20 h to or until single colonies are clearly visible.
   (b) From a single colony, inoculate 1 mL LB-Miller plus 30 μg/mL kanamycin and 12.5 μg/mL chloramphenicol in a 14 mL round-bottom Falcon tube. Incubate the culture at 32 °C with shaking at 220 RPM for ~6–8 h or until slightly turbid.
   (c) Add 10 μL of 10 % arabinose and mix gently (*see* **Note 16**). Incubate 1 additional hour at 32 °C with shaking at 220 RPM.
   (d) Dilute 2 μL of the liquid culture media into 1 mL of fresh LB-Miller in a 1.5 mL eppendorf tube and mix gently. Pipette 100 μL of LB-Miller directly on to an LB/Agar plus12.5 μg/mL chloramphenicol plate and add 2 μL of the diluted media. Spread the mixture evenly on the plate and incubate at 32 °C for ~20 h or until single colonies are clearly visible (*see* **Note 17**).
4. Identify clones with successful NEO removal by colony PCR screening following the general procedures outlined in step 2(a–d) above but using an appropriate primer pair to detect the excision of the NEO cassette. The forward primer recognizes the sequence near the end of the BGHpA (upstream of the first FRT site) and the reverse primer binds immediately downstream of the BoxB homology arm. The size of the PCR product depends on the size of the BoxB homology arm (*see* **Note 18**). Furthermore, when growing overnight cultures for glycerol stocks and BAC miniprep DNA following NEO removal, the kanamycin selection is omitted (only add 12.5 μg/mL chloramphenicol). Store modified BAC miniprep DNA at 4 °C until further analysis.

3.7 Analysis of Sequential BAC Recombineering Steps by PFGE

To perform a comparative DNA analysis at various stages of the recombineering process, follow the general procedures outlined in Sect. 3.3 using the unmodified BAC miniprep DNA and the BAC miniprep DNA collected after each modification step (e.g., homologous recombination, NEO removal). The most useful analysis is by restriction digestion using enzymes that produce BAC DNA fragments that are of moderate size (20–50 kb) and that are well spaced. In many cases the enzyme selected for fingerprint analysis of the unmodified BAC clones will also be suitable for this analysis (*see* **Note 19**). Comparison of multiple samples at each stage of the recombineering is recommended in order to identify and avoid BAC clones with unintended rearrangements, although such events are very rare.

3.8 Large-Scale BAC DNA Preparation

Once the final BAC modifications have been completed and verified, it is necessary to prepare BAC DNA of high purity and integrity, ideally with a concentration of at least 100 ng/μL or higher. This is best accomplished using the NucleoBond BAC 100 Kit from Clontech (*see* **Note 20**). The BAC 100 Kit uses proprietary silica-based anion exchange column technology together with folded filters for cellular lysate clarification, thereby circumventing the need for toxic chemicals or high-speed centrifugation steps which can shear large DNA species such as BACs. The kit is designed specifically for purification of large DNA constructs up to ~300 kb and incorporates the recommended higher buffer volumes that facilitate higher DNA yield.

1. Briefly remove the glycerol stock of the modified BAC clone in EL250 cells from the –80 °C freezer and streak a fresh LB/Agar plate plus 12.5 μg/mL chloramphenicol. Incubate the plate at 32 °C for ~20 h or until colonies are clearly visible.
2. Inoculate 400 mL of LB (Miller) plus 12.5 μg/mL chloramphenicol with a single colony from the BAC plate (*see* **Note 21**). Incubate at 32 °C for ~20 h in a shaking incubator at 220 RPM.
3. Transfer the bacteria into a 500 mL centrifuge container and harvest the bacterial by centrifugation at 5,000 RPM for 15 min at 4 °C. Discard the supernatant.
4. Follow the manufacturer's recommended protocol for the NucleoBond BAC 100 Kit under "low-copy plasmid purification (Maxi/BAC)." We recommend using clear 50 mL polycarbonate Oak Ridge tubes for isopropanol precipitation and subsequent centrifugation steps (*see* **Note 22**). For the final reconstitution step, we use 400 μL of MilliQ water to dissolve the BAC DNA pellet with gentle rocking and tapping. Do not overdry the DNA pellet. Do not vortex or re-pipette the DNA solution. Typical BAC DNA yields range from 100–400 ng/μL. If the yield of BAC DNA is below 100 ng/μL, it is recommended to repeat the culture and DNA purification steps.

3.9 Purification of Linearized BAC DNA

Circular BAC DNA can be used directly for pronuclear injection. However, the available evidence indicates a higher success rate for injection of linearized BAC DNA [40], and linearization avoids random breakage of the BAC that occurs following injection of circular BAC DNA. On the other hand, the purification of high concentration linearized BAC DNA that is of very high integrity can be a formidable task. Despite exercising great care in handling samples, it is inevitable that BAC DNA will degrade to some appreciable extent. The goal is to limit degradation to a minimum to ensure the highest chance of successful integration of the intact BAC into the mouse genome following pronuclear injection. A few simple principles for curtailing degradation include the omission of any high-speed centrifugation steps that are likely to shear BAC DNA, the use of high-purity reagents in preparing microinjection buffer components, and minimizing BAC DNA storage time prior to injection. However, the most rigorous method for isolation of intact linearized BAC DNA (at the exclusion of degraded BAC DNA) is to excise the linearized BAC fragment directly from a preparative PFG. This is critical since BAC DNA that is completely sheared will still appear as a single tight band by conventional 1 % agarose gel electrophoresis but can be easily detected by PFGE. Therefore, the excision and purification of intact BAC DNA directly from a PFG can lead to a notably higher success rate in the generation of BAC transgenic mice and rats, as has been reported previously [41, 42].

1. Linearize ~10–15 μg of modified BAC DNA by restriction digestion with 10 μL of NotI high-fidelity enzyme in a 100 μL reaction volume (*see* **Note 23**). Incubate at 37 °C for 2–4 h.
2. Tape together several wells in the middle of the gel comb to make one long continuous well and cast a preparative 1 % agarose gel in 0.5× TBE buffer. Leave at least two single wells on each side for loading DNA ladders and reference BAC DNA samples.
3. Fill the PFGE chamber with 2 L of 0.5× TBE buffer. Turn on the power module, drive module, controller module/pump (set to 70), and chiller module (set to 14 °C). Remove the comb and load the solidified gel into the chamber and pre-run the system for 30 min (*see* **Note 6** for run parameters). Ensure that the system is operating properly and the buffer is circulating with no obstructions or air bubbles.
4. Pause the drive module, stop the pump, and turn off the chiller. Add 6× DNA loading buffer to the DNA sample and load the majority of the digested modified BAC DNA into the main well. Add a small sample (~5 μL) of the remaining sample into the flanking reference wells. Load the MidRange I PFG Marker and 2.5 kb Molecular Ruler into the outside wells. Restart the drive module and run the gel for 20 min before turning back on the pump and chiller (*see* **Note 7**).
5. At the end of the run, remove the gel and cut away the entire section containing the main sample lane and temporarily place it to the side. Stain the flanking gel segments with 0.5 mg/mL ethidium bromide solution for ~30 min. If necessary, destain the gel by gentle agitation in MilliQ water for 20 min. Image the flanking gel segments on a UV transilluminator system

to visualize DNA bands and mark the position of the upper BAC insert on both sides using a razor blade. Turn off the UV light and place the unstained main gel segment in between the stained segments to reassemble the full gel. Using the physical markings on the flanking segments as a guide, excise the long gel band containing the linearized BAC DNA (Fig. 5) (*see* **Note 24**). Save the gel slice at 4 °C.

6. Perform electroelution of linearized BAC DNA. A diagram of the electroelution setup is shown in Fig. 5.
   (a) Set up a ring stand with clamps on your benchtop. Secure the electroelution column to the stand with the clamps. The column should be perfectly vertical. Attach the IV drip reservoir to the stand about 8 in. above the top of the column. Place a P200 pipette tip on the end of the drip line tubing for better control of droplet size. Fill the drip reservoir with 100 mM caproic acid. Allow the acid to flow into the line all the way to the end (no air bubbles in the line), then clamp the line shut.
   (b) Place several 1.0 cm × 1.0 cm sections of dialysis membrane in a petri dish filled with MilliQ water. It is convenient to place the petri dish directly under the column to catch the water flowing through the column.
   (c) Fill the column with MilliQ water. As the water drains through the column, force out any air bubbles that could impede flow using a rubber air bulb. Just before the water completely drains out of the column, pack the bottom of the column with ~1 mL of G50 Sephadex beads in 1× TE. Add more G50 Sephadex beads to completely pack the column to the top of the narrow part of the column.
   (d) Gently fill the upper part of the column with 1× TE. Allow the 1× TE to flow through the column by gravity until the level reaches about 1 cm above the top of the G50 beads. As the level approaches this line, lift a piece of the rinsed dialysis membrane (using forceps to grip the edge) and secure the membrane over the bottom opening of the column using the end cap (*see* **Note 25**). The cap should fit securely to prevent leakage of fluid from the column. If necessary, a rubber band can be placed around the outer ring of the end cap and stretched over the clamp to provide additional support and ensure the cap stays firmly in place.
   (e) After the end cap is secured, wait 5 min before proceeding to ensure there is no leak in the column. If the level of the fluid in the column decreases, remove the cap and dialysis tubing and go back to step "d."

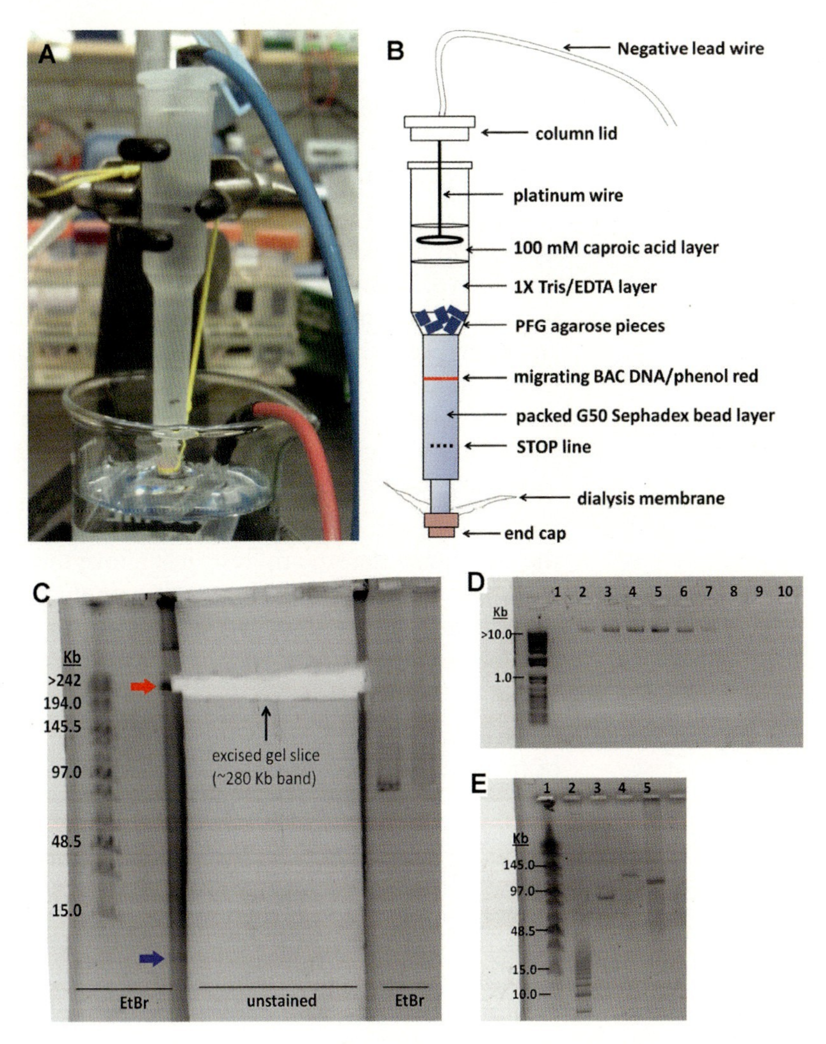

Fig. 5 Purification of linearized BAC DNA. (**a, b**) Photograph and corresponding diagram of the custom electroelution column. (**c**) Excision of intact BAC DNA fragments from a 1 % PFG agarose gel (*red arrow*), effectively separating the BAC DNA from the vector (*blue arrow*). The flanking gel segments were stained with ethidium bromide (EtBr) to visualize and mark the position of the target BAC DNA band. A gel slice was excised from the center unstained portion of the gel using the markings on the flanking gel segments as a guide. (**d**) Following electroelution, the sequential fractions are analyzed on a 1 % agarose gel using standard electrophoresis. The highest fractions (samples 3–6) are pooled and then subjected to spot dialysis against microinjection buffer. (**e**) Verification of linearized BAC DNA integrity following the electroelution and dialysis procedures. *Lane 1*: MidRange I PFG Marker. *Lane 2*: 2.5 kb marker. *Lanes 3–5*: analysis of three different linearized BAC DNA samples. Note the slight BAC DNA degradation for the sample in *lane 5*

(f) Gently lower the column into a 250 mL beaker filled with 1× TE. The membrane and end cap should just break the surface of the TE. Ensure that there are no air bubbles trapped on the submerged tip of the column as this will impede the electroelution.

(g) Connect the positive lead wire to the power supply. Submerge the platinum wire end of the positive lead into the beaker containing the 1× TE (*see* **Note 26**). Connect the negative lead wire to the power supply. Do NOT attach the column cap at this time.

(h) On a clean piece of parafilm, mince the excised gel slice containing the BAC DNA with a clean razor blade. Push the finely minced pieces together in a tight ball. Add 0.5 μL of the concentrated phenol red stock to the ball and allow the dye to spread into the agarose/DNA pieces. Gently transfer the ball containing the BAC DNA into the column. The ball should stay together as much as possible and should rest right on top of the bed of G50 Sepharose beads and under the layer of 1× TE.

(i) Gently place the P200 tip attached to the IV drip line into the column so that the tip rests just above the 1× TE layer and at the side wall of the column. Open the drip line with slow flow (about 1 drip per second) so that the caproic acid fills the uppermost part of the column without disturbing the agarose pieces.

(j) Gently place the negative lead wire with cap onto the top of the column. The platinum wire is shaped in a ring and should be centered in the column submerged within the caproic acid layer only.

(k) Turn on the power supply to ~120 V setting. The current should read 1–2 mA. After about 10 min you should see a tight red band forming within the top part of the G50 bead layer. This band will migrate to the bottom of the column. The BAC DNA runs just in front of the red band (*see* **Note 27**).

(l) When the tightly formed red band reaches the mark near the bottom of the column, turn off the power supply. Remove the lid of the column. Carefully raise the column and move the beaker out of the way.

(m) With a 1 mL pipette, carefully remove all liquid down to the bead layer. Some beads may become lodged in the pipette tip, so it is bet to start around the edges.

(n) Prepare a tube rack with (10) 1.5 mL eppendorf tubes labeled 1–10. Open the caps to facilitate collection of the droplets from the column.

(o) Slowly pipette in 1 mL of 1× TE into the upper part of the column. Remove the rubber band being sure not to dislodge the end cap. Wipe the excess wetness around the rim of the end cap (but not touching the membrane), then remove the cap and quickly position the tube rack with open eppendorf tubes under the column.

(p) Collect two drips per eppendorf tube (in sequential order from tube 1–10) for analysis of the fractions. We find that when the column is prepared correctly and the power is turned off at the correct time, the highest fraction of BAC DNA normally elutes in the second through fifth drips. When the phenol red is visible in the drips, most of the BAC DNA has already eluted.

7. Prepare a standard 1 % agarose gel to analyze the fractions. Load 5 μL of each fraction sequentially to the gel wells. Run the gel at 100 V for 30–40 min and then analyze the gel image to determine the fractions with the most concentrated BAC DNA (Fig. 5d). Combine the 3 or 4 highest fractions into one tube, which should yield at least 200–400 μL total volume of electroeluted, linearized BAC DNA.

8. Perform dialysis of electroeluted BAC DNA by carefully pipetting the sample using a P1000 tip onto the center of a 25 mm, 0.025 μm filter spot dialysis disc floating shiny side up in 50–100 ml of microinjection buffer (*see* **Note 28**). Let the dialysis proceed undisturbed at 4 °C overnight. The extended dialysis step will result in a desirable concentration of the BAC DNA by approximately 10–20 %. Collect the dialyzed BAC DNA using a P1000 tip and taking care not to mix the sample DNA with the reservoir of microinjection buffer. The sample should now be clear with no traces of phenol red. Store the injection-ready BAC DNA at 4 °C. Polyamines are added to the BAC DNA sample exactly one week prior to the scheduled injection date (*see* **Note 29**).

3.10 Final PFGE Analysis of Injection-Ready BAC DNA

If the injection-ready, linearized BAC DNA has been successfully prepared at a concentration of at least 2 ng/μL, it should be possible to visualize the product by PFGE (*see* **Note 30**). It is ideal to verify the BAC DNA size and integrity by PFGE immediately before the sample is used for pronuclear injection. This ensures that the DNA was intact with minimal degradation leading up to the injection procedure (Fig. 5e). If desired, the unused sample can be saved and further analyzed by PFGE soon after the pronuclear injection procedure has concluded to verify that the sample has not degraded during the injection procedure (*see* **Note 31**). To estimate the concentration of the injection-ready BAC DNA, compare the sample to linearized BAC DNA of known concentrations prepared by NotI digestion and dilution with MilliQ water. Highly concentrated BAC DNA prepared using the BAC 100 Kit should be used for preparing the standards to ensure sufficient accuracy. It is practical to use BAC DNA standards covering the range 0.5–5.0 ng/μL and load 20 μL of each sample per well.

3.11 Pronuclear Injection and Identification of Transgenic Founder Lines

To produce BAC transgenic founders, the purified and linearized BAC DNA is injected at a concentration of 0.5–2.0 ng/μL into the male pronucleus of fertilized eggs at the one-cell stage. The surviving embryos are then transplanted into pseudopregnant female surrogate mice. In our studies these procedures were carried out by the Duke Neurotransgenic Laboratory according to our previously outlined pronuclear injection protocol [43]. Further information and detailed protocols pertaining to standard pronuclear injection techniques can be found elsewhere [44, 45]. Resultant offspring were genotyped to identify transgenic founders by tail PCR using the HotSHOT method [46]. PCR-positive founders were crossed to pure C57Bl/6J mice to verify germline transmission of the transgene to the offspring and establish the line. A minority of founder lines created in this manner exhibit mosaicism and do not transmit the transgene through the germline at the expected frequency. In some cases it is possible to establish germline transmission after extensive breeding and PCR genotyping of multiple litters produced by the mosaic founder, assuming that the transgene is present in at least some proportion of the germ cells.

With the methods we have described for preparing high-integrity BAC DNA, we routinely find that 10–15 % of pups born following pronuclear injection are PCR positive for the transgene. Low numbers of pups born may indicate toxicity of the transgene or toxicity from injection of BAC DNA that was too highly concentrated (>5 ng/μL). Conversely, excessively high numbers of pups born with a very low positive ratio may indicate an insufficient amount of injected BAC DNA.

4 Overview of Additional BAC Recombineering Methods

The majority of BAC transgenic lines developed to date have utilized BAC clones that contain one or more genes other than the gene that was the explicit target for manipulation. This issue has been largely overlooked and underreported given that the goal of BAC transgenesis is first and foremost to ensure a desirable pattern of gene expression based on the large regulatory elements contained in the BACs. However, the presence of "extra" genes can be a major confound in many experimental contexts. Some extra genes may only be expressed in various tissues that do not share overlap with the expression profile of the target transgene but that may still alter the phenotype of the animal. In some BAC transgenic lines, the extra genes may be overexpressed to a considerable extent in the same tissue as the transgene of interest, which poses the largest potential confound. This was convincingly demonstrated for multiple GENSAT BAC transgenic mice derived from the Drd1a spanning BAC clone RP23-47M2 and the Drd2 spanning BAC clone RP23-161H15, respectively [26, 47]. In Drd1a-BACTRAP mice, the extra gene Sfxn1 is overexpressed in Drd1a-expressing striatal medium spiny neurons, while in Drd2-BACTRAP mice, the extra gene Ttc12 is overexpressed in Drd2-expressing striatal medium spiny neurons [26]. Although not unexpected based on the BAC clones selected for development of these transgenic mouse lines, it is important to point out that these findings were derived from an unbiased gene expression profiling approach. In contrast, few BAC transgenic studies have directly investigated the overexpression of extra genes and the contribution of extra genes to an observed phenotype [48] or made strong inferences about the likely contribution of extra genes to an unexpected phenotype [49, 50]. To circumvent this underappreciated problem, we have employed methods for eliminating undesirable "extra" genes from BAC clones. These strategies may be broadly useful in developing improved BAC transgenic lines with fewer potential confounds. Here we briefly discuss the strategies we have devised and overview critical steps for implementing these strategies. Further details regarding these BAC recombineering strategies have been recently described in a separate study [51].

4.1 Removal of Potentially Confounding "Extra" Genes by BAC Trimming

The most straightforward approach for eliminating extra genes contained in BAC clones is to remove segments from either end of the BAC insert, a method known as BAC trimming or shaving. One method for accomplishing this task takes advantage of the strategically placed incompatible loxP and mloxP sites at either end of the BAC vectors used to create the RP23 and RP24 libraries. By inserting a compatible site (loxP or mloxP) at a defined location of the BAC by homologous recombination, one can then remove end segments by expression of Cre recombinase in the bacteria. However, this strategy may be undesirable as it requires two sequential steps and will result in the removal of restriction sites used to liberate the BAC vector from the modified insert (unless added back during the recombineering steps). Failure to remove the vector sequence harboring the loxP sites will result in mouse lines that are not compatible for use with Cre/loxP studies and renders this approach unsuitable for developing Cre-inducible BAC transgenic lines.

A more versatile method for BAC trimming has been described in which small antibiotic selection cassettes are used to replace BAC DNA segments up to 60 kb using a single recombineering step [52, 53]. The selection cassettes are PCR amplified using long primers that add ~50 bp homology arms to either end [37] then used for recombineering according to the general procedures outlined in Sect. 3.6. The homology arms define the exact BAC region that will be deleted by the insertion of the selection cassette, and the selection cassette enables subsequent identification of properly modified BAC clones. In addition, many different selection cassettes can be utilized for BAC trimming (e.g., ampicillin resistance-bla, kanamycin resistance-neo, chloramphenicol resistance-cam, blasticidin resistance-bsd, and streptomycin resistance-rpsL), providing options for sequential rounds of BAC trimming and transgene targeting using different selection markers. To further generalize this strategy for deleting fragments from any BAC clone in the RP23 and RP24 library, one of the homology arms can be designed on the end of the BAC vector such that only the second homology arm needs to be customized for each unique BAC clone to define the deletion region (the region containing undesirable extra genes). In addition, we include a NotI restriction site (or other suitable restriction site) located adjacent to the unique homology arm but before the selection cassette. Once the deletion of the BAC segments has been confirmed by restriction digestion and PFGE as outlined in Sect. 3.3, the BAC insert can be liberated from the BAC vector plus selection cassette by NotI digestion.

We have used this method to successfully delete precisely defined BAC DNA segments >80 kb as demonstrated for the DAT BAC clone RP24-269I17 (Fig. 6a, b). It is preferable to perform the BAC trimming step first to generate a glycerol stock of the trimmed BAC clone lacking extra genes. This enables the use of the trimmed BAC clone for introducing any transgene cassette of interest without the need for repeating the BAC trimming steps each time. However, if desired, it is also possible to perform BAC trimming and insertion of the transgene cassette simultaneously and with high efficiency by using two different antibiotic-resistance markers and double selection. It is worth noting that other methods have been described for removal of extra genes from BAC clones to produce BAC transgenic mouse lines. In one study a modified BAC clone was subjected to restriction digestion and PFGE, and, fortuitously, a fragment could be excised that contained suitably large 5′ and 3′ flanking regions but excluded several extra genes originally present in the BAC clone [54]. In another study a spontaneous deletion was identified in a single clone during the recombineering steps, and this deletion serendipitously removed the only extra gene present in the BAC clone [55]. However, the recombineering-based method we have outlined for BAC trimming affords greater reliability and precision for routine deletion or extra genes.

4.2 Inactivation of Potentially Confounding "Extra" Genes by Minimal Insertion Mutations

Although BAC trimming can be a versatile and powerful method for deleting both small and large segments of BAC DNA, in some cases this method is not permissible for the removal of extra genes. For example, when an extra gene is located within the critical 5′ and 3′ flanking regions of the targeted gene, deletion of these flanking regions would likely negatively affect the pattern of gene expression. We have successfully trimmed BAC clones to within 15–20 kb of 3′ flanking sequence with respect to the targeted gene region (and noting that the ~50 kb 5′ flanking region was not modified) without altering the expected pattern of transgene expression in the brain. Therefore, BAC trimming may still be the

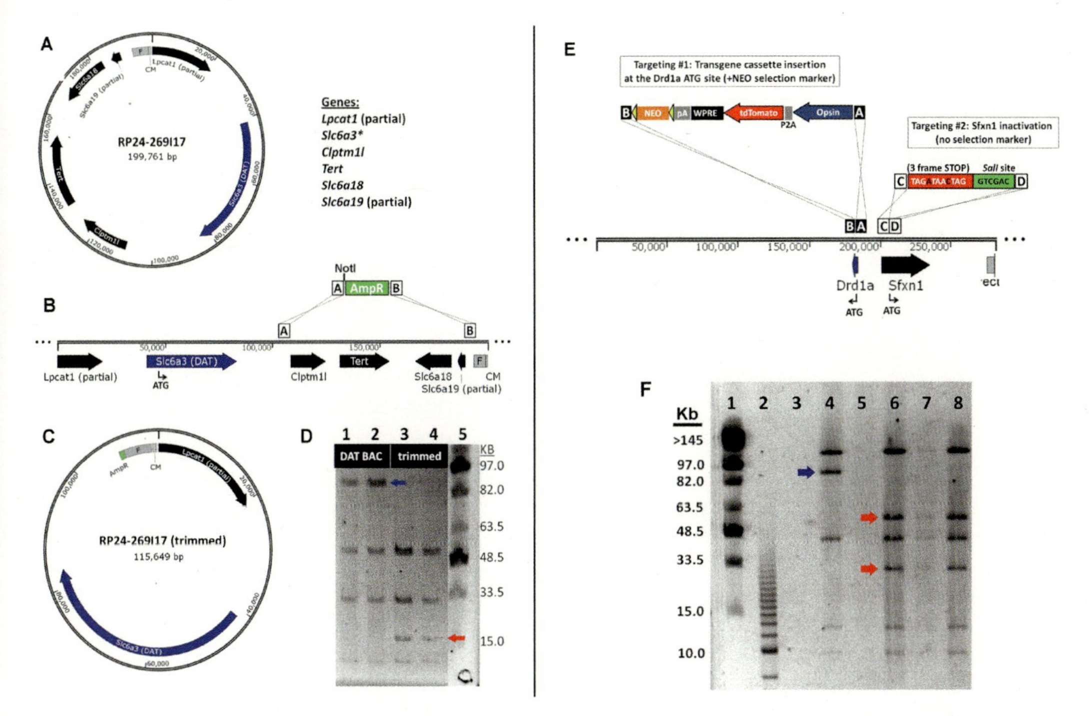

Fig. 6 BAC recombineering strategies for avoiding expression of extra genes. (**a**) Strategy for deletion of extra genes by BAC trimming. A diagram of the *DAT* (*Slc6a3*)-spanning BAC clone RP24-269I17 reveals the presence of many extra genes including *Clptm1l*, *Tert*, and *Slc6a18* and the truncated genes *Lpcat1* and *Slc6a3*. (**b**) The extra genes *Clptm1l*, *Tert*, *Slc6a18*, and *Slc6a19* located in the 3′ flanking region of the *DAT* gene are deleted by homologous recombination. The recombination event replaces the ~84 kb BAC region with a 1 kb ampicillin resistance (AmpR) cassette. (**c**) Diagram of the "trimmed" *DAT* BAC clone that can now be used for insertion of a transgene cassette at the *DAT* gene by BAC recombineering. (**d**) Verification of DAT BAC trimming by diagnostic restriction digestion with SalI and PFGE. *Lanes 1* and *2*: modified DAT BAC DNA. *Lane 3* and *4*: trimmed DAT BAC DNA. *Lane 5*: MidRange I PFG Marker. In successfully trimmed BAC clones, the 91 kb DNA fragment (*blue arrow*) is reduced to 17 kb (*red arrow*). In addition, a 6 and 4 kb fragment are deleted but are too small to be visualized on the PFG. (**e**) Strategy for inactivation of extra genes by dual targeting. The *Drd1a*-spanning BAC clone RP23-47M2 contains the gene *Sfxn1* in the promoter region of the *Drd1a* gene. In the dual-targeting strategy, two simultaneous homologous recombination events must occur. The first targeting event is the insertion of the transgene cassette (including the NEO selection marker) at the start site of the *Drd1a* gene. The second targeting event is the insertion of a triple frame stop mutation (and SalI restriction site) immediately downstream of the *Sfxn1* start codon. Double recombinant clones are identified by kanamycin resistance. (**f**) Verification of Sfxn1 gene inactivation in modified Drd1a BAC DNA by diagnostic restriction digestion with SalI and PFGE. *Lane 1*: MidRange I PFG Marker. *Lane 2*: 2.5 kb marker. *Lane 4*: Drd1a BAC DNA (+*Sfxn1*). *Lanes 6–8*: Drd1a BAC DNA following *Sfxn1* inactivation. For successfully modified BAC clones, the ~85 kb DNA fragment (*blue arrow*) is cleaved into 60 and 35 kb fragments (*red arrows*). Note that all other bands remain unaltered

method of choice for eliminating extra genes that are located >15 kb from the 3′ end of the targeted gene region. If the extra genes are located <15 kb on the 3′ flanking region or at any position within the 50 kb 5′ flanking region, alternative strategies must be employed to minimize the alterations to the DNA sequence. We devised a strategy to inactivate extra genes nested within 5′ flanking regions of the primary target gene by inserting stop codons in all three reading frames to interrupt the coding region of the extra gene. The targeting vector contains the three frame stop mutation and a carefully selected diagnostic restriction site flanked by ~1 kb homology arms to target the insertion event immediately downstream of the start codon for the extra gene. In this way the inactivation of an extra gene can be carried out with minimal insertions as small as 15–20 bp (Fig. 6d).

The targeting vector for gene inactivation does not contain any selection marker and therefore must be used in conjunction with a second targeting vector that does contain a selection marker, such as targeting vectors constructed using iTV1 as described in Sect. 3.4. The two targeting vectors are jointly used for homologous recombination steps in EL250 cells containing the target BAC clone as outlined in Sect. 3.6. This dual-targeting strategy allows for simultaneous insertion of a transgene cassette and inactivation of an extra gene by selecting for EL250 cells that have undergone successful homologous recombination at both locations within the BAC DNA sequence. In order to ensure a high likelihood that the antibiotic selection appropriately isolates double-recombinant clones, the concentrations of each linearized targeting vector are carefully proportioned. The amount of linearized targeting vector for gene inactivation should be roughly 20× higher than the linearized iTV1-based targeting vector. This is best accomplished by diluting the iTV1-based vector 1:20 and using 1 μL of each in the electroporation step. In addition, relatively long homology arms (~1 kb) are intentionally used for the gene inactivation targeting vector to ensure a high efficiency of recombination. When the efficiency of recombination and proportion of targeting vectors is optimized, it is generally possible to identify one or more double-recombinant clones by PCR screening of <50 colonies (*see* **Note 32**). Because the insertion includes a new restriction site, the successful gene inactivation targeting event can be readily verified by restriction digestion and PFGE fingerprint analysis of the modified BAC DNA (Fig. 6e). Following establishment of the BAC transgenic line, successful gene inactivation can also be verified in various tissues by analysis of mRNA or protein levels.

5 Notes

1. The specific lengths of 5′ and 3′ flanking DNA for selection of an appropriate BAC clone are recommendations based on our experience and successful outcomes in creating BAC transgenic mice from the literature. However, the extent of 5′ and 3′ flanking regions necessary for recapitulating endogenous gene expression patterns is gene specific, and in some cases it

may be possible that important regulatory elements are located more distal to the coding region. Nonetheless, in the majority of cases, it is both practical and sufficient to select BAC clones with at least 50 kb of 5′ and 50 kb of 3′ flanking sequence.

2. Searching for BAC end sequencing data on the UCSC Genome Browser: visit the website at http://genome.ucsc.edu/cgi-bin/hgGateway. Under the genome pulldown menu, select "mouse." For assembly select "July 2007" since the latest assembly for Dec 2011 does not seem to support searching for BAC clones. Enter the BAC clone ID in the search term box and submit. If the database has information on this clone, the genome alignment will automatically appear with the specified BAC clone highlighted in the left-hand column. If there is no information on the specified clone, you will receive an error message at the top of the page. For successful searches, click on the BAC clone ID in the left-hand column to display the "genomic alignment of BAC ends" data. Links are provided that contain the primer information (T7 or SP6) and the sequencing reads. This information can be used to verify the start, end, and orientation of the BAC insert relative to the BAC vector sequence.

3. Creating a digital sequence file of selected BAC clones: If the orientation of the BAC clone is not available through online databases, construct the digital sequence with random orientation. This can be verified and corrected at a later time using BAC end sequencing with T7 or SP6 or by PCR-based methods.

4. Growing cultures of *E. coli* harboring BAC DNA: BAC clones replicate very slowly compared to the propagation of high-copy-number plasmids in bacteria. It is important to grow BAC clones at 32 °C to avoid unwanted rearrangements or recombination of the BAC DNA, especially when BAC DNA is propagated in the EL250 strain. When streaking bacteria containing BAC clones from a stab culture onto LB plates, it may take >20 h of incubation at 32 °C or longer before clear individual colonies can be observed.

5. If no PCR products can be obtained, it is likely that the ends of the BAC DNA do not match to the expected sequence. In this situation it is best to screen a different BAC clone rather than spend time to track down the true BAC ends. However, another strategy is to prepare high concentration, highly pure BAC DNA for use in end sequencing, although in our experience there is a high failure rate with this approach. Alternatively, several new primer pairs can be designed to PCR amplify ~500 bp regions of DNA at variable distances from the predicted BAC ends. In this way it is possible to determine if the BAC clone covers the wrong genomic region or if the ends are merely slightly offset from the predicted ends. We have used

this strategy to identify BAC ends that were both longer and shorter than the predicted ends, in some cases by as much as 20 kb.

6. The PFGE run parameters can vary greatly depending on the size range of BAC DNA fragments that need to be resolved. Here is a list of recommended parameters for adequate resolution of common DNA size ranges:

   For separation from 10 to 275 kb:

   Volts/cm: 6 V (highest setting)

      Initial switch time: 5 s

      Final switch time: 15 s

      Pre-run time: 30 min

      Total run time: 20 h

   For separation from 5 to 150 kb:

      Volts/cm: 6 V (highest setting)

      Initial switch time: 0.5 s

      Final switch time: 10 s

      Pre-run time: 30 min

      Total run time: 16 h

   For separation from 5 to 100 kb:

      Volts/cm: 6 V (highest setting)

      Initial switch time: 0.5 s

      Final switch time: 5 s

      Pre-run time: 30 min

      Total run time: 16 h

   For separation from 1 to 50 kb:

      Volts/cm: 6 V (highest setting)

      Initial switch time: 0.1 s

      Final switch time: 4 s

      Pre-run time: 30 min

      Total run time: 16 h

7. When loading the BAC DNA samples into the wells, avoid unnecessary re-pipetting for sample mixing with the loading buffer to prevent possible shearing of the BAC DNA. Some protocols advocate the use of cutoff pipette tips for loading BAC DNA; however, we find that this is not necessary if care is taken in gently pipetting and loading the samples. It is important to pause the run and turn off the pump and chiller while loading the BAC DNA samples into the wells to avoid sample loss. In addition, it is critical to run the program for 20 min before restarting the pump and chiller to allow the DNA to enter the

agarose gel before circulating the buffer to avoid sample loss from the wells.

8. When designing the cloning strategy for insertion of the BoxA homology arm, it is important to take into consideration the cloning sites to be used in the subsequent steps for inserting the cDNA for the transgene of interest. In some cases it may be desirable to incorporate additional restriction sites in the BoxA reverse primer in order to facilitate the later insertion of the cDNA for the transgene of interest.
9. If the vector-only control plate has a similar number of colonies as the vector plus insert plate, it may be necessary to perform a PCR screen of the colonies (with replicate plate) in order to identify successfully ligated (positive) clones. This can be carried out with the T7 forward primer and a reverse primer that recognizes the BoxA (or in the later step, BoxB) sequence.
10. The DNA vectors we have used such as iTV1 are available upon request. The ChR2(H134R)-EYFP construct is available through Addgene (www.addgene.org, plasmid# 20940). Particular attention should be given to the junction between the BoxA homology arm and the transgene cDNA. Avoid inclusion of unnecessary vector sequence as this may have undesirable effects on transgene expression. In some cases it may be possible to design a cloning strategy in which the junction between the BoxA homology arm and the cDNA of the transgene (the sequence immediately upstream of and including the ATG) exactly matches the endogenous sequence for the gene that was replaced. This design may require more complicated cloning procedures, such as fill-in or chew-back reactions (using T4 polymerase, klenow, or mung bean nuclease) and less efficient blunt cloning. It is expected that the creation of a "perfect" junction at this critical site will yield the highest likelihood of success.
11. In our work we have employed a two homology arm targeting strategy. Alternative strategies including single homology arm targeting designs using shuttle vectors have been utilized with excellent success [56]. The single homology arm strategy is particularly efficient for high-throughput transgenic mouse line development given that the time required to generate BAC targeting vectors can be much shorter. Indeed this strategy has proven highly effective as the central strategy employed in the GENSAT project. We favor the two homology arm strategy because it allows for flexibility in terms of defining how much (if any) of the BAC sequence will be replaced by the insertion of the transgene cassette during the homologous recombination step, given that the BoxB homology arm can theoretically be located at any position after the ATG site up to the location of the stop codon for the particular gene that was replaced.

12. It is absolutely critical that the targeting vector DNA is completely linearized in preparation for subsequent homologous recombination steps. A small amount of undigested vector will ruin the efficiency of targeting and result in plasmid contamination of the EL250 cells harboring the BAC clone of interest. To ensure complete enzymatic digestion, it is strongly recommended to use no more than 1 μg of DNA per reaction and the maximal amount of restriction enzyme allowable for the reaction volume. In extreme cases it may be necessary to perform a second round of restriction digestion and gel purification to eliminate all traces of undigested plasmid.
13. This protocol specifies the use of LB-Lennox rather than LB-Miller for preparing electrocompetent EL250 cells. In our experience the efficiency of BAC DNA transformation by electroporation is dramatically reduced when using LB-Miller, which differs only in the amount of NaCl.
14. From this point on the EL250 cells should always be kept on ice to ensure the best results. If a tabletop centrifuge with refrigeration capability is not available, it is acceptable to substitute a normal tabletop centrifuge that is moved into a cold room and allowed to cool to 4 °C prior to use.
15. Heat shock of EL250 cells for 15 min induces transient expression of the phage genes required for efficient homologous recombination. Extended heat treatments can result in cell death and should be avoided.
16. The arabinose is necessary to activate Flp recombinase expression in EL250 cells.
17. The same plating step can be performed with an LB/Agar plate plus 30 μg/mL kanamycin to check for the efficiency of Flp-mediated NEO excision. If the efficiency is near 100 %, there will be no colonies on this control plate. In our experience the efficiency of NEO removal is generally >90 %.
18. The NEO removal screening PCR is designed to amplify a product of reduced size (generally ~600–800 bp) for BAC DNA clones that have undergone successful Flp-mediated excision of the FRT-NEO-FRT cassette as compared to clones that retain the FRT-NEO-FRT cassette. Depending on the PCR reaction conditions, clones that retain the NEO may also yield a product of larger size (~2.2–2.4 kb). In rare cases some clones may show partial excision and yield both the large and small PCR products, and these clones should be discarded.
19. Certain restriction enzymes tend to have few restriction sites in BAC sequences and are generally useful for fingerprint analysis of many different BACs. Some examples include PvuI, BssHII, MluI, SalI, AgeI, and SbfI, many of which are available as "high-fidelity" versions that are more desirable for digestion of BAC DNA.

20. We have had similar success using the Clontech NucleoBond Xtra MAXI Kit. The main difference is the integrated filter within the anion exchange column, as opposed to the separate column and folded filters that come with the BAC 100 Kit. The choice is merely a matter of preference.
21. It is very important to pick a colony from a freshly streaked plate in order to get the best possible yield of BAC DNA. At this particular stage it is not recommended to store the bacterial plates at 4 °C before use.
22. The use of the clear polycarbonate Oak Ridge style tubes is helpful to visualize the BAC DNA pellet after centrifugation. In some cases the highly pure BAC DNA pellet may also have a clear and glossy appearance. For this reason it is highly recommended to mark the orientation of the tubes each time they are loaded into the centrifuge to denote the expected location of the DNA pellet. Extreme care should be taken to avoid dislodging and discarding the DNA pellet in the ethanol wash step.
23. The use of high-fidelity restriction enzymes is recommended whenever possible to ensure a rapid and complete linearization.
24. If desired, stain the leftover gel segment and reassemble the gel (minus the excised band) for viewing under UV light to verify that the correct portion of the gel was collected. This method of gel excision without staining and direct visualization avoids intercalation of ethidium bromide into the BAC DNA and UV exposure that could compromise the integrity of the BAC DNA sample.
25. The column cap can be made from the cutoff plastic end of an 18 G 1½ needle. The inside rim must be smoothened to ensure a proper fit and prevent leakage. This piece is crucial to ensure that the dialysis membrane is securely fastened at the end of the column and that the sample will not leak out during the electroelution procedure.
26. To construct the negative lead wire, the end of the wire is stripped and soldered to a 6 cm length of platinum (or iridium platinum) wire which will contact the solution in the upper portion of the column. The end of the platinum wire is shaped into a ring for a distributed electric field through the column. To construct the positive lead wire, the end of the wire is stripped and soldered to a 3 cm length of platinum (or iridium platinum) wire which will submerged in the 1× TE solution that comes in contact with the column end cap. The cutoff end of a plastic Pasteur pipette is used to fashion a guard to cover the bare platinum of the positive lead wire and secured in place with electrical tape. Multiple rectangular windows are cut into the plastic so as not to impede current flow.

27. The total time necessary for the band to reach the bottom can vary depending on the packing of the column and the tightness of the end cap. In general it should take about 30 min. If it takes significantly longer, there may be air bubbles in the column. If it moves too fast then the column is probably leaking and you will note the red band appears diffuse rather than tight. In either case you will lose your sample and need to start over. It is important to carefully monitor the column during this procedure. If a problem arises, it may be possible to correct if identified quickly.
28. It is highly recommended to use high-purity "water for embryo transfer" to make the microinjection buffer. If desired, the buffer can be stored at 4 °C for up to 6 months. However, it is absolutely critical to ensure that the pH of the microinjection buffer is 7.4 ± 0.05 units and to ensure the sterility of the solution at the time of use. Any variation can adversely affect the viability of the injected zygote embryos.
29. It was reported that the addition of polyamines to the BAC DNA one week prior to pronuclear injection can increase the incidence of multiple founder lines resulting from a single BAC DNA injection [4]. It is presumed that the effect is due to condensation of the BAC DNA in the presence of polyamines, which seems to provide a dual benefit of facilitating successful microinjection of the intact BAC DNA as well as prevent degradation of BAC DNA during extended storage at 4 °C [40]. This step can increase the chance of successfully generating a useful BAC transgenic mouse line by providing larger numbers of founders to analyze for functional transgene expression. The 1,000× polyamine stock is diluted 1:10 in fresh microinjection buffer to yield a 100× stock. Filter the 100× stock solution through a 0.2 μm syringe filter. Add polyamines to the injection-ready BAC DNA to a final concentration of 1× (30 μM spermine and 70 μM spermidine) at exactly 1 week prior to the scheduled injection date.
30. In our experience the minimal amount of linearized BAC DNA that can be reliably visualized with this method (PFGE and post staining with ethidium bromide) is approximately 40 ng (load 20 μL of 2 ng/μL sample plus DNA loading buffer). It is best if the BAC DNA concentration falls within the range of 10–15 ng/uL, which is routinely achieved with when the BAC DNA band is cut directly from the PFG and purified. If the BAC DNA is in this higher range, load approximately 100 ng in the well. In some rare cases, we have found that certain very large BAC DNAs could not be purified at concentrations >2 ng/μL; however, as indicated above, this was still sufficient for visualization by PFGE.

31. By confirming the size and integrity of the BAC DNA both immediately prior to and following pronuclear injection, it is possible to largely exclude the quality of the BAC DNA as a potential cause for failed injections. Note that this verification strategy is unable to address BAC DNA integrity problems caused by improper injection procedures, e.g., shearing of BAC DNA during microinjection.
32. Failure in this procedure is most often manifest as a complete lack of colonies on the plates. This is likely the result of too little iTV1-based vector (the vector providing the selection marker) in the electroporation step. Repeat the procedure using a lower dilution (e.g., 1–2 μL of a 1:10 dilution). Caution should be observed in preparing the linearized targeting vectors to ensure a complete restriction digestion and to avoid uptake of contaminating plasmid DNA.

Acknowledgments

We wish to acknowledge Dr. Bernd Gloss for providing the iTV1 plasmid and for significant contributions on the BAC recombineering and electroelution procedures described in this work. We thank Dr. Karl Deisseroth for providing the original ChR2(H134R)-EYFP plasmid. The EL250 strain used for BAC recombineering was generously provided by Dr. Neal Copeland. This work was supported in part by an American Recovery and Reinvestment Act grant from the US National Institute of Mental Health to G.F. (RC1-MH088434), a National Alliance for Research on Schizophrenia and Depression: The Brain and Behavior Research Foundation Young Investigator Award to J.T.T., and US National Institutes of Health Ruth L. Kirschstein National Research Service Awards to J.T.T. (F32-MH084460).

References

1. Gong S et al (2007) Targeting Cre recombinase to specific neuron populations with bacterial artificial chromosome constructs. J Neurosci 27(37):9817–9823
2. Yang XW, Model P, Heintz N (1997) Homologous recombination based modification in Escherichia coli and germline transmission in transgenic mice of a bacterial artificial chromosome. Nat Biotechnol 15(9):859–865
3. Gong S, Kus L, Heintz N (2010) Rapid bacterial artificial chromosome modification for large-scale mouse transgenesis. Nat Protoc 5(10):1678–1696
4. Gong S et al (2003) A gene expression atlas of the central nervous system based on bacterial artificial chromosomes. Nature 425(6961): 917–925
5. Kozorovitskiy Y et al (2012) Recurrent network activity drives striatal synaptogenesis. Nature 485(7400):646–650
6. Witten IB et al (2011) Recombinase-driver rat lines: tools, techniques, and optogenetic application to dopamine-mediated reinforcement. Neuron 72(5):721–733
7. Wan Y, Feng G, Calakos N (2011) Sapap3 deletion causes mGluR5-dependent silencing

of AMPAR synapses. J Neurosci 31(46): 16685–16691
8. Chen M et al (2011) Sapap3 deletion anomalously activates short-term endocannabinoid-mediated synaptic plasticity. J Neurosci 31(26): 9563–9573
9. Higley MJ et al (2011) Cholinergic interneurons mediate fast VGluT3-dependent glutamatergic transmission in the striatum. PLoS One 6(4):e19155
10. Witten IB et al (2010) Cholinergic interneurons control local circuit activity and cocaine conditioning. Science 330(6011):1677–1681
11. Ding JB et al (2010) Thalamic gating of corticostriatal signaling by cholinergic interneurons. Neuron 67(2):294–307
12. Lobo MK et al (2010) Cell type-specific loss of BDNF signaling mimics optogenetic control of cocaine reward. Science 330(6002): 385–390
13. Kravitz AV et al (2010) Regulation of parkinsonian motor behaviours by optogenetic control of basal ganglia circuitry. Nature 466(7306): 622–626
14. Hnasko TS et al (2010) Vesicular glutamate transport promotes dopamine storage and glutamate corelease in vivo. Neuron 65(5): 643–656
15. Gittis AH et al (2010) Distinct roles of GABAergic interneurons in the regulation of striatal output pathways. J Neurosci 30(6): 2223–2234
16. Kim JC et al (2009) Linking genetically defined neurons to behavior through a broadly applicable silencing allele. Neuron 63(3):305–315
17. Gertler TS, Chan CS, Surmeier DJ (2008) Dichotomous anatomical properties of adult striatal medium spiny neurons. J Neurosci 28(43):10814–10824
18. Kreitzer AC, Malenka RC (2007) Endocannabinoid-mediated rescue of striatal LTD and motor deficits in Parkinson's disease models. Nature 445(7128):643–647
19. Lobo MK et al (2007) Genetic control of instrumental conditioning by striatopallidal neuron-specific S1P receptor Gpr6. Nat Neurosci 10(11):1395–1397
20. Day M et al (2006) Selective elimination of glutamatergic synapses on striatopallidal neurons in Parkinson disease models. Nat Neurosci 9(2):251–259
21. Zhao S et al (2010) Fluorescent labeling of newborn dentate granule cells in GAD67-GFP transgenic mice: a genetic tool for the study of adult neurogenesis. PLoS One 5(9):e12506
22. Yu-Taeger L et al (2012) A novel BACHD transgenic rat exhibits characteristic neuropathological features of Huntington disease. J Neurosci 32(44):15426–15438
23. Cannon JR et al (2013) Expression of human E46K-mutated alpha-synuclein in BAC-transgenic rats replicates early-stage Parkinson's disease features and enhances vulnerability to mitochondrial impairment. Exp Neurol 240: 44–56
24. Bateup HS et al (2008) Cell type-specific regulation of DARPP-32 phosphorylation by psychostimulant and antipsychotic drugs. Nat Neurosci 11(8):932–939
25. Doyle JP et al (2008) Application of a translational profiling approach for the comparative analysis of CNS cell types. Cell 135(4): 749–762
26. Heiman M et al (2008) A translational profiling approach for the molecular characterization of CNS cell types. Cell 135(4):738–748
27. Schmidt EF et al (2012) Identification of the cortical neurons that mediate antidepressant responses. Cell 149(5):1152–1163
28. Zhao S et al (2011) Cell type-specific channelrhodopsin-2 transgenic mice for optogenetic dissection of neural circuitry function. Nat Methods 8(9):745–752
29. Halassa MM et al (2011) Selective optical drive of thalamic reticular nucleus generates thalamic bursts and cortical spindles. Nat Neurosci 14(9):1118–1120
30. Guo ZV et al (2014) Flow of cortical activity underlying a tactile decision in mice. Neuron 81(1):179–194
31. Hull C, Regehr WG (2012) Identification of an inhibitory circuit that regulates cerebellar Golgi cell activity. Neuron 73(1):149–158
32. Ren J et al (2011) Habenula "cholinergic" neurons co-release glutamate and acetylcholine and activate postsynaptic neurons via distinct transmission modes. Neuron 69(3):445–452
33. Hagglund M et al (2010) Activation of groups of excitatory neurons in the mammalian spinal cord or hindbrain evokes locomotion. Nat Neurosci 13(2):246–252
34. Ciotta G et al (2011) Recombineering BAC transgenes for protein tagging. Methods 53(2): 113–119
35. Schwartz DC, Cantor CR (1984) Separation of yeast chromosome-sized DNAs by pulsed field gradient gel electrophoresis. Cell 37(1): 67–75
36. Liu P, Jenkins NA, Copeland NG (2003) A highly efficient recombineering-based method

for generating conditional knockout mutations. Genome Res 13(3):476–484

37. Yu D et al (2000) An efficient recombination system for chromosome engineering in Escherichia coli. Proc Natl Acad Sci U S A 97(11): 5978–5983
38. Zhang Y et al (1998) A new logic for DNA engineering using recombination in Escherichia coli. Nat Genet 20(2):123–128
39. Muyrers JP et al (1999) Rapid modification of bacterial artificial chromosomes by ET-recombination. Nucleic Acids Res 27(6): 1555–1557
40. Van Keuren ML et al (2009) Generating transgenic mice from bacterial artificial chromosomes: transgenesis efficiency, integration and expression outcomes. Transgenic Res 18(5): 769–785
41. Abe K et al (2004) Establishment of an efficient BAC transgenesis protocol and its application to functional characterization of the mouse Brachyury locus. Exp Anim 53(4): 311–320
42. Takahashi R et al (2000) Generation of transgenic rats with YACs and BACs: preparation procedures and integrity of microinjected DNA. Exp Anim 49(3):229–233
43. Feng G, Lu J, Gross J (2004) Generation of transgenic mice. Methods Mol Med 99: 255–267
44. Ittner LM, Gotz J (2007) Pronuclear injection for the production of transgenic mice. Nat Protoc 2(5):1206–1215
45. Vintersten K, Testa G, Stewart AF (2004) Microinjection of BAC DNA into the pronuclei of fertilized mouse oocytes. Methods Mol Biol 256:141–158
46. Truett GE et al (2000) Preparation of PCR-quality mouse genomic DNA with hot sodium hydroxide and tris (HotSHOT). Biotechniques 29(1):52, 54
47. Chan CS et al (2012) Strain-specific regulation of striatal phenotype in Drd2-eGFP BAC transgenic mice. J Neurosci 32(27):9124–9132
48. Yang XW et al (1999) BAC-mediated gene-dosage analysis reveals a role for Zipro1 (Ru49/Zfp38) in progenitor cell proliferation in cerebellum and skin. Nat Genet 22(4):327–335
49. Lin C et al (2009) Construction and characterization of a doxycycline-inducible transgenic system in Msx2 expressing cells. Genesis 47(5):352–359
50. Kramer PF et al (2011) Dopamine D2 receptor overexpression alters behavior and physiology in Drd2-EGFP mice. J Neurosci 31(1):126–132
51. Ting JT, Feng G (2014) Recombineering strategies for developing next generation BAC transgenic tools for optogenetics and beyond. Front Behav Neurosci 8:111
52. Testa G et al (2004) BAC engineering for the generation of ES cell-targeting constructs and mouse transgenes. Methods Mol Biol 256: 123–139
53. Hill F et al (2000) BAC trimming: minimizing clone overlaps. Genomics 64(1):111–113
54. Chao HT et al (2010) Dysfunction in GABA signalling mediates autism-like stereotypies and Rett syndrome phenotypes. Nature 468(7321): 263–269
55. Belforte JE et al (2010) Postnatal NMDA receptor ablation in corticolimbic interneurons confers schizophrenia-like phenotypes. Nat Neurosci 13(1):76–83
56. Gong S et al (2002) Highly efficient modification of bacterial artificial chromosomes (BACs) using novel shuttle vectors containing the R6Kgamma origin of replication. Genome Res 12(12):1992–1998

Chapter 10

Engineered Rabies Virus for Transsynaptic Circuit Tracing

Jennifer Selever and Benjamin R. Arenkiel

Abstract

Transsynaptic tracing using modified rabies virus (RV) is a powerful new technology in neuroscience that allows for visualization of targeted neurons and their synaptic connections. Here, we describe how a genetically engineered version of RV can be used for transsynaptic tracing studies of mammalian neuronal cells by providing protocols for viral isolation, propagation, pseudotyping, and concentration. The resulting genetically modified RV shows neuronal infectivity both in vitro and in vivo. Once the target neuron has been infected, the RV replicates and "jumps" presynaptically to connected neurons to provide a visual map of synaptic connectivity.

Key words Neurons, Neural tracing, Rabies virus, Pseudotyped rabies virus

1 Introduction

A longstanding goal for both basic and clinical neuroscience has been to elucidate the wiring diagrams of intact mammalian neural circuits. An emergent method to address this challenge has been to utilize genetically engineered viral vectors to label patterns of synaptic connectivity. With the advent of viral vectors and modified RV tracers, cell type-specific genetic manipulation of neurons and their connections is possible. Viral vectors are useful tools to deliver genetic material in a highly specific manner and allow the study of functional connectivity by facilitating in vivo image reconstruction and recording of neuronal activity [1, 2]. Furthermore, virally packaged genetic material can be administered to a broad range of species, including those that are not easily genetically engineered such as primates and humans. Moreover, virus administration can be performed at any time point and can be used to selectively target desired cells based on either injection site or promoter expression.

Benjamin R. Arenkiel (ed.), *Neural Tracing Methods: Tracing Neurons and Their Connections*, Neuromethods, vol. 92, DOI 10.1007/978-1-4939-1963-5_10, © Springer Science+Business Media New York 2015

In the past, viral-mediated transsynaptic tracing [3–6] was accomplished using wild-type viral particles. Two types of viral vectors that have been implemented for this purpose include RV and herpes. Herpes viruses belong to a family of double-stranded DNA viruses, while RV is in the family of negative-strand RNA viruses [7]. Interestingly, although they are significantly different in their genetic makeup, they are both endowed with the unique ability to selectively infect neuronal cells. This cell type-specific infectivity derives from their enveloped coat particles, which consist of both host membrane and virally encoded glycoproteins. The composite envelope mediates neuronal membrane binding and subsequent neuron-to-neuron transsynaptic spread.

Two common strains of herpes virus used for neuronal tracing include herpes virus simplex 1 (HSV-1) [8] and pseudorabies virus (PRV) [9]. Both types of viruses predominantly spread presynaptically in a retrograde direction and each has been used to elucidate patterns of synapse and circuit connectivity in the rodent brain [10]. However, one major drawback of using herpes viruses for circuit analysis is polysynaptic spread and the challenge to decipher weak direct connections from strong indirect ones. Recently, Dr. Ed Callaway's group has developed a monosynaptic RV tracing system to overcome the limitations of previous transsynaptic tracers [11, 12]. The notable features of this system are as follows: (1) RV can be genetically targeted in a cell type-specific manner, (2) RV shows robust retrograde transfer, and (3) RV can be engineered to label cells connected to a single source cell in a monosynaptic manner. The method utilizes a deletion-mutant RV in which a portion of the viral genome encoding a viral glycoprotein, which is normally required for viral replication and spread, is absent. Instead, this gene is provided in *trans* by a targeted neuron of interest to reconstitute the viral genome and produce infectious particles. Reconstituted RV shows retrograde infection, therefore labeling neurons presynaptic to the target neuron. Because the engineered RV on its own lacks the gene encoding its wild-type glycoprotein, and the presynaptically labeled neurons do not express the viral glycoprotein to provide it in *trans*, it is unable to produce more infectious particles and transsynaptic spread comes to a halt. This is the key factor allowing for monosynaptic tracing using genetically engineered RV. To provide selectivity of infection to distinct neuronal subsets, the RV can be pseudotyped to contain an envelope protein that is engineered to only infect a specific neuronal subtype that expresses its cognate receptor, which is normally not found on the mammalian neuronal membrane [12].

1.1 Components of the Rabies Virus

RV is a neurotropic negative-strand RNA virus with a cylindrical, bullet-like morphology and belongs to the family *Rhabdoviridae* [13]. It is an enveloped, single-stranded RNA virus packaged as a ribonucleoprotein complex. The 11–12 kb viral genome is composed of five genes: nucleoprotein (N), phosphoprotein (P), matrix protein (M), glycoprotein (G), and viral RNA polymerase (L) [14, 15]. RV has a round architecture on one end, is planar on the other, and is covered on all sides with glycoprotein G spikes. The membrane composed of the matrix protein lies underneath the envelope and surrounds the viral core, which contains the helically oriented ribonucleoprotein.

1.2 Rabies Virus Infection

The virus utilizes its glycoprotein G spikes during endosomal transport to infect the host and bind to the neuronal membrane. Once within the host cell, the RV takes over the host cell machinery and the host begins transcribing and replicating the viral genome. Transcription and replication of the RV genome occur in the cytoplasmically located Negri body, which is absent in uninfected neurons. Therefore, the Negri body is a hallmark morphological characteristic of RV infection and is used for viral identification [16]. Once transcribed, the viral components are then packaged to make new infectious viral particles, which then retrogradely spread to presynaptically connected neurons, eventually infecting the entire nervous system. Infected neurons will ultimately die, so there is an approximate 10-day time window in which experimentation can be performed before cytopathic changes are evident [17, 18].

1.3 Rabies Virus Tracing

RV tracing utilizes a mutant viral genome in which the Glycoprotein G is deleted and replaced with enhanced Green Fluorescent Protein (EGFP). This vector is designated SADΔG-EGFP and requires to be typed and/or pseudotyped in vitro. The glycoprotein G is then introduced in *trans* in the initially infected neurons, which are designated "source cells" for transsynaptic tracing studies. With G provided *in trans* by the host cell, viral transcription, replication, and packaging occurs within the source cell. Once packaged, the RV will jump one time retrogradely to mark presynaptically connected neurons. Upon transsynaptic transfer to newly infected neurons, the RV is unable to propagate, since G is absent in these cells. In this scheme, both the source cell and the neurons presynaptic to it will be labeled with EGFP. The benefits associated with RV tracing over other tracers are the exceptional infection efficiency, and the retrograde transfer. Alongside EGFP, G-deleted RV can also be engineered to express any number of genetic elements and/or reporters, including rescue cDNAs, shRNA, synaptic silencers, calcium indicators, or light-gated and ligand-dependent ion channels [19–26].

Another potent attribute of RV vectors for labeling neuronal circuits is the ability to target infection to desired neuronal subtypes. This can be done through pseudotyping the virus with foreign ligands and expressing the cognate receptors to the ligand in targeted neurons. The best described components for this exploit the EnvA/TVA and EnvB/TVB systems, which normally exist in avian species and are absent from mammalian cells [11, 12, 27–29]. To implement this system, targeted cells are first marked for initial infection by expressing an avian sarcoma leukosis virus receptor, ASLV-A (TVA) or ASLV-B (TVB), on their membrane [10, 12, 30, 31]. There are two methods by which TVA can be anchored to the membrane: by attachment with a phospholipid anchor (Tva800) or a transmembrane domain (Tva950). For Tva800, multiple ligand-bound receptors are necessary for viral entry, but only one ligand-receptor complex is required for infection through Tva950 [32]. To utilize the full potential of the targeted system, the G-deleted, EGFP-expressing RV is encapsulated in an envelope ASLV-A glycoprotein, EnvA. EnvA directs the virus to cells that express the cognate TVA receptor. If G is expressed within these cells *in trans*, RV propagates and jumps presynaptically to connected neurons. If the next neuron lacks G, the deletion-mutant RV can no longer self-assemble and retrograde transfer stops, allowing for monosynaptic tracing. Through creative applications of this technology, it is possible to study connected neuronal networks morphologically, electrophysiologically, and on the systems level.

1.4 Limitations and Benefits of Rabies Tracing

RV is an RNA virus, which never synthesizes DNA, so it cannot express promoter-driven DNA transgenes. However, depending on the application, this might be considered an advantage, since the vector does not integrate into the genome and is incapable of producing undesired mutations. Other popular retroviruses such as HIV and MoMLV insert into the host genome and can potentially interfere with the host genome integrity, thus leading to potential gain or loss of function [19]. Another limitation is that RV is cytotoxic to infected neurons. Given the extremely high levels of gene expression that underlie normal cell functions, there is only a narrow window (7–14 days with current vectors) in which infected neurons can be studied before cell death occurs.

1.5 Protocol to Produce G-Deleted Rabies Virus

Notes: Many of the cell lines and viral starter stocks are available from the Callaway laboratory [19]. Addgene distributes the plasmids necessary for G-deleted rabies viral production and pseudotyping.

The G-deleted RV production is a multi-step protocol, consisting of vector construction, isolation of viral particles, viral

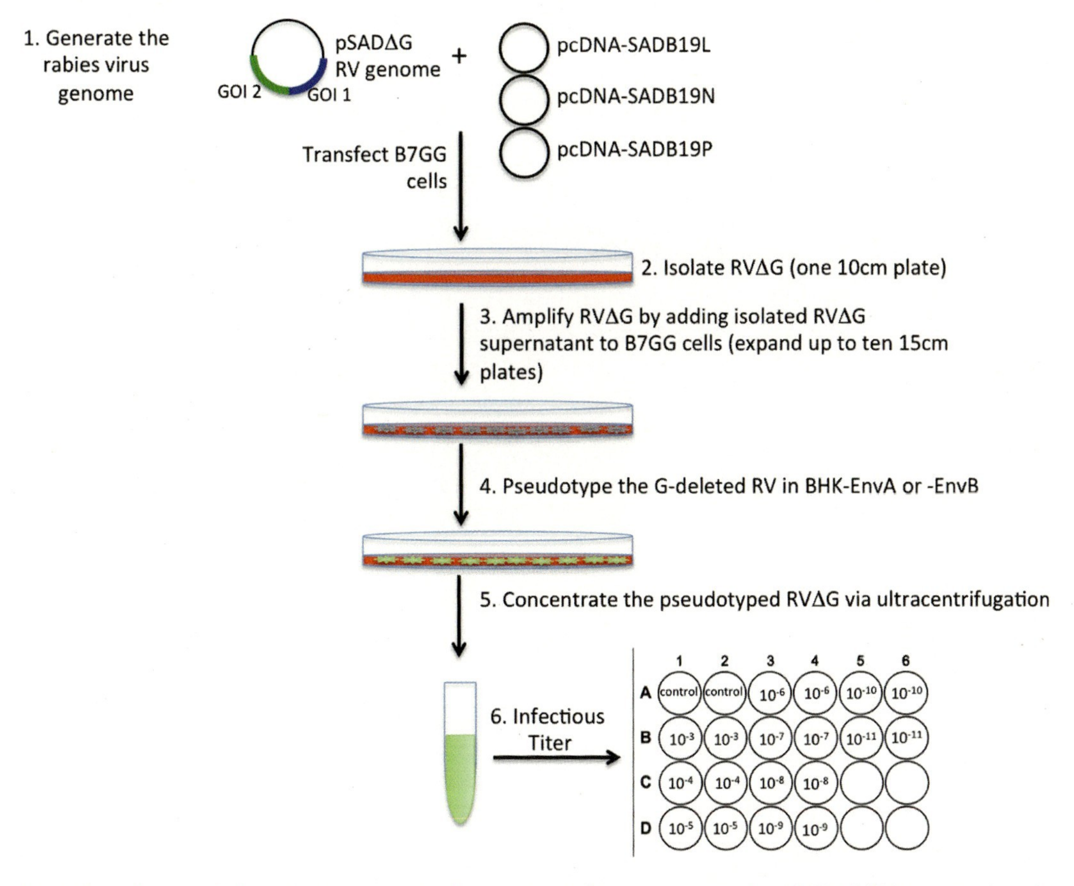

Fig. 1 Flow diagram of the rabies virus production protocol. *Step 1* generates the pSADΔG RV genome containing "gene of interest 1 and 2" (GOI 1 and GOI 2) in place of wild-type glycoprotein G. B7GG cells expressing T7 polymerase and rabies virus glycoprotein are transfected with the pSADΔG RV genome and helper plasmids expressing the G, L, N, and P components necessary to produce infectious rabies particles. In *step 2*, new G-deleted RV is isolated from transfected cells. *Step 3* amplifies the rabies virus, hence increasing the titer. The unpseudotyped RV can be used once completing *step 3*. However, proceed to *step 4* to pseudotype, if the goal is to only infect cells expressing TVA or TVB receptors. *Step 4* consists of infecting BHK-EnvA or BHK-EnvB cells with unpseudotyped RV. RV generated and packaged in the BHK-Env cells will be encapsulated in EnvA or EnvB. The resulting virus will be capable of infecting cells expressing TVA or TVB receptors, respectively. Proceed to *step 5* to concentrate the pseudotyped G-deleted RV by two rounds of ultracentrifugation. *Step 6* tests the infectious titer of the produced RV. Cells are plated in a 24-well plate and infected with a serial dilution of RV from 10^{-3} to 10^{-11}. Pseudotyped RV needs to be titered in HEK 293T and HEK 293T-TVA800 (EnvA) or HEK 293T-TVB (EnvB) to ensure pseudotyped RV is not contaminated with unpseudotyped RV

amplification in vitro, pseudotyping (if desired), concentrating, and titering (Fig. 1). If plasmids are being purchased from Addgene, the vector construction and isolation steps are not necessary. However, if generation of a novel vector is desired, the first two steps are required.

2 Materials

Cell Lines

B7GG cells: These cells are used to make the RV. They express rabies glycoprotein G for packaging, T7 RNA polymerase to transcribe the RNA genome, and a histone-tagged GFP for identification.

Pseudotyping cell line: If it is desired to pseudotype the RV with either EnvA or EnvB, BHK-EnvA or BHK-EnvB cells are available. The G-deleted, EGFP-expressing rabies virus is encapsulated in an envelope ASLV-A (or -B) glycoprotein, EnvA or EnvB. The envelope hones the virus to cells expressing the TVA or TVB receptor.

Titer Cell Lines

HEK 293T cells: These cells are used to titer RV particles that are not pseudotyped with EnvA or EnvB.

HEK 293T-TVA800 cells: This cell line expresses TVA and can be used to titer G-deleted virus pseudotyped with EnvA.

HEK 293T-TVB cells: HEK 293T-TVB cells express TVB and can be used to titer G-deleted virus pseudotyped with EnvB.

Plasmids
Attain the following plasmids from Addgene or the viral supernatant from the Callaway laboratory:

pSADΔG-F3

pcDNA-SADB19G

pcDNA-SADB19L

pcDNA-SADB19N

pcDNA-SADB19P

To generate G-deleted rabies virus, follow the protocol established in Osakada et al., Neuron 2011 [29].

3 Methods

3.1 Step 1: Generate pSADΔ G RV Genome

1. Engineer "your transgene of interest" within the pSADΔG-F3 plasmid. This may require multiple cloning steps or formation of an intermediate shuttle vector. Nonetheless, it is imperative to minimize the insert size. The rabies virus packaging size is limited to total transgene inserts of less than 3.7 kb [19], so it is best to remove any unnecessary components. If only inserting one element, it is advisable to cut with a single restriction enzyme in the

first multiple cloning site (MCS) and one in the second MCS to directionally insert the transgenic fragment. For successful ligation, backbone to insert ratio should be 1:3 or better.

2. Transform 1–2 μl of ligation product into thawed chemically competent cells.
3. Incubate DNA/cells on ice for 10 min.
4. Heat shock the cells at 42 °C for 1 min, then cool on ice for an additional 2 min.
5. To the cells, add 250 μl of SOC medium and incubate in a 37 °C shaking incubator for 1 h.
6. Spread various concentrations of transformed cells on 100 μg/ml carbenicillin LB plates to attain single colony isolates.
7. Incubate inverted plates overnight at 37 °C.
8. The next morning, pick at least eight single colonies and grow in liquid LB with carbenicillin.
9. Miniprep plasmid DNA from cells using a purification kit per the manufacturer's protocol.
10. Verify proper insert via restriction digest of the isolated plasmid DNA. It may be necessary to verify directionality of the insert.
11. Grow a maxi prep of one positive clone and extract plasmid DNA using a kit following the manufacturer's protocol.
12. Sequence the DNA to ensure no mutations are present in the clone. If a mutation is present, isolate another clone and sequence it.

3.2 Step 2: Transfect and Isolate RVΔG

It will take approximately 10 days to recover rabies virus from cDNA. It is important to note that B7GG cells grow best at subconfluent amounts. When splitting, collect all cells, resuspend in media to a single cell state, and reseed at a 1:5 ratio. Too low or too high cell density is not advised for optimal viral production:

1. Plate one 10 cm dish with three million B7GG cells at 1:5 ratio to ensure adequate number of cells for transfection on day 2. Culture cells at 37 °C and 5 % CO_2 overnight.
2. The next day, 30 μg rabies virus genomic vector (pSADΔG-F3), 15 μg pcDNA-SADB19N, 7.5 μg pcDNA-SADB19P, 7.5 μg pcDNA-SADB19L, and 5 μg pcDNA-SADB18G are transfected into B7GG cells prepared in step 1. Lipofectamine 2000 reagent is used for the transfection per manufacturer's protocol.
3. Allow each tube, one with OPTI-MEM and plasmid and one with OPTI-MEM and Lipofectamine 2000, to incubate at room temperature for 5 min prior to mixing them together.

4. Thoroughly mix the plasmids into the Lipofectamine solution and let incubate at room temperature for 15 min.
5. Administer 2.5 ml of the plasmid/Lipofectamine mixture drop wise over the plated cells. Return the plate to the 37 °C and 5 % CO_2 incubator for 6 h.
6. Remove medium from the cells, wash with 10 % FBS/DMEM and reapply 10 ml fresh 10 % FBS/DMEM. Return the plate to the same incubation conditions for one overnight cycle.
7. The next day, remove the plate from 37 °C and transfer to 35 °C and 3 % CO_2 incubation conditions overnight.
8. The next day, first wash the cells with PBS, then detach with 4 ml of 0.25 % trypsin-EDTA. Allow the cells to detach in 37 °C for 5–10 min.
9. Add 10 % FBS/DMEM and gently pipet cells enough times to achieve single cell suspension. Collect all medium and cells, and transfer to a conical tube.
10. Pellet cells at 300 × *g* for 5 min and aspirate off supernatant. Carefully resuspend cell pellet in 10 ml of 10 % FBS/DMEM and transfer to a 15 cm plate. Incubate the cells overnight at 35 °C and 3 % CO_2. *Note*: Culture conditions for all successive steps are 35 °C and 3 % CO_2. These conditions foster low cellular multiplicity and elevated viral production compared to 37 °C and 5 % CO_2.
11. The following day, remove the medium and replace it with a new 20 ml aliquot of 10 % FBS/DMEM. Return the plate to the 35 °C incubator for 2 days.
12. Add 5 ml of 10 % FBS/DMEM to the plate and incubate an additional 2 days at 35 °C.
13. Collect and filter the supernatant. Use a 0.45 μm filter to allow for the viral capsid to remain intact post filtration.
14. Feed the cells with an additional 20 ml of 10 % FBS/DMEM and culture at 35 °C conditions for 2 days. Monitor the cells daily and visualize the spread of virus via fluorescent EGFP marker.
15. Administer 5 ml medium to the cells and return to 35 °C incubation conditions for 2 more days.
16. Once the virus has spread to nearly 100 % of cells, filter the supernatant through a 0.45 μm filter.
17. Mix the supernatant from both collection days together.
18. Proceed to step 3, amplification of the virus.

3.3 Step 3: Expand the RVΔG

The purpose of step 3 is to propagate the virus in B7GG cells in order to expand the amount of packaged virus. This step can be completed subsequent to the previous two steps, or as the initial step when starting with a frozen viral supernatant aliquot.

If beginning with the virus recovered from step 2:

1. Plate B7GG cells on three 15 cm dishes to 60 % confluency.
2. Aspirate the medium off, wash with PBS, and split the collected viral supernatant evenly among the three dishes. Add 7 ml of 10 % FBS/DMEM. Culture the cells at 35 °C and 3 % CO_2.

If beginning with frozen viral stock:

1. Plate B7GG cells at 60 % confluency in a 15 cm dish.
2. Remove the medium, wash with PBS, and add viral supernatant and 10 % FBS/DMEM at multiplicity of infection (MOI) rate of 0.1–0.3. Culture cells at 35 °C and 3 % CO_2.

From this point on, follow the same protocol for both.

3. Approximately 6 h later, change the medium to 24 ml of 10 % FBS/DMEM. Allow the cells to incubate for 4 days. Always ensure the medium maintains a red appearance. If the medium begins to yellow, add 5 ml fresh medium.
4. Collect viral supernatant and filter through a 0.45 μm filter, aliquot, and freeze at −80 °C. If it will be used immediately, the supernatant can be stored in 4 °C for 24–36 h.
5. Add an additional 24 ml of 10 % FBS/DMEM, and culture for 2 more days. Monitor the viral spread by EGFP expression. It is best to collect the final aliquot of supernatant when EGFP is expressed in nearly all cells.
6. Collect, filter, and aliquot the RV supernatant. Store at −80 °C. Depending on the application, it may be necessary to increase the number of plates ensuring adequate quantity of RV to complete all anticipated experiments.

3.4 Step 4: Pseudotyping the Rabies Virus

The goal of step 4 is to pseudotype the G-deleted RV with either EnvA or EnvB. The procedure is the same for either envelopes; however, utilize BHK-EnvA or BHK-EnvB cells accordingly. The EnvA- or EnvB-pseudotyped rabies virus will only be able to infect cells expressing TVA or TVB cell surface receptors, respectively.

1. Plate BHK-EnvA cells in 15 cm dishes to 60 % confluency.
2. Once cell density is at 60 %, remove medium and replace with unpseudotyped G-deleted rabies virus supernatant at a MOI of 0.5 and incubate for 6 h.
3. Remove the medium containing the unpseudotyped virus and wash the plate twice with PBS. It is important to remove all unpseudotyped virus because it will contaminate the pseudotyped virus.

4. Trypsinized plates with warm 0.25 % trypsin and allow cells to detach in 37 °C for 5 min.
5. Add equal volume of 10 % FBS/DMEM to trypsinized cells and pipet. Collect all cells in a tube. Wash the plates with additional medium, and add to the tube.
6. Pellet the cells by centrifugation at 200 × *g* for 3 min. Aspirate and discard medium.
7. Evenly reseed the pelleted cells to ten 15 cm plates. Total volume should be 20–25 ml on each plate. Incubate overnight.
8. The next day, aspirate off all possibly contaminated medium and replace with fresh 10 % FBS/DMEM. Culture cells for 2 days.
9. Collect and filter (0.45 μm) the supernatant.

At this point, the supernatant can be aliquoted and frozen at −80 °C for future usage; however, if concentrated virus is necessary, proceed to step 5.

3.5 Step 5: Concentrate the Pseudotyped Rabies Virus

Concentration of the virus will require the use of an ultracentrifuge and rotors capable of handling large (35 ml per tube) and small volumes (5 ml per tube) at 70,000 and 50,000 × *g*, respectively. Important note: Compatible tubes are necessary to ensure proper centrifugation at high rpms. All tubes should be properly counterbalanced when centrifuging. Before filling the tubes, they should be cleaned with 70 % ethanol, rinsed with autoclaved water, and allowed to air dry in the sterile tissue culture hood.

1. Evenly fill six tubes with supernatant from step 4.
2. Centrifuge in a Beckman SW28, or equivalent swing bucket rotor, at 70,000 × *g* for 2 h at 4 °C.
3. Post centrifugation, under the hood, pour off the supernatant and allow the tubes to dry upside down. A faint pellet may be visible.
4. To each tube, add 300 μl of HBSS, cover tubes with parafilm, and lightly vortex to resuspend the pellet.
5. Load 2.5 ml of a 20 % sucrose/HBSS solution into the bottom of each small tube (tube needs to be suitable for use in SW55 rotor, or equivalent).
6. Add the resuspended virus, dropwise, to the tubes containing the sucrose cushion. Do not disturb the cushion or introduce air bubbles.
7. Centrifuge in a SW55, or equivalent swing bucket rotor, at 50,000 × *g* for 2 h at 4 °C.

8. Post centrifugation, under the hood, pour off the supernatant and allow the pellet to dry upside down. A pipet may be used to aspirate off any residual liquid.
9. Each viral pellet is resuspended in 100 μl of HBSS by gentle vortexing. To allow for complete resuspension, let the tubes rest in the refrigerator for an hour before aliquoting and freezing the concentrated virus.
10. To ensure that the viral solution is homogenous, pipet the pellet in the HBSS, and then aliquot virus to 0.2 μl tubes at a volume of 3–5 μl each. These tubes may be frozen at −80 °C for an extensive timeframe. However, the virus cannot undergo multiple freeze/thaw cycles.

3.6 Step 6: Titer Virus

G-deleted "wild-typed" RV can be titered by infecting HEK 293T cells. In contrast, the pseudotyped rabies virus requires titering with both HEK 293T and HEK 293-TVA or TVB cells, depending on the envelope used. If the pseudotyped virus is pure, it will only infect the HEK 293-TVA or TVB cells. However, if the HEK 293T cells are positive for infection, then the virus is contaminated with wild-type G-deleted virus. Use of contaminated virus will confound experimental results.

1. Plate appropriate cells in 24-well dishes seeded at 1.5×10^5 cells per well in 500 μl DMEM with 10 % FBS.
2. The next day, infect cells with serial dilutions of virus spanning 10^{-3}–10^{-11}. 250 μl of each dilution is plated in duplicate. Important note: Very small volumes of virus will skew the titer results, so tips need to be changed prior to each dilution.
3. Incubate plates at 3 % CO_2 and 35 °C until GFP is visible, normally 24–36 h.
4. Determine the viral titer by counting infectious units per milliliter at the dilution resulting in 10–100 GFP positive cells. Calculate the average. Deduce the stock titer from this average.

4 Important Considerations

1. The pSADΔG genome can receive a maximum insert size of 3.7 kb [19, 29].
2. During the pseudotyping (step 4), it is imperative to remove all unpseudotyped virus after the initial infection by extensive washing with PBS, trypsinizing, centrifuging, resuspending in a new aliquot of medium, plating, and finally changing the medium once more. Following this extensive plating protocol will minimize the presence of unpseudotyped RV from the desired pseudotyped RV [19].

3. It is imperative to have a low passage stock of the B7GG cells.
4. For ease of producing G-deleted virus in the future, production of a seed stock of G-deleted rabies virus is helpful.

In conclusion, the RV tracing system is a powerful new tool for the advancement of neuroscience. Artfully crafted rabies viruses expressing a myriad of constructs and reporters such as Cre, FLP, Channelrhodopsin-2, GCaMP, and many more allow genetic manipulation of neurons based on circuit connectivity. The power of this technique is evident in vitro (Fig. 2) and in vivo (Fig. 3). In light of the many published manuscripts using this new technology, there are still limitations to the system. As the expression of RV accumulates in neurons, lethality is inevitable. In the future, it will be imperative to limit the lethality of this vector and thus improve upon this technology.

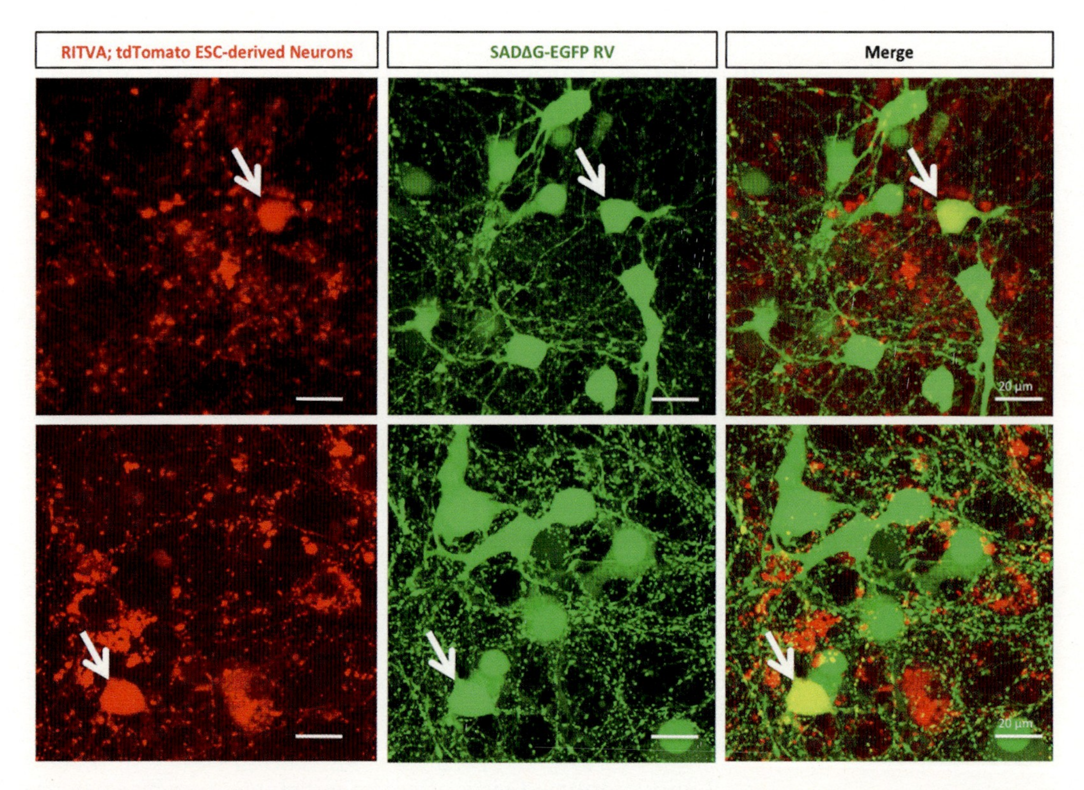

Fig. 2 In vitro RV tracing. Neurons produced by differentiating embryonic stem cells expressing Rabies-G-IRES-Neo-IRES-TVA and tdTomato reporter gene in the ROSA26 locus [17, 33, 34] and coseeded with primary cortical neurons. The cultured neurons are infected with EnvA-pseudotyped SADΔG-EGFP RV. The RV infects the TVA [35] expressing neurons (*yellow*, *arrow*) and moves synaptically to presynaptic cortical neurons (*green*)

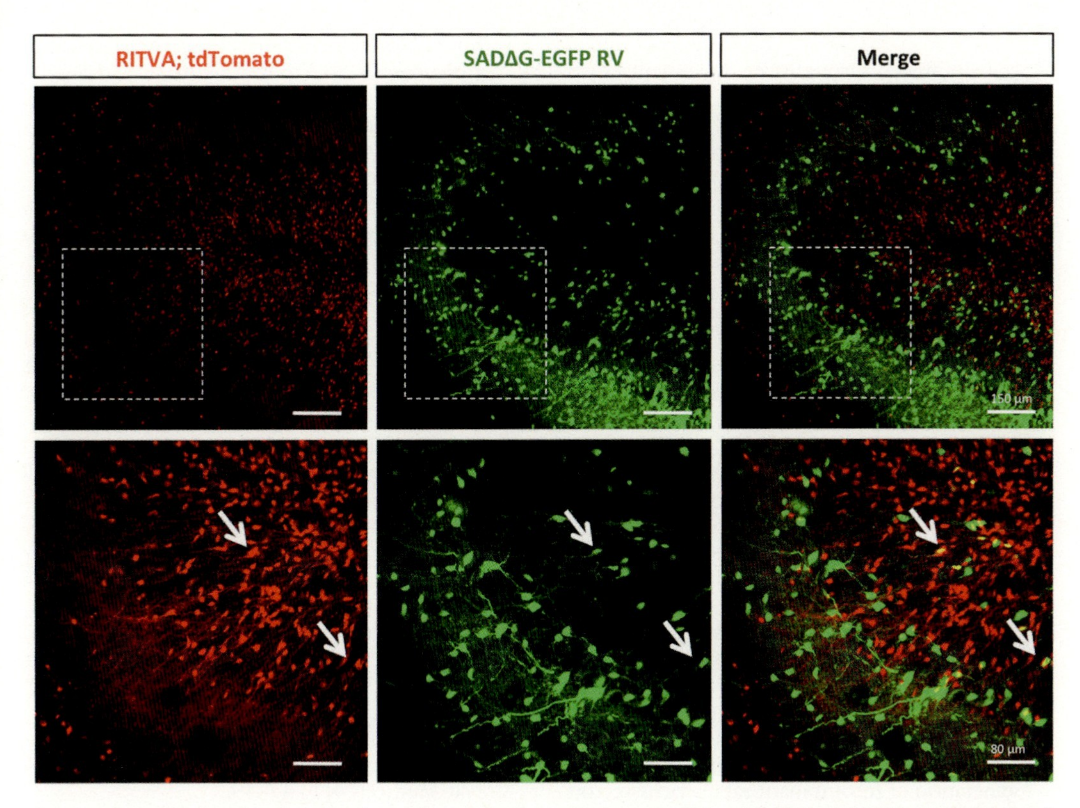

Fig. 3 In vivo RV tracing. Genetically engineered mice express an allele of CAG-loxP-STOP-LoxP-Rabies-G-IRES-TVA (RITVA) and an allele of loxP-STOP-loxP-tdTomato in the ROSA26 locus [36]. Upon Cre recombination in the olfactory bulb, the RITVA expressing cells are labeled *red* via tdTomato expression. These cells can be infected by the EnvA-pseudotyped SADΔG-EGFP RV (*yellow*, *arrows*) and package and synaptically release the SADΔG-EGFP RV to presynaptic neurons. Those neurons turn green since they express EGFP. However, the presynaptic neurons are incapable of packaging the RV because they do not express wild-type G and mono-synaptic RV tracing has occurred

References

1. Arenkiel BR, Ehlers MD (2009) Molecular genetics and imaging technologies for circuit-based neuroanatomy. Nature 461(7266): 900–907
2. Luo L, Callaway EM, Svoboda K (2008) Genetic dissection of neural circuits. Neuron 57(5):634–660
3. Kuypers HG, Ugolini G (1990) Viruses as transneuronal tracers. Trends Neurosci 13(2): 71–75
4. Ugolini G (1995) Specificity of rabies virus as a transneuronal tracer of motor networks: transfer from hypoglossal motoneurons to connected second-order and higher order central nervous system cell groups. J Comp Neurol 356(3):457–480
5. Aston-Jones G, Card JP (2000) Use of pseudorabies virus to delineate multisynaptic circuits in brain: opportunities and limitations. J Neurosci Methods 103(1):51–61
6. Chen S et al (1999) Characterization of transsynaptic tracing with central application of pseudorabies virus. Brain Res 838(1–2): 171–183
7. Voyles BA (1993) The biology of viruses, 1st edn. William C. Brown, Boston, MA, p 386
8. Lilley CE, Branston RH, Coffin RS (2001) Herpes simplex virus vectors for the nervous system. Curr Gene Ther 1(4):339–358
9. Enquist LW (2002) Exploiting circuit-specific spread of pseudorabies virus in the central nervous system: insights to pathogenesis and circuit tracers. J Infect Dis 186(Suppl 2):S209–S214
10. Callaway EM (2008) Transneuronal circuit tracing with neurotropic viruses. Curr Opin Neurobiol 18(6):617–623

11. Wall NR et al (2010) Monosynaptic circuit tracing in vivo through Cre-dependent targeting and complementation of modified rabies virus. Proc Natl Acad Sci U S A 107(50): 21848–21853
12. Wickersham IR et al (2007) Monosynaptic restriction of transsynaptic tracing from single, genetically targeted neurons. Neuron 53(5): 639–647
13. Schnell MJ et al (2010) The cell biology of rabies virus: using stealth to reach the brain. Nat Rev Microbiol 8(1):51–61
14. Conzelmann KK et al (1990) Molecular cloning and complete nucleotide sequence of the attenuated rabies virus SAD B19. Virology 175(2):485–499
15. Finke S, Conzelmann KK (2005) Replication strategies of rabies virus. Virus Res 111(2): 120–131
16. Albertini AA et al (2008) Structural aspects of rabies virus replication. Cell Mol Life Sci 65(2):282–294
17. Garcia I et al (2012) Tracing synaptic connectivity onto embryonic stem cell-derived neurons. Stem Cells 30(10):2140–2151
18. Ugolini G (2010) Advances in viral transneuronal tracing. J Neurosci Methods 194(1):2–20
19. Osakada F, Callaway EM (2013) Design and generation of recombinant rabies virus vectors. Nat Protoc 8(8):1583–1601
20. Scanziani M, Hausser M (2009) Electrophysiology in the age of light. Nature 461(7266):930–939
21. Miyawaki A (2005) Innovations in the imaging of brain functions using fluorescent proteins. Neuron 48(2):189–199
22. Tian L et al (2009) Imaging neural activity in worms, flies and mice with improved GCaMP calcium indicators. Nat Methods 6(12): 875–881
23. Armbruster BN et al (2007) Evolving the lock to fit the key to create a family of G protein-coupled receptors potently activated by an inert ligand. Proc Natl Acad Sci U S A 104(12):5163–5168
24. Lechner HA, Lein ES, Callaway EM (2002) A genetic method for selective and quickly reversible silencing of mammalian neurons. J Neurosci 22(13):5287–5290
25. Magnus CJ et al (2011) Chemical and genetic engineering of selective ion channel-ligand interactions. Science 333(6047): 1292–1296
26. Tan EM et al (2006) Selective and quickly reversible inactivation of mammalian neurons in vivo using the Drosophila allatostatin receptor. Neuron 51(2):157–170
27. Choi J, Young JA, Callaway EM (2010) Selective viral vector transduction of ErbB4 expressing cortical interneurons in vivo with a viral receptor-ligand bridge protein. Proc Natl Acad Sci U S A 107(38):16703–16708
28. Marshel JH et al (2010) Targeting single neuronal networks for gene expression and cell labeling in vivo. Neuron 67(4):562–574
29. Osakada F et al (2011) New rabies virus variants for monitoring and manipulating activity and gene expression in defined neural circuits. Neuron 71(4):617–631
30. Sena-Esteves M et al (2004) Optimized large-scale production of high titer lentivirus vector pseudotypes. J Virol Methods 122(2): 131–139
31. Wickersham IR et al (2007) Retrograde neuronal tracing with a deletion-mutant rabies virus. Nat Methods 4(1):47–49
32. Gray ER et al (2011) Binding of more than one Tva800 molecule is required for ASLV-A entry. Retrovirology 8:96
33. Garcia I, Kim C, Arenkiel BR (2012) Genetic strategies to investigate neuronal circuit properties using stem cell-derived neurons. Front Cell Neurosci 6:59
34. Garcia I, Kim C, Arenkiel BR (2013) Revealing neuronal circuitry using stem cell-derived neurons. Curr Protoc Stem Cell Biol Chapter 2:Unit 2D 15
35. Maguire AM et al (2008) Safety and efficacy of gene transfer for Leber's congenital amaurosis. N Engl J Med 358(21):2240–2248
36. Takatoh J et al (2013) New modules are added to vibrissal premotor circuitry with the emergence of exploratory whisking. Neuron 77(2): 346–360

Chapter 11

Genetic Labeling of Synapses

Carlos Lois and Wolfgang Kelsch

Abstract

A major challenge in neuroscience is to unravel how the synaptic contacts between neurons give rise to brain circuits. A number of techniques have been developed to visualize the synaptic organization of neurons. In this chapter, we focus on genetic methods to mark specific types of synapses so that synaptic sites can be visualized throughout the entire dendritic or axonal arbor of single neurons. Genetic synaptic labeling can be achieved by cell-type-specific viral or transgenic delivery of synaptic proteins tagged by fluorescent proteins. Sparse genetic labeling of neurons permits semiautomated quantification of the distribution and densities of selected types of synapses in segregated domains of the axonal and dendritic trees. These approaches can reduce the complexity and ambiguity of attributing synaptic sites to unravel principles of the synaptic organization of identified neuronal types in the circuit.

Key words Synaptic organization, Synaptic markers, Neurons, Genetic synaptic labeling, Synaptophysin, PSD-95, Retroviruses

1 Introduction

A central hurdle toward understanding brain function is the complex organization of synaptic contacts between neurons that form circuits. A plethora of techniques have been developed to solve this challenging task of deciphering the synaptic organization of single neurons and mapping synaptic connectivity in neuronal circuits. In this chapter, we will focus on recently developed genetic markers to identify synaptic contacts and discuss their caveats and limitations.

Traditionally, three main approaches, morphological correlates of synapses, antibody staining, and electron microscopy, had been employed to describe or infer synaptic contacts of neurons:

1. In some cases, synapses can be identified on the basis of their association with neuronal structural specializations. For instance, many excitatory input synapses are located in dendritic spines, in which case spines may be used as a morphological proxy for synapses. One limitation of this method is that a large proportion of synapses, such as excitatory input synapses on cell somata

Benjamin R. Arenkiel (ed.), *Neural Tracing Methods: Tracing Neurons and Their Connections*, Neuromethods, vol. 92, DOI 10.1007/978-1-4939-1963-5_11, © Springer Science+Business Media New York 2015

and inhibitory synapses, are not associated with spines [1, 2]. In addition, morphological methods are not useful to quantify output synapses, which are mostly located in axon terminals (Ref?). It is easy to identify axon terminals, but it is not possible to accurately quantify the density and measure the size of output synapses by simple morphological analyses. Similarly, the extent of overlap between axons and postsynaptic dendrites has been used to infer the presence of synaptic connectivity based on the fact that synapse formation requires a physical proximity. However, ultrastructural studies reveal that axons and dendrites frequently approach each other without making synapses. For example, in several well-studied systems like C. elegans, the rat hippocampus, or the mouse retina, only 15–25 % of physical contacts are synaptic [3–5]. Thus, close proximity between neurons cannot be used to reliably identify synapses.

2. Antibody labeling against synaptic markers is a powerful method to label synapses in cultured neurons. However, this method is suboptimal in brain sections because the large number of synapses present severely complicates the attribution of synapses to individual new neurons.
3. Electron microscopy (EM) represents the "gold standard" assay for identifying synapses but its main disadvantages are that it is labor intensive, can only be applied to fixed (nonliving) tissue, and cannot be used to trace long-range connections (*see* also below section "Electron Microscopy").

The limitations of these methods have driven the ongoing development of tools to genetically label the synapses of identified neurons.

Recently developed genetic approaches to visualize synapses fall into two broad categories depending on the experimental demand:

- Visualizing specific types of synapses (i.e., glutamatergic input synapses or gabaergic output synapses) in single neurons.
- Visualizing the synaptic contacts between neurons by genetic labeling of both the pre- and postsynaptic neurons.

These approaches reduce the complexity and ambiguity of attributing synaptic sites by genetically targeting a very small subset of neurons in the examined tissue. In this chapter we will discuss these two genetic approaches considering their potential applications and respective limitations.

2 Genetic Labeling of the Synaptic Organization of Single Neurons

Labeling synapses with genetically encoded markers addresses some of the limitations of the abovementioned techniques and significantly simplifies the quantification of synaptic organization and development in neurons [6–9]. The visualization of pre- and postsynaptic terminals can be achieved via expression of fluorescent proteins fused to proteins specifically located in synapses.

2.1 Molecular Targets for Genetic Labeling of Synapses

Few synaptic proteins have been extensively tested for their use as genetic synaptic markers. We will particularly focus here on two synaptic proteins fused to fluorescent proteins that have been successfully applied in us and others. The first one, PSD-95, is a scaffolding protein restricted to clusters in the postsynaptic density of most glutamatergic synapses [9–13]. The second protein, synaptophysin, can be used to identify release sites on axon terminals as it is selectively localized at presynaptic terminals [14].

Postsynaptic Targets for Excitatory Synapses

To visualize glutamatergic input synapses, we and others have expressed a PSD-95-GFP fusion protein. PSD-95 is a scaffolding protein that localizes to the postsynaptic density of glutamatergic synapses [12] and has been extensively used as a postsynaptic marker of glutamatergic synapses [9–11, 13] both for confocal imaging in fixed tissue and in vivo imaging.

PSD-95-GFP-positive clusters overlap with endogenous PSD-95 expression. PSD-95-GFP-positive clusters are contacted by the presynaptic marker bassoon and are concentrated at asymmetric synapses at the ultrastructural level [6, 7]. Furthermore PSD-95 is already highly expressed at birth [15]. Thus, PSD-95-GFP is useful to follow synaptic development as neurons start to differentiate because it appears early during assembly of the postsynaptic density. For example, expression of PSD-95-GFP fusion protein in progenitor cells with retroviruses was a useful method to investigate the dynamics of synapse formation in newly generated neurons produced in the brain of adult mice [16]. Other postsynaptic proteins like SAP-102 fusion proteins can complement PSD-95-GFP to monitor synapse formation, as they are expressed with a different temporal profile during maturation of glutamatergic synapses.

Finally, retroviral expression of PSD-95-GFP did not change the strength and number of glutamatergic synapses in cultured neurons [6]. The absence of a direct effect of the synaptic marker on synapse number and stability is critical for many experiments, since it has been reported in cultured hippocampal neurons that five- to tenfold overexpression of PSD-95 by transient transfection led to strengthening or increase in the number of AMPA receptor-mediated mEPSCs [17]. Our experiments revealed that the modest level of expression achieved with retroviral expression did not change the strength or number of AMPA receptor-mediated mEPSCs in vitro, further supporting the idea that PSD-95-GFP can be used to genetically label postsynaptic sites. Thus, in all strategies utilizing fusion proteins as genetic markers of synapses, there is a critical balance to achieve, because it is necessary to obtain sufficient expression so that the fusion protein can be reliably detected, but low enough not to interfere with normal cell function.

To the best of our knowledge, other candidate fusion proteins are less established for labeling of glutamatergic synapses. For example, one could imagine visualization of specific subsets of glutamatergic neurons during development. Glutamatergic synapses are characterized by sequential addition of NMDA receptor subunits GluN2B followed by GluN2A. Both NMDA receptor subunits can be engineered as fusion proteins and cluster at the expected sites where glutamatergic synapses are regularly found [18]. However, in contrast to PSD-95-GFP fusion protein, overexpression of these fusion proteins have overt effects on neuronal maturation [18]. Hence, the effects of the expressed fusion protein itself on synapse properties like elimination or strengthening are important to consider. Certain of these effects may be mitigated by directed mutagenesis of the synaptic marker, but there is the possibility that dominant negative effects (such as sequestration of interacting proteins) could occur.

Postsynaptic Targets for Inhibitory Synapses

In contrast to the well-established marker PSD-95-GFP for glutamatergic synapses, less is known about equivalent marker for inhibitory synapses. A scaffolding protein of inhibitory synapses, gephyrin, may be useful [19], but a thorough characterization is necessary to confirm whether gephyrin only clusters at synapses or whether it also clusters at significant levels, i.e., in the cytoplasm. In addition, it has not been determined the fraction of GABAergic synapses that are actually labeled by this marker. Finally, it is not known the extent by which gephyrin labels synapses as neurons mature. Other promising candidates may be $GABA_A$-receptor subunits fused to fluorescent proteins [20]. It is however important to again consider direct effects of these receptor fusion proteins on synapse function or maintenance [21]. Similar gain-of-function problems can arise if other structural synaptic proteins, like neurexins or neuroligins, are used as marker proteins (*see* also next section on GRASP).

Presynaptic Targets for Release Sites

Presynaptic synapses (output synapses) can be labeled by a synaptophysin-GFP fusion protein. Synaptophysin is a 38 kDa synaptic vesicle glycoprotein that is expressed in virtually all neurons in the brain and spinal cord [14]. Despite its well-known interaction with the essential synaptic vesicle protein synaptobrevin, the exact function of synaptophysin remains unclear as it is not an obligatory protein for vesicle formation or fusion [22, 23]. Synaptophysin-GFP has been extensively used to study the distribution and density of presynaptic sites in neurons both in vitro and in vivo [8, 24–26]. There is currently no indication that its overexpression results in gain of additional synapses. Its expression is nearly ubiquitous in all presynaptic sites in contrast to other presynaptic marker proteins like Bassoon. Finally, it appears relatively early during synapse formation [27]. A previous study that examined synapse formation in zebra fish indicated that transient clusters are formed in the cytosol of developing axons potentially indicating transport of pre-clustered synaptic proteins to the terminals [8]. Such transient clustering should be considered when studying initial synapse formation.

2.2 Genetic Labeling of Contact Between Synaptic Partner Neurons

For certain questions the visualization of synaptic contacts between two types of neurons is critical. For this purpose, complementary genetic labeling techniques have been developed.

Transsynaptic Labeling with Replication-Deficient Rabies Vectors

A technique that has recently gained substantial attention is retrograde transsynaptic labeling using genetically modified rabies viruses, also called "monosynaptic tracing" [28]. Monosynaptic tracing is reviewed in detail in a separate chapter of this book, and here we will focus on its current limitations and opportunities. This method has currently several drawbacks that have to be considered when applying this technique in its current state [29]. Replication of rabies viruses interferes with the protein synthesis machinery of the host cell, thus limiting the time window between its initial expression until the death of the labeled neurons to a few days [30].Thus, it is important to be aware of the effects of rabies replication on the health of the neuron for any imaging and functional studies . Also, once the virus jumps to a presynaptic cell, that cell will remain labeled independently of whether the synaptic contact is lost [29]. Particularly during states of high synapse turnover such as during development, this technique can be used to sample the cumulative history of previous and current synaptic partners. Finally, it appears that rabies virus do not cross over all the synapses. For example, cortical pyramidal cells are estimated to have approx. 10,000 presynaptic partners, but transsynaptic labeling with rabies only reveals hundreds of presynaptic cells [31]. Similarly it is not clear whether rabies viruses will cross different types of synapses with the same efficiency [29]. Again, once these constraints are better understood, they could provide further insight into synaptic properties and their distribution.

GFP Reconstitution Across Synaptic Partners (GRASP)

Another approach that has been initially developed in C. elegans is "GFP Reconstitution Across Synaptic Partners" (GRASP) [32]. GRASP genetically labels synaptic partners utilizing a two-component synaptic labeling system [33]. GRASP takes advantage of a version of GFP divided into two fragments that can emit fluorescence only when the two halves of GFP are combined. To visualize exclusively synaptic contacts, the two GFP fragments need to be appended to the extracellular domains of transmembrane proteins that localize to synapses. GRASP labeling matched ultrastructural synaptic contacts in C. elegans circuits [32, 34]. GFP fluorescence requires no exogenous cofactors and therefore can be monitored in vivo like other fluorescent genetic synaptic marker proteins. GRASP has now been modified for application in C. elegans, Drosophila, and, most recently, mouse. In transgenic mice [35, 36] the GRASP system currently appears to have low sensitivity. Although signals in especially large photoreceptor synapses are robust, known contacts at smaller synapses in the inner plexiform layer of the retina, the spinal, or the cortex using other Cre transgenic lines were not detected potentially due to insufficient expression of the 2 GRASP partners [36]. In contrast, using high levels of expression via adeno-associated viral vectors, it is possible to detect GRASP signals with higher sensitivity in the mouse hippocampus and cortex [35].

Finally, an important concern regarding GRASP is the potential stabilization of otherwise transient cell-cell contacts, as the binding of the two elements of the split GFP is irreversible. Thus, the two components of the GRASP system act as a strong cell adhesion complex that will keep cell permanently attached to each other. This may become particularly important in dynamic situations such as during brain development where a large fraction of synapses is usually eliminated before achieving the eventual synaptic organization. In summary, GRASP is a promising technique to visualize synaptic connections between neurons that might become widely useful once several key technical limitations are solved.

3 Comparison with Other Techniques

3.1 Histological Approaches to Visualize the Synaptic Organization of Neurons

As already stated above conventional antibody labeling of synaptic proteins in vibratome or frozen tissue sections by light microscopy has two major drawbacks. First, it relies on antibodies against synaptic proteins. However, most antibodies against synaptic proteins that work in cultured neurons do not work well in tissue sections. Second, it is frequently impossible to see the "tree in the forest" due to the extremely high density of labeling in tissue sections, which makes it impossible to attribute or even quantify the number of synaptic sites of single neurons.

Emerging imaging techniques like "array tomography" [37] may be able to overcome some of these problems. Array tomography is a volumetric microscopy method based on physical serial sectioning. Ultrathin sections (50–200 nm) of a plastic-embedded tissue are cut using an ultramicrotome and bonded in an ordered array to a glass coverslip. Due to the ultrathin sectioning, antibody labeling is substantially improved as antibodies penetrate efficiently through the sections. Because these arrays are very effectively stabilized by the glass substrate, they can withstand many repeated cycles of staining, imaging, and elution. This permits using many antibodies serially (20 or more) to each individual section.

The resulting two-dimensional image tiles can then be reconstructed computationally into three-dimensional volume images for visualization and quantitative analysis. However, "array tomography" relies on the efficiency of antibody staining and preservation of tissue antigenicity and can result in potentially ambiguous and incomplete results.

Electron microscopy is technically more challenging than imaging by light microscopy but allows for high-resolution analysis of pre- and postsynaptic sites by visualizing synaptic vesicles and postsynaptic sites, respectively. Conventional EM using serial sections has been used, for example, to reconstruct the whole set of individual synapses between single adult-born neurons and their synaptic partners [38]. However, this kind of work is extremely labor intensive and can only be used to reconstruct a handful of neurons per experiment. Recent developments within the past few years hint at the possibility of semiautomated sectioning and imaging of large neuropil volumes [39], thus facilitating high-throughput ultrastructural analyses of synapses. However, only relatively small brain volumes can presently be imaged (on the order of $450 \times 350 \times 50$ μm, e.g., [3]), mainly because electron microscopy image acquisition and analysis remain a formidable challenge [39]. In addition, the main limitation for the EM approach is that this technique focuses on the local structure under analysis, because long-range connections from distant parts of the brain cannot be analyzed.

There are two main advantages of electron microscopy and array tomography. First, these techniques provide very high-resolution images. Second, in contrast to genetic methods, they reveal brain circuits in their "native" status, without perturbing their function by addition of extra molecules to their synapses. However, both electron microscopy and array tomography are restricted to fixed (dead) tissue and do not allow functional studies via in vivo imaging or electrophysiological recordings as is possible for genetic labeling of synapses. Due to the extensive tissue processing, e.g., serial electron microscopy, caution has to be applied concerning nonproportional changes in extracellular due to tissue shrinkage and subsequent artifacts that can distort the wiring diagram of circuits during semiautomated analysis [39].

4 Delivering Genetic Synaptic Markers into Neurons

As outlined above delivery of genes to a small number of neurons and obtaining appropriate levels of expression is critical to successfully using genetic synaptic markers. Techniques such as gene gun delivery (in slices), in vivo plasmid electroporation, and adeno-associated viruses are less explored for genetic labeling of synapses as the high copy numbers of transferred genes into single neurons with these methods may result in unpredictable effects on synapse function. In addition, there is a high variation in the copy number of transgenes delivered per cell using adeno-associated viruses, electroporation, or gene gun. Therefore, analysis of labeled synapses may be complicated by the degree of variation in brightness of clusters among neurons. Finally, above a certain level of expression, a significant level of the marker protein might be located outside of synaptic sites. In contrast, retroviral vectors provide moderate expression levels and a relative consistency in the level of expression between cells. Due to these advantages, retroviruses have been extensively used in many studies of genetic labeling of synapses. An alternative approach to achieve reproducibility of gene expression is provided by generating transgenic animal lines. We will therefore discuss here expression of synaptic markers by retroviruses and tissue-specific transgenesis.

4.1 Viral Delivery of Genetic Synaptic Markers

Retroviral vectors are the most commonly used vehicle to deliver genetic synaptic markers into neurons. Retroviral vectors fall by-and-large into two popular systems: HIV-derived lentiviruses and Moloney oncoretroviruses. Both viral systems are replication deficient, meaning that after the initial infection of a cell, they are not able to replicate and infect other target cells. A major difference between the two systems resides in the cells they can infect [40]. Whereas oncoretroviruses can only infect cells that are in the progress of dividing, lentiviruses can infect both quiescent and dividing cells. Oncoretroviruses have therefore been widely used in developmental biology as they allow birthdating of newly generated neurons and tracking their maturation. The packaging size of the inserted genes is comparable in both systems and sufficient to harbor most genetic synaptic marker and an additional gene of interest, such as those coding for ion channels, growth factors, or cell adhesion molecules. Thus, by introducing both a synaptic marker and a gene of interest, it is possible to assess the effects of various manipulations on synapse formation and dynamics. Retroviral expression under the control of different promoter fragments derived from the human synapsin, CMV, or RSV promoter provides appropriate levels of expression in mammals. Cell-type-specific promoters are however, despite intense search, still not available for most neuron types. For rodents, the CamKII promoter is a good option to label excitatory neurons, but currently there is no good candidate promoter to label inhibitory neurons. Using the abovementioned promoters and retroviral systems, we have observed a limited variability in the brightness of the fusion proteins PSD-95-GFP and synaptophysin-GFP when comparing cells in the same tissue section. Moreover, these differences of expression between cells had little effect on the analysis by confocal microscopy as the brightness of the synaptic clusters was generally saturating and the differences occurred largely in the mostly dim and diffusely distributed fusion protein fluorescence throughout neuronal processes. These low levels of diffusely distributed synaptic fusion protein do not interfere with detection of synaptic clusters, but can be very helpful as, in the case of in vivo imaging studies, it helps to visualize the neuronal tree along which synaptic clusters form [11, 41]. In case of confocal analysis in fixed tissue, the low levels of diffusely distributed GFP in the neuronal processes can be rendered detectable by amplifying its signal with immunofluorescent staining in a different color (for instance, Texas red antibodies to amplify the PSD-95-GFP diffuse signal) that substantially facilitates the reconstruction of finer neuronal processes.

4.2 Generation of Transgenic Animals Expressing Genetic Synaptic Markers

Another potentially attractive approach is the generation of transgenic lines expressing genetic synaptic markers in selected subsets of neurons. This approach generally can have two advantages. First, transgenic animals provide reproducible levels of gene expression for a given cell type, eliminating the cell-to-cell variability discussed with viral delivery methods. Second, and more importantly, cell-type specificity can be obtained by using knock-in-techniques or large promoter region fragments (>100 kb) for conventional transgenesis. The downside of most cell-type-specific mouse line is that the high density of labeled cells may make it difficult to reconstruct and attribute synaptic clusters to single neurons. This limitation could be overcome using transgenic strategies that result in stochastic and sparse expression of genetic synaptic markers, such as incomplete activation of inducible Cre (Nathans, Luo paper in PLOS ONE). For some of these model systems, the use of genetic synaptic markers, PSD-95-GFP and synaptophysin-GFP, has been successfully applied in combination with in vivo imaging to study synaptic development of single neurons [8].

4.3 Different Ways of Attributing Synapses to Specific Neurites

For most experimental question, genetically labeled synapses need to be attributed to neuronal processes of single neurons. Here, different approaches exist:

1. Immunohistochemical methods.
   - (a) For imaging the synaptic organization of single neuron's visualization of the dendritic or axonal tree that contain the genetically labeled synapses in fixed tissue, a simple method based on immunofluorescence antibody labeling can be used. To attribute PSD-GFP-positive clusters to a particular neuron, we took advantage of the presence of low levels of diffuse PSD-95-GFP protein in the cytoplasm, which were not detectable by its endogenous fluorescence. However, this diffuse PSD-95-GFP protein can be easily visualized by amplifying its signal with antibodies raised against GFP (followed by staining with a secondary antibody coupled to a red or blue fluorophore to distinguish it from the intrinsic green fluorescence of PSD positive clusters) and allowed attribution of PSD-GFP clusters to a dendritic arbor belonging to a particular neuron.

2. Co-expression with membrane-tagged XFPs.

   (a) An alternative approach is the expression of both the genetic synaptic marker and a membrane-tagged fluorescent protein of another color. Different approaches exist to co-express two genes with a single retrovirus.

   IRES: internal ribosomal entry sites can be engineered to obtain two different proteins from single mRNA (bicistronic expression). However, this strategy is problematic in many cases as it is usually found that the second protein is expressed at much lower levels than the first protein. In some cases it has been reported that the protein located in behind the IRES may be expressed at levels tenfold lower than the first protein.

   2A linkers: The use of T2A-linker sequences derived from picornaviruses is a relatively recent strategy to obtain bicistronic expression. In this strategy the first gene is engineered without its stop codon, followed by a 2A sequence (about 18 amino acids), and finally followed by the complete coding sequence of the second gene. When the ribosome reaches the 2A sequence, it will release the first protein plus the 2A sequence and will start translating the second protein. In contrast to IRES, the two genes separated by 2A sequences are expressed at similar levels. However, there are a few caveats for bicistronic cassettes with 2A sequences. First, in some situations instead of two independent proteins, a large fusion protein will be produced that includes the open reading frames of the first and second proteins plus the intervening amino acids of the 2A sequence. Second, the first protein will have an extra "tail" consisting of the 2A sequence at its C-terminus, which in some cases can affect its function. We recently observed that the T2A linker provides sufficient levels of expression of synaptophysin-GFP and membrane-tagged dimeric tomato protein in a retroviral vector (unpublished observations, Fig. ???). This approach may be particularly interesting for in vivo imaging that aims at attributing synaptic clusters to processes of specific neurons.

   (b) Bicistronic expression of a synaptic genetic marker plus the recombinase Cre can be used to combine genetic synaptic labeling with conditional mouse or viral transgenic tools. XFP reporter lines are available that displayed high levels of fluorescent protein expression upon Cre recombination. Cre is a highly efficient recombinase, and having a low expression of it (e.g., following an IRES) is not problematic because it is sufficient to induce recombination of the loxP sites. Thus, in this strategy, the XFP would be driven by a strong promoter in a loxP-dependent reporter transgenic mouse or virus.

This approach can be further expanded combining multiple conditional mouse lines and/or viral conditional gene delivery that is Cre dependent.

4.4 Technical Considerations of Genetic Labeling

Two main points have to be considered when designing genetic synaptic markers:

1. *Expression level: not too low and not too high (influence of synapse strength).*
   When introducing these genetically encoded markers, it is critical to ensure that only modest levels of overexpression are achieved because excessive levels of these proteins may interfere with synaptic development or function [17]. Fortunately, retroviral vectors, which deliver single copies of transgenes into their target cells, produce modest levels of expression that are sufficient to detect fluorescent synaptic marker proteins, while at the same time leave synapse number and strength unperturbed [6].
2. *Influence on synapse stability*
   There are concerns to consider in this respect. First, expression of these genetic synaptic markers may influence the number or strength of synapses. For example, neuroligins are known to directly induce synapse formation even with nonneuronal targets [42] and thus have to be carefully tested before using them as fusion proteins to study synaptic development or reorganization as discussed above. Second, labeling approaches based on the transneuronal interaction of transmembrane proteins may interfere with synapse turnover. For example, an important drawback of GRASP is that the reconstitution of the split GFP creates a strong cell-cell adhesion that can stabilize otherwise transient synapses. Thus, it is critical to keep in mind that using proteins to label synapses could corrupt the normal synaptic development and distribution.

5 Imaging and Quantification of Genetic Synaptic Markers

5.1 Two-Photon Microscopy

Two-photon microscope technology has taken off considerably in recent years and is still the only technique that allows for synapse imaging in vivo. This technique is extremely useful for observing real-time changes to experimental manipulations and allows investigators to visualize synapse dynamics.

Neurons up to 800 μm below the brain's surface can be imaged [43]. In some cases it becomes even possible to image even deeper brain structures at high resolution such as superficial dendrites of neurons in CA1 of the hippocampus by removing part of the neocortex and white matter above the hippocampus [44]. Finally, there has been great interest in using two-photon microscopy associated with endoscope lenses to image deep within the brain, but several technical obstacles to obtain a sufficient spatial resolution need to be solved before this method can be routinely used [45].

5.2 Confocal Microscopy

As deep structures of the brain are beyond the depth limitation of two-photon microscopy in vivo, in vitro time-lapse confocal imaging of brain slices is sometimes carried out to study the dynamics of synapse formation [38, 46]. The main concern about this technique is that the integrity of cultured adult brain slices is dramatically perturbed over long time periods with currently available culture techniques [47], as well as the possibility of abnormal synapse rearrangement due to fluctuations in culture conditions [48]. Confocal imaging of fixed slices is much more commonly used to study synaptic organization, especially because this method of observation is technically straightforward and enables investigators to analyze many neurons simultaneously. Time course experiments can be performed to observe spine formation over days and months, but because only a snapshot of the synapses can be obtained in fixed slices, this technique cannot be used to analyze the short-term dynamics of synapse formation in real time.

5.3 Data Analysis and Quantification

Genetic synaptic markers provide substantial advantages when it comes to the quantification of densities of synapses in dendritic or axonal domains. With genetically encoded markers, one can, in principle, analyze the complete set of a neuron's excitatory input synapses and output synapses, including those not associated with structural specializations such as spines or axon terminals.

Genetically labeled synapses appear as discrete bright clusters that can be semiautomatically detected and measured upon thresholding of images using freely available software packages (e.g., ImageJ-based MacBiophotonics). The obtained data about synapse organization can be combined with reconstruction of the morphology of individual neurons to obtain the spatial distribution of synapses along neuronal trees and measures of local cluster density. Most of the current analysis is limited to two-dimensional projections of 3D image stack. With the advancement of existing software tools, analysis in 3D may soon be amenable.

The main challenge of these approaches is the still highly labor intense reconstruction of neuronal processes, because it is essential to attribute genetically labeled synapses to specific neuronal domains such as axons or dendrites. One may imagine that reconstruction of dendritic and axonal trees of sparsely labeled neurons is an easy task to be automated. This however turned out to be a formidable challenge [49]. As a first step in this process, accurate semiautomated reconstruction software in 3D (e.g., [50]) could facilitate neuronal reconstruction to then attribute synaptic organization to neuronal trees in 3D.

6 Future Directions: Combination of Synaptic Labeling and Transneuronal Tracing

Genetic synaptic markers provide powerful tools that can help answer many open questions on the development and synaptic wiring of brain circuits when considering their currently existing limitations and caveats. With the continuous efforts to improve the existing tools, validation of novel marker proteins for specific types of synapses, these tools become increasingly more flexible for specific research questions. They may thus help to find the "tree in the forest" among the complex meshwork of neuronal processes and synapses in brain circuits. We believe that for many circuits understanding the typical wiring diagram of neurons may help to build realistic, unifying models of their function without necessarily having to know each detail of the entity of each individual's circuit under study.

Finally the strength of genetic tools might come into full action when genetic synaptic markers and transneuronal tracing will be combined. One example for such a dual approach shall be outlined here. Once a presynaptic neuron has been labeled at a given time point via transneuronal labeling with rabies, it will remain so independently of whether the synaptic contact is lost. In contrast to this permanent labeling (a trace of the neuron's connectivity history), labeling with genetic synaptic markers, e.g., with PSD-95-GFP, will be lost immediately when the synapse is lost. Hence, a combination of both techniques can serve to identify the history of cells that had been presynaptic to the target neuron at one point and the current state of synaptic contacts between the neurons.

7 Appendix: Gene Delivery of Genetic Synaptic Markers with Retroviruses and Quantification of Synapses

This appendix describes some specific procedures to deliver retroviruses carrying genetically encoded synaptic markers into the brain of rodents. General details about production and titration of lentiviral and retroviral vectors can be found elsewhere.

8 Injection of Viruses into the Brain

Viral prep: It is critical that the viral suspension is very clean. During the preparation of the viruses, there will be some cellular debris that can be strongly autofluorescent. To eliminate this debris it is useful to centrifuge the viral prep with a 20 % sucrose cushion.

Stereotaxic injection: It is critical to minimize the damage to the brain during injection. In particular, bleeding associated with the injection will cause very high levels of autofluorescence that will make quantification of synapses very difficult. To minimize damage it is useful to use thin borosilicate pipettes pulled to an outer diameter of approx. 15–20 μm. It is not advisable to inject through metallic needles as this will cause severe tissue damage on the injection site. Similarly, it is advisable to inject the viral prep slowly, at a rate of approx. 1 μl over 5 min. Rapid injection can severely damage and distort the tissue. Regarding the timing of imaging after injection, for lentiviral vectors, the expression of the transgene peaks as early as 3 days, but there will likely be some acute damage in the injection area at this early time. Thus, it is advisable to wait at least a week so that the autofluorescence due to damage is resolved before perfusion of the animal.

9 Acquisition and Analysis of Genetically Labeled Synapses

This section describes the procedure that we optimized to visualize the synaptic organization of single genetically labeled neurons, which can be easily modified for individual experimental needs. The procedure is divided into three main steps, and technical issues are highlighted that are critical in our experience: preparation of the tissue ("Preparation of Tissue"), image acquisition by confocal microscopy ("Image Acquisition"), and semiautomated image analysis ("Quantification of Synaptic Clusters").

9.1 Preparation of Tissue

1. The protocol is described for small rodents, but can be easily adapted to other species. Animals are transcardially perfused initially with phosphate buffered saline (PBS, 1×) for 10–15 s, followed by 4 % paraformaldehyde (PFA) for 3–5 min. The animal should become rigid within the first 30 s to 1 min of perfusion with PFA. It is optimal to use an overdose of an injectable anesthetic drug (such as Ketamin/Xylazin) and to start perfusion when the heart is still beating. PBS should be infused at a pressure such that the liver becomes pale within 10–15 s and clear PBS flows out of the right atrium after this period. It is equally important to perfuse with relatively low pressure, because at high perfusion pressure, the small capillaries in the brain will break and perfusion will not be homogeneous throughout the brain. PBS should be set to pH 7.0–7.4 and pre-warmed to 32–37 °C to prevent contraction of smaller blood vessels in the brain. Following PBS, perfusion should be immediately switched to room-temperature PFA. Incorrect perfusion leads to delayed fixation with PFA, which results in beaded structures of dendrites and loss of genetically labeled synaptic clusters. After perfusion is complete, the brains are extracted from the skull and post-fixed in 4 % PFA overnight at 4 °C.

2. After preparation of floating sections with a vibratome (e.g., 50 μm sections), tissue can be incubated (overnight at 4 °C) with primary antibody raised against the fluorescent protein tagged to the synaptic marker. The following day sections are rinsed in PBS and stained with a secondary fluorescent antibody for two hours at room temperature. Sections are then washed with PBS and mounted with an aqueous mounting medium that preserved fluorescent molecules. This procedure allows for the visualization of the neuronal tree and to attribute synaptic clusters to a neuron and specific dendritic domains. Blocking solutions (PBS containing 1 % bovine serum albumin or related serum proteins) for antibody incubation usually contain a detergent, i.e., Triton X-100, to permeabilize the tissue. We keep the procedure and times as constant as possible to avoid introducing additional variability, i.e., by differentially affecting the brightness of the fluorescence of the synaptic clusters.

9.2 Image Acquisition

Neurons expressing synaptic marker proteins can be conveniently imagined using confocal laser scanning microscopy. In most experimental conditions, it is advantageous to image sections that are sparsely labeled, where individual neurons are clearly separated from each other. In this case it is easy to analyze the full dendritic arbor of a single neuron without having to deal with neurites that could belong to neighboring cells. Confocal stacks are acquired at high magnification (with a 60–63× oil immersion objective) to efficiently capture emitted light from the clustered fluorescent proteins. The pixel size should be sufficiently small to obtain high intensity of all the pixels that are grouped in individual synaptic clusters, and to easily distinguish them from the occasionally observed noise that may result in random isolated pixels with high intensity. As most synaptic clusters have a size around 1 μm, we found a cluster size between 0.2×0.2 and $0.3 \times 0.3\ \mu m^2$ most reliable for subsequent analysis. Laser excitation intensities should be set to levels that result in little or no obvious bleaching of the clusters. This can be easily tested by imaging the same neuron twice in the same day and comparing the clusters among the two images. A reference section containing neurons with good synaptic cluster intensity should be used to guarantee comparable acquisition conditions over time. The sensitivity of the photomultipliers (PMT) should be set to a level that clusters just saturate but low enough that individual clusters do not become confluent due to overexposure. Similarly, the pinhole size should be kept in the recommended range [51] for the chosen magnification. Once the settings are initially defined with a test sample, the conditions should be kept constant throughout the different imaging session. Upon acquisition of confocal stacks, maximum density projections are prepared for further image analysis. Two-dimensional projections are generally used for analysis, as current version of most image processing software cannot handle 3D data for quantification.

9.3 Quantification of Synaptic Clusters

Analysis of densities and distribution of genetically labeled synapses can be semiautomated. Fully automated analysis is currently limited by the still challenging task for computers to reconstruct neuronal trees due to overlap of labeled neurons and discontinuities in the processes deriving from histological processing and incomplete filling with fluorescent proteins. Thus, reconstruction of processes has to be performed individually or at least be supervised.

Contrary to reconstruction of dendritic trees, analysis of clusters can be fully automatized provided the original image quality has a good signal-to-noise ratio. Signal-to-noise ratio for these experiments means high intensity of fluorescence in the synaptic cluster and low levels of autofluorescent background outside of synaptic sites. Similarly, it is important that there is a low level of fluorescence originating from diffusely distributed XFP in the cells' processes outside the synapses. Another potential source of "contamination" with artifactual autofluorescent clusters can be due to lipofuscin granules observed in some brain structures and species. The appearance of these autofluorescent granules is difficult to predict. For example, we have found them in mouse olfactory bulb and dentate gyrus, but not in the rat olfactory bulb or mouse neocortex. These autofluorescent artifacts can be easily diagnosed as they are excited by all wavelengths. In contrast, real genetic synaptic markers containing XFPs can only be detected at a specific wavelength (e.g., 550 nm for GFP). In addition, these autofluorescent granules can usually be excluded from analysis as they are mostly present in cell bodies.

Given these considerations the analysis is relatively straightforward using different analysis software packages. We will describe the different steps of analysis and particularly refer to the ImageJ-based MacBiophotonics software (www.macbiophotonics.ca/). Similar procedures can be performed in Metamorph software from Molecular Probes.

Steps:

1. Open maximum density projection (*File>Open*).
2. Define pixel size for subsequent distance measurements (*Analyze>Set scale*).
3. Split color channels (*Image>Color>Split channels*).
4. Choose the channel that displays the fluorescent synaptic clusters.
5. Set inclusive threshold so that only clusters are included (*Image>Adjust>Threshold*). The threshold value should be kept constant throughout the analysis. Therefore, it proves useful to use a reference as described in the acquisition part to set the threshold.
6. Draw a contour using the freehand selection tool to define a region of interest to measure a specific dendritic domain and exclude neighboring neurons.

7. Perform region measurement (*Analyze>Analyze particles*). Desired parameter data can be set in the results window (*Analyze>Set Measurements*) and copied to a data sheet of a given statistics program.

Measure the length of the neuronal processes in the region of interest using the freehand line tool and *Analyze>Measure*.

References

1. Price JL, Powell TP (1970) The synaptology of the granule cells of the olfactory bulb. J Cell Sci 7:125–155
2. Woolf TB, Shepherd GM, Greer CA (1991) Serial reconstructions of granule cell spines in the mammalian olfactory bulb. Synapse (NY) 7:181–192
3. Briggman KL, Helmstaedter M, Denk W (2011) Wiring specificity in the direction-selectivity circuit of the retina. Nature 471:183–188
4. White JG, Southgate E, Thomson JN et al (1986) The structure of the nervous system of the nematode Caenorhabditis elegans. Phil Trans R Soc Lond B Biol Sci 314:1–340
5. Mishchenko Y, Hu T, Spacek J et al (2010) Ultrastructural analysis of hippocampal neuropil from the connectomics perspective. Neuron 67:1009–1020
6. Kelsch W, Lin C-W, Lois C (2008) Sequential development of synapses in dendritic domains during adult neurogenesis. Proc Natl Acad Sci U S A 105:16803–16808
7. Livneh Y, Feinstein N, Klein M et al (2009) Sensory input enhances synaptogenesis of adult-born neurons. J Neurosci 29:86–97
8. Meyer MP, Smith SJ (2006) Evidence from in vivo imaging that synaptogenesis guides the growth and branching of axonal arbors by two distinct mechanisms. J Neurosci 26:3604–3614
9. Niell CM, Meyer MP, Smith SJ (2004) In vivo imaging of synapse formation on a growing dendritic arbor. Nat Neurosci 7:254–260
10. Ebihara T, Kawabata I, Usui S et al (2003) Synchronized formation and remodeling of postsynaptic densities: long-term visualization of hippocampal neurons expressing postsynaptic density proteins tagged with green fluorescent protein. J Neurosci 23:2170–2181
11. Gray NW, Weimer RM, Bureau I et al (2006) Rapid redistribution of synaptic PSD-95 in the neocortex in vivo. PLoS Biol 4:e370
12. Sheng M (2001) Molecular organization of the postsynaptic specialization. Proc Natl Acad Sci U S A 98:7058–7061
13. Washbourne P, Bennett JE, McAllister AK (2002) Rapid recruitment of NMDA receptor transport packets to nascent synapses. Nat Neurosci 5:751–759
14. Südhof TC, Jahn R (1991) Proteins of synaptic vesicles involved in exocytosis and membrane recycling. Neuron 6:665–677
15. Shu F, Ohno K, Wang T et al (2001) Developmental changes in PSD-95 and Narp mRNAs in the rat olfactory bulb. Dev Brain Res 132:91–95
16. Sassoé-Pognetto M, Utvik JK, Camoletto P et al (2003) Organization of postsynaptic density proteins and glutamate receptors in axodendritic and dendrodendritic synapses of the rat olfactory bulb. J Comp Neurol 463:237–248
17. El-Husseini AE, Schnell E, Chetkovich DM et al (2000) PSD-95 involvement in maturation of excitatory synapses. Science (NY) 290: 1364–1368
18. Kelsch W, Li Z, Eliava M et al (2012) GluN2B-containing NMDA receptors promote wiring of adult-born neurons into olfactory bulb circuits. J Neurosci 32:12603–12611
19. Chen JL, Villa KL, Cha JW et al (2012) Clustered dynamics of inhibitory synapses and dendritic spines in the adult neocortex. Neuron 74:361–373
20. Chen L, Wang H, Vicini S et al (2000) The γ-aminobutyric acid type A (GABAA) receptor-associated protein (GABARAP) promotes GABAA receptor clustering and modulates the channel kinetics. Proc Natl Acad Sci 97:11557–11562
21. Jacob TC, Bogdanov YD, Magnus C et al (2005) Gephyrin regulates the cell surface dynamics of synaptic GABAA receptors. J Neurosci 25:10469–10478
22. Pennuto M, Bonanomi D, Benfenati F et al (2003) Synaptophysin I controls the targeting of VAMP2/Synaptobrevin II to synaptic vesicles. Mol Biol Cell 14:4909–4919
23. Valtorta F, Pennuto M, Bonanomi D et al (2004) Synaptophysin: leading actor or walk-on role in synaptic vesicle exocytosis? BioEssays News Rev Mol Cell Dev Biol 26:445–453
24. Nakata T, Terada S, Hirokawa N (1998) Visualization of the dynamics of synaptic vesicle

and plasma membrane proteins in living axons. J Cell Biol 140:659–674
25. Li Z, Murthy VN (2001) Visualizing postendocytic traffic of synaptic vesicles at hippocampal synapses. Neuron 31:593–605
26. Kaether C, Skehel P, Dotti CG (2000) Axonal membrane proteins are transported in distinct carriers: a two-color video microscopy study in cultured hippocampal neurons. Mol Biol Cell 11:1213–1224
27. Bergmann M, Lahr G, Mayerhofer A et al (1991) Expression of synaptophysin during the prenatal development of the rat spinal cord: correlation with basic differentiation processes of neurons. Neuroscience 42:569–582
28. Wickersham IR, Lyon DC, Barnard RJO et al (2007) Monosynaptic restriction of transsynaptic tracing from single, genetically targeted neurons. Neuron 53:639–647
29. Wickersham IR, Feinberg EH (2012) New technologies for imaging synaptic partners. Curr Opin Neurobiol 22:121–127
30. Scott CA, Rossiter JP, Andrew RD et al (2008) Structural abnormalities in neurons are sufficient to explain the clinical disease and fatal outcome of experimental rabies in yellow fluorescent protein-expressing transgenic mice. J Virol 82:513–521
31. Marshel JH, Mori T, Nielsen KJ et al (2010) Targeting single neuronal networks for gene expression and cell labeling in vivo. Neuron 67:562–574
32. Feinberg EH, Vanhoven MK, Bendesky A et al (2008) GFP reconstitution across synaptic partners (GRASP) defines cell contacts and synapses in living nervous systems. Neuron 57:353–363
33. Cabantous S, Terwilliger TC, Waldo GS (2005) Protein tagging and detection with engineered self-assembling fragments of green fluorescent protein. Nat Biotechnol 23:102–107
34. Park J, Knezevich PL, Wung W et al (2011) A conserved juxtacrine signal regulates synaptic partner recognition in Caenorhabditis elegans. Neural Dev 6:28
35. Kim J, Zhao T, Petralia RS et al (2012) mGRASP enables mapping mammalian synaptic connectivity with light microscopy. Nat Methods 9:96–102
36. Yamagata M, Sanes JR (2012) Transgenic strategy for identifying synaptic connections in mice by fluorescence complementation (GRASP). Front Mol Neurosci 5:18
37. Micheva KD, Smith SJ (2007) Array tomography: a new tool for imaging the molecular architecture and ultrastructure of neural circuits. Neuron 55:25–36
38. Toni N, Teng EM, Bushong EA et al (2007) Synapse formation on neurons born in the adult hippocampus. Nat Neurosci 10:727
39. Briggman KL, Denk W (2006) Towards neural circuit reconstruction with volume electron microscopy techniques. Curr Opin Neurobiol 16:562–570
40. Sanes JR (1989) Analysing cell lineage with a recombinant retrovirus. Trends Neurosci 12: 21–28
41. Livneh Y, Mizrahi A (2012) Experience-dependent plasticity of mature adult-born neurons. Nat Neurosci 15:26–28
42. Scheiffele P (2003) Cell-cell signaling during synapse formation in the CNS. Annu Rev Neurosci 26:485–508
43. Helmchen F, Denk W (2005) Deep tissue two-photon microscopy. Nat Methods 2:932–940
44. Mizrahi A, Crowley JC, Shtoyerman E et al (2004) High-resolution in vivo imaging of hippocampal dendrites and spines. J Neurosci 24: 3147–3151
45. Flusberg BA, Cocker ED, Piyawattanametha W et al (2005) Fiber-optic fluorescence imaging. Nat Methods 2:941–950
46. Galimberti I, Gogolla N, Alberi S et al (2006) Long-term rearrangements of hippocampal mossy fiber terminal connectivity in the adult regulated by experience. Neuron 50:749–763
47. Berdichevsky Y, Sabolek H, Levine JB et al (2009) Microfluidics and multielectrode array-compatible organotypic slice culture method. J Neurosci Methods 178:59–64
48. Kirov SA, Petrak LJ, Fiala JC et al (2004) Dendritic spines disappear with chilling but proliferate excessively upon rewarming of mature hippocampus. Neuroscience 127:69–80
49. Donohue DE, Ascoli GA (2011) Automated reconstruction of neuronal morphology: an overview. Brain Res Rev 67:94–102
50. Brown KM, Barrionuevo G, Canty AJ et al (2011) The DIADEM data sets: representative light microscopy images of neuronal morphology to advance automation of digital reconstructions. Neuroinformatics 9:143–157
51. Hanrahan O, Harris J, Egan C (2011) Advanced microscopy: laser scanning confocal microscopy. In: O'Driscoll L (ed) Gene expression profiling. Humana, Totowa, NJ, pp 169–180

Chapter 12

Genetic Pathways to Circuit Understanding in Drosophila

Jennifer J. Esch, Yvette E. Fisher, Jonathan C.S. Leong, and Thomas R. Clandinin

Abstract

Genetic tools enable a diverse array of experimental approaches to dissect the relationships between brain function and behavior. Many genetic tools have been developed for use in the fruit fly, *Drosophila melanogaster*, making it a powerful model system due to its rich behavioral repertoire on one hand and, on the other, the accessible size and scale of its highly stereotyped nervous system. The stereotypy of the fly brain, in particular, makes it possible to interrogate, in essence, the same cell types—genetically, morphologically, physiologically, molecularly, and behaviorally—across individual flies. In this chapter, we describe how genetic tools can be used to target specific cell types in the fly. We then illustrate how these tools can be used to measure or manipulate a cell type's activity, thereby shedding light on that cell type's function within an intact circuit. Finally, we describe how tools that directly manipulate genes in a cell-type-specific manner can be used to identify the molecular and cellular mechanisms underlying each cell's function.

Key words Neuron, *Drosophila melanogaster*, Binary expression system, Cell type, Silencing methods, Calcium imaging, Gene manipulation

1 Introduction

Thanks to the rapid development of new technologies and ever-growing interest, neuroscience has begun to link phenomenological understanding of neuronal computation and organismic behavior to mechanistic, biological underpinnings. Given its wealth of powerful genetic tools, the fruit fly, *Drosophila melanogaster*, has emerged as an apt model system for the study of neural circuits and their roles in behavior. Previous studies of the fly have made many fundamental contributions to neuroscience. Among these was the demonstration that individual genes can play central roles in shaping behavior and in specifying the development of a complex nervous system. The aim of this chapter is to describe the use of modern techniques to study the function of neuronal circuits, extending from the level of single neurons to the organismic level, behavior. As we will discuss, many of the techniques available to

Benjamin R. Arenkiel (ed.), *Neural Tracing Methods: Tracing Neurons and Their Connections*, Neuromethods, vol. 92, DOI 10.1007/978-1-4939-1963-5_12,

the *Drosophila* neuroscientist owe much to the rich history of *Drosophila* neurogenetics but are now evolving to answer old questions as well as pose new ones. A long-term goal of this effort is to establish causal relationships between the functioning of individual neurons and the computations in the rest of the brain.

The fly brain contains on the order of 100,000 neurons distributed over two hemispheres and organized into a series of neuropils. In flies, like in other organisms, one can group neurons that have similar morphology, functional response properties, or expression of molecular markers. Neurons that are grouped by any one feature commonly share their other defining features. Moreover, the fly brain is highly stereotyped from individual to individual, with each fly containing nearly the same number of neurons of each identified type [17, 30, 103]. The consequence of this stereotypy is that it is possible to study the same cell in every individual fly. Fly neuroscientists aim to identify and gain genetic access to cell types of interest and then use a wealth of tools to probe their response properties, connections, and behavioral functions.

Beyond this exquisite stereotypy of cell types, the fly has several additional attractive properties that have led to its adoption as a prominent model system for neuroscience. First, flies possess a diverse repertoire of behaviors, including courtship, locomotion, and learning. Second, like vertebrate neurons, fly neurons are connected by a mixture of chemical and electrical synapses, and many neurons are spiking. Finally, flies exemplify the best qualities of genetic model organisms: facile genetics, a wealth of molecular tools for genomic manipulation, short generation time, and simple laboratory culture. Taken together, these advantages set an exciting trajectory for understanding how groups of neurons compute.

To understand the computational role of particular neurons, one must first answer three critical questions. How does the activity pattern of a single cell type correlate with specific sensory stimuli or behavior? How does altering this pattern affect the activity of other neurons? How does this manipulation affect behavior? Answering these questions requires techniques that allow for cell-type-specific control and observation of neural activity, tools that are the focus of the first part of this chapter. Once a cell type has a demonstrated role in a computation, one may begin to ask more subtle questions about the circuit- and cell-level mechanisms that enable this function. What are the types of inputs that this cell receives? How are these inputs integrated? What signal does this cell transmit to downstream circuits? To answer these questions, it is helpful to manipulate genes in a cell-type-specific manner. Techniques for gene manipulation will be covered in the second part of this chapter.

2 Defining the Role of a Neuron

2.1 Achieving Cell-Type-Specific Genetic Labeling

It is a common experimental goal to obtain reproducible genetic access to a specific cell type of interest. There are two ways of gaining genetic access to specific cell types. The first takes advantage of the stereotyped sequence of cell divisions that gives rise to specific neural types in a defined temporal order. Expressing a DNA recombinase at specific times during development can give genetic access to cells that were born at those times. Specifically, the recombinase can cause a mitotic recombination event that changes the genetic makeup of a cell lineage. In this way, specific neurons can be genetically labeled and manipulated. A specific instantiation of this approach, called MARCM (mosaic analysis with a repressible cell marker), is outlined in greater detail later in the chapter [63].

The second approach to gaining genetic access to specific cell types takes advantage of developmental programs that construct cell-type-specific patterns of gene expression. Cell types, as defined by their morphology, location, number, and response properties, differentially express specific sets of genes. The specific expression of these genes is regulated by the combinatorial binding of transcription factors at specific binding sites within gene regulatory elements, termed enhancers. Thus, one may gain genetic control over a cell type by co-opting endogenous enhancer elements to express heterologous genes. Since most genes are not uniquely expressed in a single cell type, using the regulatory region of any single gene does not typically provide single cell-type specificity. Depending on one's experiment, the genetic manipulation of additional cell types may be inconsequential. Otherwise, additional genetic techniques can be used to achieve further cell-type specificity.

Introduction to Binary Expression Systems

One method for co-opting endogenous enhancer elements to express heterologous genes is to directly fuse enhancers to a coding sequence of interest and to introduce this transgene into the genome. The enhancers confer cell-type specificity to expression of the coding sequence of interest, while the coding sequence is chosen for its biological effect. Often, one would like to manipulate the same group of cells in different ways by expressing different proteins. Binary expression systems allow cell-type specificity of expression to be uncoupled from the identity of the expressed protein. This modularity is accomplished by placing expression of a heterologous transcriptional activator under the control of the enhancer of interest (the "driver") and then, in parallel, placing the expression of "effector" proteins under the control of regulatory elements recognized by the heterologous factor. Critically, the heterologous transcriptional activator and its cognate regulatory element are designed to be largely independent of endogenous gene regulation. Typically, the coincidence of driver and effector is

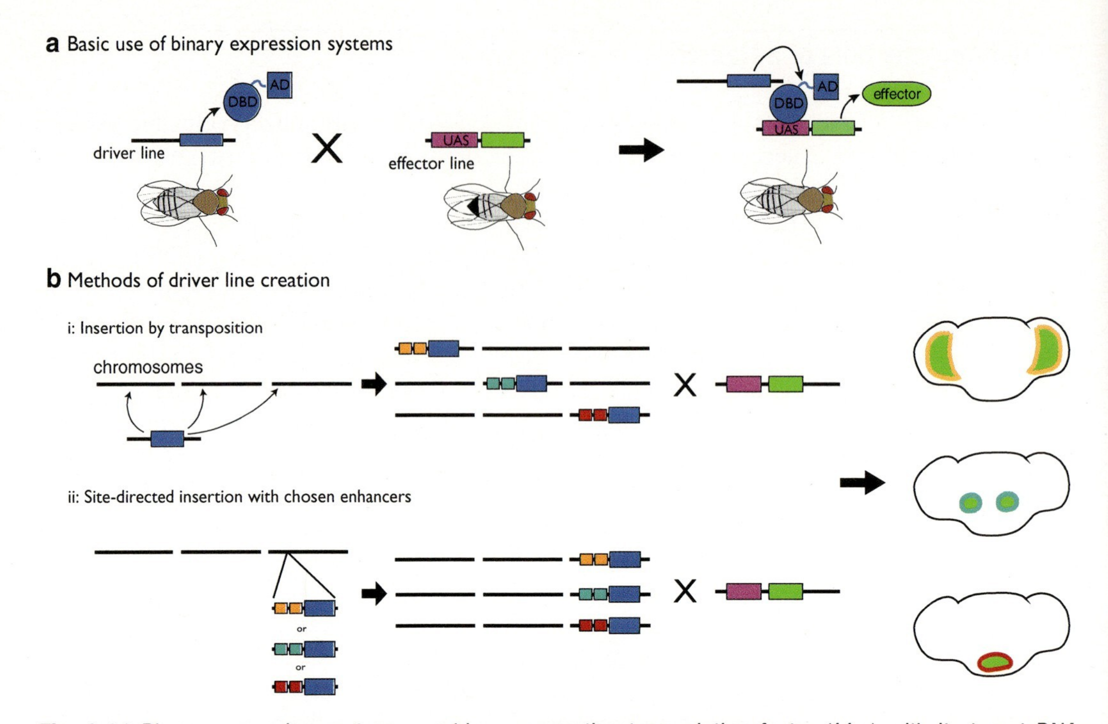

Fig. 1 (**a**) Binary expression systems combine a nonnative transcription factor (*blue*) with its target DNA sequence (upstream activating sequence (UAS), *magenta*) to control the expression of an effector protein. In cells that express the transcription factor, binding of the transcription factor with the factor's DNA-binding domain (DBD) to its cognate sequence positioned upstream of the coding sequence drives effector expression through the action of the factor's activation domain (AD). Flies that carry a binary system transcription factor, "driver lines," can be crossed to any appropriate "effector" line to drive expression of the effector in only the specific cells and tissues that have transcription factor expression. (**b**) *i*: Transposons, DNA sequences that can be mobilized using a transposase, can be used to insert DNA fragments throughout the genome. When this DNA includes the coding sequence of a nonnative transcription factor (*blue*), one can capitalize on the diverse positions of insertion events that capture the activities of different local enhancers (*orange*, *red*, *turquoise*), causing the transcription factor to be expressed under the control of the local regulatory sequences. These driver lines can then direct the expression of an effector sequence (*green*) that is just downstream of the nonnative transcription factor's cognate binding sequence (*magenta*). *ii*: By using specific landing sites already in the fly genome, combined with plasmid-based expression of the phiC31 integrase, one can insert the DNA fragment of choice into a specific genomic location. DNA of choice can include presumptive local enhancers from a gene of interest (*orange*, *red*, *turquoise*), a minimal promoter, and the sequence for a nonnative transcription factor (*blue*), to generate a driver line that should mimic the expression of the gene of interest. These driver lines can then direct the expression of an effector sequence (*green*) that is just downstream of the nonnative transcription factor's cognate binding sequence (*magenta*)

achieved simply by mating the fly strains that separately contain the driver and the effector transgenes (Fig. 1a).

There are three such binary systems in common use, each with different regulatory target sequences. The binary system that is used most widely is based on the yeast transcription factor GAL4 and its regulatory target sequence, UAS [10]. Two additional binary systems include the LexA-*lexAop* system and the QF-*QUAS* system [61, 94]. These binary systems can be used in the same fly to independently target two or more groups of cells. Because they are modular, the use of multiple copies of driver and effector transgenes provides semiquantitative control of effector dosage. In addition, many of these systems can be repressed by the expression of other specific factors. Finally, as we will discuss further below, these transcriptional binary systems can also be combined with recombinase-based binary systems to achieve further modularity.

Methods for Generating Cell-Type-Specific Drivers

There are two types of approaches to generating cell-type-specific driver lines. The first approach, known as enhancer trapping, requires a morphological description of a cell type of interest but requires no prior knowledge of gene expression patterns. Here, a binary system transcription factor is expressed in a pattern determined by a particular set of endogenous genomic enhancers. This ubiquitous tool is created by inserting a transgene encoding the transcription factor into the genome using transposition (Fig. 1bi). Transposition can be applied on large scale to produce libraries of fly strains bearing insertions at thousands of distinct insertion sites. Expression of each insertion will be influenced by its local genomic landscape. The resulting diversity of expression patterns can then be screened for expression within a cell type of interest by cell morphology (*see* below). Expression patterns can be diffuse, labeling many different neuron types distributed across the brain, but can also be restricted to single cell types. The two most commonly used transposons for making enhancer traps, P elements and PiggyBac elements, have different biases for insertion sites and thus tend to produce expression patterns that are more or less broad, respectively [9, 33, 86, 89]. Using driver lines created from transposable element insertions has been useful in both defining cell types and in generating tools for manipulating them.

Driver lines can also be constructed in one of two directed ways to recapitulate known patterns of gene expression. First, homologous recombination can be used to insert the DNA encoding the transcription factor directly into the genomic locus of a gene of interest, capturing its expression pattern. Such recombination can be done either into the endogenous locus, thereby replacing part of the gene's coding sequence with the transcription factor, or it can be done to an additional copy of the locus introduced on a bacterial artificial chromosome [14, 71, 104, 126]. Driver lines generated using this method are believed to recapitulate the expression pattern of the gene of interest, subject to some modest deviations [104, 126]. Second, a more commonly used approach is to fuse putative regulatory elements of a gene of interest directly to the coding sequence for the transcription factor DNA and then to insert this transgene into the genome, either using transposons [10, 47] (Fig. 1bi) or using integrases, such as phiC31, which target specific landing sites [89] (Fig. 1bii). While enhancer fusions were first introduced into the genome using transposable elements, directed insertion using integrases is now generally preferred. Nonetheless, as we do not fully understand all the factors that determine how enhancer elements control expression patterns, inserting transcription factors via homologous recombination into endogenous loci is likely the best strategy to precisely recapitulate a gene's expression patterns.

Characterizing Driver Lines at the Single-Cell Level

As many driver lines exhibit broad expression patterns, an experimental necessity is to determine which cell types are targeted by a driver. Given the stereotypy of the fly brain, characterizing the morphologies of all of the neurons labeled by a driver provides a useful catalog of the targeted cell types. Morphological characterization is typically accomplished by using the driver line to express a cellular label such as a membrane-tagged fluorescent protein. Given the small physical dimensions of the fly brain, entire cells, including all dendrites and axons, can be visualized in a single brain. Also owing to the compact size and stereotypy of the fly brain, many *Drosophila* neurons have been previously described and named, allowing driver lines to be mapped onto identified neurons [17, 30, 103].

The overlap of the processes of multiple cells in space can make it difficult to resolve the morphology of single cells optically, especially when an expression pattern labels many cell types in close proximity to one another. A number of recombinase-based genetic techniques have been developed to address this problem. These so-called "Flp-out" or "mosaic" techniques rely on the use of a recombinase, typically the yeast recombinase Flp, to stochastically change the genotype of individual cells, in this way sparsening their labeling [35, 51, 57, 63, 120, 126]. An alternative to labeling sparse subsets of cells targeted by a driver is to label each of the cells with a different effector using strategies inspired by the mouse "Brainbow" [37, 42, 68, 115].

Refining Driver Line Expression Patterns

It is rare for any single driver to target exactly one cell type, but it is relatively common to obtain a number of drivers with overlapping cell-type specificity. From overlapping drivers, it is often possible to create a new expression pattern that is more specific than that seen in either parent line. For example, a Boolean AND gate can be constructed genetically to target only the cell types common to both drivers. It is also possible to construct a Boolean NOT gate, whereby only the cell types unique to a single parent line are targeted. While several different molecular implementations are possible for a given logical operation, all of these operations rely on functional interactions of two or more proteins that are expressed under the control of different parent driver lines. Here we outline three commonly used strategies for refining driver line expression patterns.

One kind of genetic AND gate exploits the fact that binary system transcription factors like GAL4 or LexA contain two functional domains, a DNA-binding domain (DBD), which specifically targets the cognate activation sequence, and a separate transcriptional activation domain (AD), which is necessary for effector expression (Fig. 2a). The expression of each of these two domains can be placed under the control of distinct enhancer elements and the domains themselves engineered so that they will dimerize to reconstitute a fully functional transcription factor when expressed in the same cell. In this so-called "split" system, cells that express only one half of the transcription factor have no effector expression. Thus, by placing each half of the transcription factor under the control of different enhancer elements, the effector will be expressed only in the cell types in which both enhancers are active, creating a genetic intersection.

A second kind of a genetic AND gate makes effector expression contingent on both a DNA recombinase and a binary system transcription factor. A "stop" sequence, flanked by recombinase sites, is placed between the activation sequence recognized by the transcription factor and coding sequence of the effector (Fig. 2b), thereby preventing expression. In the presence of the recombinase, the stop sequence is removed, and the transcription factor drives effector expression. In this way, effector expression can be limited to the intersection of the expression patterns of a transcription factor driver (i.e., GAL4, LexA, QF) and a recombinase driver [4, 91, 126]. The utility of this method is often limited by the availability and efficiency of the recombinase driver line that expresses in the neurons of interest.

The most commonly used Boolean NOT gates use transcriptional repressors. For example, in the GAL4-UAS system, expression of the transcriptional repressor GAL80 blocks GAL4 activity [63]. Thus, when GAL4 and GAL80 are expressed in distinct but overlapping sets of cells, the effector is expressed in cells that express GAL4 but not GAL80. A temperature-sensitive variant of Gal80 affords additional temporal control. In the QF-QUAS system, the transcriptional repressor QS blocks QF activity (Fig. 2c) [94]. Furthermore, QS-mediated suppression can be inhibited by the small molecule quinic acid. When flies are fed quinic acid, QF activity is derepressed, allowing for additional temporal control.

Genetic intersections are intended to give reproducible refinement of cell-type-specific expression in every animal. Alternatively, it can be advantageous to refine driver line expression patterns stochastically such that a different subset of cells is labeled in each animal. In these methods, the activity of a binary system is modified by the variable expression of a recombinase, using either Flp-out or mosaic techniques, to create stochastic Boolean AND or NOT gates. These methods work well when the goal is to target a

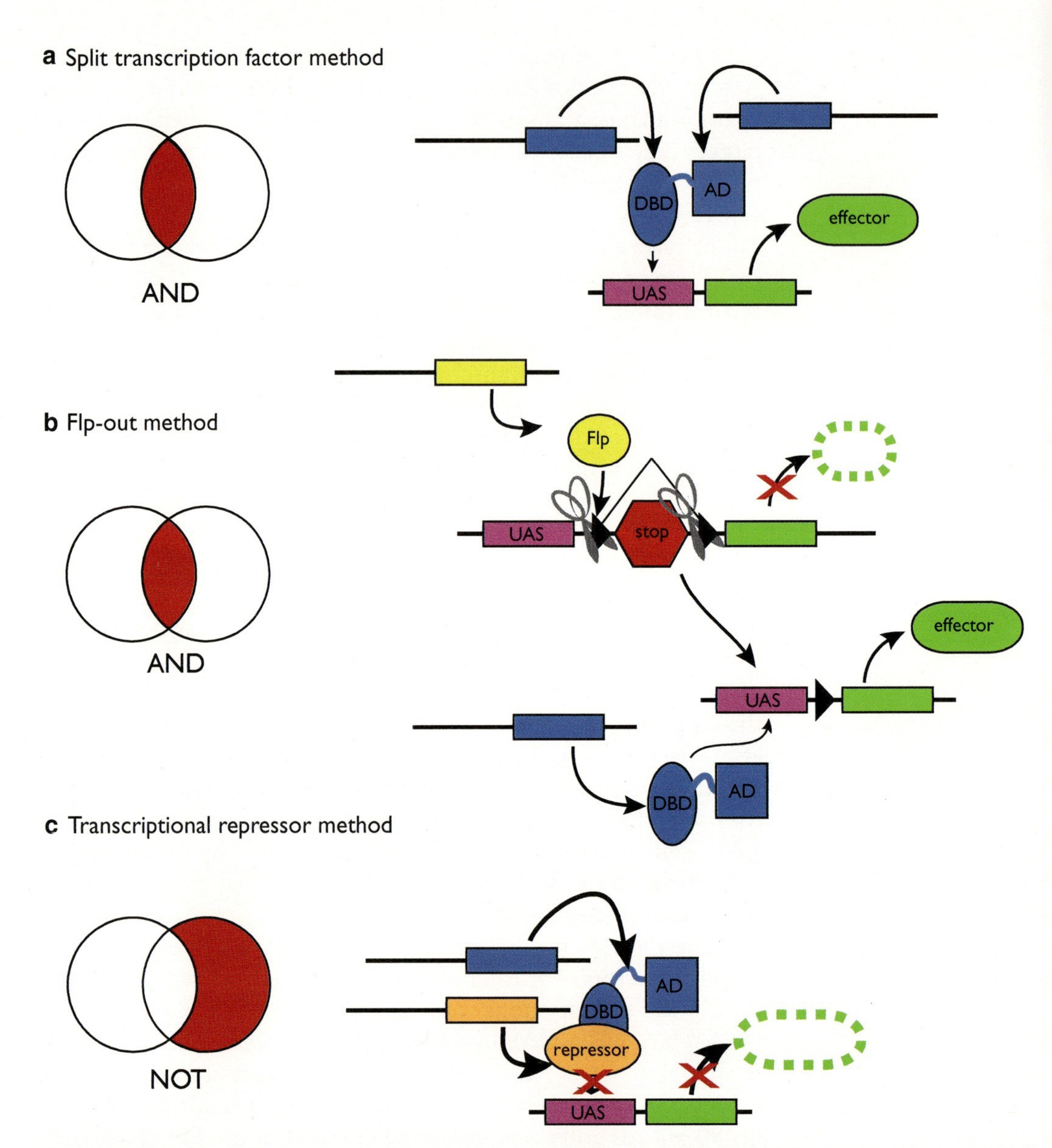

Fig. 2 (**a**) Expression of an effector is limited to cells that express both halves of a split driver system, a logical "AND" gate. One half of the split system includes the part of the transcription factor that binds to the specific activation sequence upstream of the effector sequence (DNA-binding domain, DBD). The other half of the system includes the machinery to activate transcription (activation domain, AD). Each component in this split system has a leucine zipper that allows dimerization when they are both expressed in the same cell, reconstituting an active transcription factor. (**b**) Expression of an effector is limited to cells that express both a recombinase (Flp) and a driver. A stop sequence between the activating sequence and the coding sequence of the effector prevents expression of the effector until the stop sequence is excised. Once FLP excises the stop sequence, expression of the effector can be driven as usual by the transcription factor. (**c**) Expression of an effector is limited to cells that have expression of a driver and do not have expression of a transcriptional repressor. Thus, the overall pattern of expression is of cells in which expression is driven by one line (the driver line) and not the other (the repressor line), a logical "NOT" gate

very small number of cells or even single neurons [35, 57]. When a cell type of interest is numerous, stochastic methods may only target a fraction of the population [80]. Depending on the redundancy of the individual cells within the cell type as well as specific experimental goals, stochastic labeling may enable some experiments but preclude others.

Choosing a Driver

At present, more than ten thousand driver lines exist, likely targeting the vast majority of cell types in the fly brain. Many of these driver lines enable repurposing of the same regulatory regions to drive different components, allowing for their use in intersectional strategies. In the FlyLight fly collection, noncoding and intronic regions of neurally expressed genes are fused to the GAL4 coding sequence and inserted in specific genomic landing sites [89]. By using specific landing sites, the same enhancer elements can be fused to different coding sequences and reinserted into the same site, allowing Gal80, split Gal4 components, or other sequences to be expressed in the same pattern [66, 69, 89, 99, 115]. Using this approach it is also possible to regenerate a fly line from the cloned DNA fragment. The landing sites used by FlyLight were chosen to maximize the expression of the inserted DNA sequences with minimal influence from the flanking genomic sequences. However, no genomic landing site is completely isolated from the surrounding chromosomal environment, and it is problematic to insert two different gene regulatory regions into the same landing site [74]. As a result, FlyLight-based intersectional approaches require multiple landing sites, each of which can differentially modify the expression pattern predicted from the characterization of the initial component enhancers.

Another method to generate intersection-ready drivers is based on the InSITE system (integrase swappable in vivo targeting element) [33]. This enhancer trap system allows for the in vivo replacement of a GAL4 element with any other component of a binary expression system (GAL80, LexA, QF, etc.) regardless of chromosomal location. Like all enhancer trap lines, InSITE drivers inherit their pattern of expression from the genomic regulatory regions surrounding their insertion site (Fig. 1bi). To exchange Gal4 for other DNA sequences, the InSITE system uses a series of recombinase- and integrase-mediated reactions that can be carried out in vivo through genetic crosses. As with FlyLight, care must be taken to confirm that newly created lines retain the expression pattern of the original tool. Different elements may differ in their expression pattern even when inserted into the same genomic location.

2.2 Making Measurements

Experimental biologists have two overarching goals: measuring biological systems and manipulating them. Highly specific genetic access, as afforded by binary systems and as refined by intersectional strategies, aids in both of these aims. Neurons transmit and receive numerous electrical and chemical signals across vast, intricate networks. From this perspective, an understanding of brain function at the level of a single cell type begins with a description of the cell type's synaptic connectivity and patterns of activity. Correlating patterns of electrical and chemical activity to sensory inputs, to the activity of other cell types, and to the behavior of the animal seed hypotheses about causal relationships between these phenomena.

Genetic Tools for Studying Cell Morphology and Connectivity

Synaptic connectivity is constrained by cell morphology, as neuronal processes must be within nanometer scale proximity in order to make synaptic contacts. While light microscopy cannot directly resolve synapses, cell-type-specific genetic access can be used to express fluorescent proteins and thereby reveal cell morphology [80]. In flies, the vast majority of morphological cell types have been catalogued, named, and registered to a reference brain [11, 17, 30, 73, 80, 87, 103, 105, 109, 111, 115, 126]. Such anatomical atlases generate predictions of pre- and postsynaptic partners for a cell type of interest, adding to our understanding of circuit architecture [7, 8].

In flies, genetically targeted fluorescent proteins can be imaged in living or fixed tissue. Confocal and multiphoton fluorescence imaging enables optical sectioning of the entire fly brain, largely eliminating the need for physical sectioning [22, 78, 114]. Fixed and cleared tissue provides higher quality images than living tissue, making fixed tissue imaging the preferred method for characterizing fine-scale morphology. Super-resolution imaging below the diffraction limit is only now being applied to the fly nervous system, opening future avenues for studying the subcellular localization of molecules critical to circuit function [92].

The ease of transgenesis in flies has catalyzed many efforts to improve molecular reagents, such as fluorescent proteins, for different applications. When a cytosolic fluorescent protein, such as GFP, is expressed in a cell of interest, the spatial extent of its expression is concentrated at the cell body and proximal processes. However, as fly neurons often have processes that are a fraction of a micron in diameter yet extend for microns, cytosolic labels fail to capture these minute processes. To solve this problem, fly neurobiologists have modified transgenes encoding fluorescent proteins to increase expression or to change subcellular localization. Different 3′ and 5′ untranslated regions (UTRs), as well as splicing signals, can result in transcripts of differing stability and lifetime, thus affecting protein levels [90]. Similarly, binary systems have also been optimized to increase transgene expression [89]. Additionally, fusion effector proteins, which combine the fluorescence of a cytosolic fluorescent protein with localization domains pirated from another protein, have been used extensively to target distal compartments of neurons. For example, fragments of the transmembrane immunoglobulin superfamily members CD2, CD4, and CD8 have been used to localize fluorescent proteins to the cell membrane [43, 63]. Another way to achieve membrane localization is to add a myristoylation tag [75, 82]. Similar strategies have also been used to generate fluorescent proteins with diverse localization patterns, targeting, for example, the nucleus [21], presynaptic sites [25, 128], mitochondria [19], and motor proteins [88, 93].

Finally, while levels of fluorescent marker expression that are too low may prevent visualization of a cell type of interest, excessive expression can lead to adverse effects, for example, protein aggregation, which can impede cell function, alter morphology, or even kill cells. Nonetheless, these caveats can be readily circumvented given the extensive molecular genetic toolkit available to the *Drosophila* neurobiologist. Thus, characterization of cell morphology is both fundamental and straightforward.

Photoactivatable and Photoconvertible Fluorescence

Photoactivatable and photoconvertible proteins are a class of fluorescent proteins that can be expressed in cell types of interest to ask specific questions about connectivity and morphology. Photoactivatable proteins, such as photoactivatable GFP (paGFP), are initially very dim and become brighter once exposed to an activating light pulse. Spatial restriction of the activating light using two-photon excitation allows these tools to highlight the morphology of those cells whose processes pass through the illuminated volume. While the use of these tools is limited to the description of projection patterns, not synaptic connectivity, projection tracing can be done in either anterograde or retrograde fashion by illuminating either axon terminals or dendrites, respectively. Photoactivation has been used to trace neuronal circuits from early olfactory processing centers deep into the brain [20, 31, 98]. Less widely used photoconvertible proteins, such as Kaede, shift their emission spectrum once exposed to a converting light pulse [15]. The advantage of photoconvertible proteins is that they allow the simultaneous visualization of both photoconverted and unphotoconverted cells, which can be used to segregate two cell types labeled by the same driver [36]. While photoactivatable and photoconvertible proteins have many exciting potential applications, technical limitations restrict the use of the current generation of reagents. Finally, even more recently developed photoswitchable proteins, whose emission spectra can be reversibly changed, give access to additional applications, particularly super-resolution microscopy [97]. These reagents will significantly expand the *Drosophila* neuroscientist's tool kit in the future.

Split-GFP

While fluorescent labeling of entire cells places a first-order constraint on synaptic connectivity, additional methods can be employed to identify synaptic contacts between cells. One method takes advantage of the fact that GFP can be split into two fragments that are individually inactive but reconstitute a functional fluorophore when in close proximity [32]. Thus, when such split-GFP partners are expressed as extracellular fusions with cell-surface proteins in separate cell populations, a fluorescent signal is obtained only when the two cells are in close apposition. This fluorescence, which may not always be unique to synaptic contacts, has been used to identify putative synaptic partners [35, 55]. Versions of split GFP that specifically localize the GFP components to pre- and postsynaptic sites seem to improve synaptic fidelity [26, 27].

Genetic Tools for Studying Neural Activity

The majority of genetic tools for measuring neuronal activity report calcium concentration, an indirect correlate of changes in membrane potential and chemical synaptic transmission. Owing to the ease of transgenesis in flies, nearly all existing genetically encoded calcium indicators (GECIs) are available as binary system effector lines. Protein engineering has also been directed toward developing voltage-sensitive fluorescent proteins and reporters of other forms of neuronal activity. Here we navigate the ever-growing stock of activity reporters and discuss specific considerations that direct the experimental applications of these tools.

Calcium Indicators

When expressed in cell types of interest, calcium sensors transduce changes in local calcium concentration into changes in fluorescence. A host of cellular mechanisms regulate intracellular calcium concentration. In the axon terminal, calcium signals serve as relatively direct reporters of synaptic transmission, as calcium influx drives synaptic vesicle fusion. Elsewhere in the cell, the relationship between calcium signals and neuronal activity can be more complex, reflecting postsynaptic currents, action potentials, and other conductances. In the fly, some neurons fire action potentials, while other neurons exhibit less stereotyped excitatory regenerative events, collectively termed spikes. Yet other neurons, especially those in early sensory neuropils, exhibit graded potentials. GECIs have been used to observe these types of activity [28, 95, 117].

Thus far, GECIs have been derived from non-calcium-sensitive fluorescent proteins using two different strategies. In one approach, two distinct fluorescent proteins are fused together via a calcium-sensitive linker domain [45, 70, 79, 112]. Fluorescence excitation and emission spectra are chosen so as to permit a Forster resonance energy transfer (FRET) between the fused fluorescent proteins, conditional upon their appropriate close steric apposition, which is in turn dependent on a calcium-sensitive linker. Early sensors of this variety included the cameleon-type sensors, as well as TN-XXL [70]. Since FRET donor fluorescence changes inversely with FRET acceptor fluorescence, FRET-based sensors give ratiometric readouts of calcium concentration. Ratiometric measurements are advantageous because the ratio signal magnitude is independent of sensor concentration and the light-scattering properties of the specimen, all of which can vary with heterogeneous samples. Thus, ratiometric measurements immediately allow for a quantitative measurement of intracellular calcium. For in vivo imaging applications, movement artifacts can easily contaminate recorded neuronal activity and become a significant challenge to data acquisition and interpretation. These problems are prominent in awake, behaving preparations, when motor behaviors can cause large, sporadic movement artifacts. Ratiometric indicators mitigate the contamination of the neural signal by movement, since a true ratiometric change is distinguished by anti-correlated fluorescence changes, whereas a movement artifact would more likely produce correlated fluorescence changes. Despite the many advantages of ratiometric calcium indicators, many groups prefer non-ratiometric alternatives due to their large signal sizes.

The second class of GECIs is based on the fusion of a circularly permuted version of GFP with a calcium-sensing domain [1–3, 83, 106, 113, 118, 129]. Calcium binding results in a structural change within the fusion protein causing fluorescence to increase. The most commonly used GECI in this class is called GCaMP. In recent years, a number of groups have improved the utility of GCaMPs through structural investigation into their function and through directed evolution. As a result, a vast array of GCaMP variants has been described, each with different fluorescence properties and calcium-binding characteristics.

From a practical standpoint, GCaMPs differ along just a few important axes: calcium affinity, onset and decay kinetics of calcium binding and unbinding, magnitude of fluorescence change in response to calcium, baseline fluorescence intensity, and excitation and emission wavelengths. Calcium affinity should be matched to intracellular calcium concentration to ensure that the sensor's dynamic range is matched to the levels of calcium that will be observed. Similarly, sensor kinetics should be matched, when possible, to the kinetics of the signals being measured. Current indicators have sufficiently large signals and rapid kinetics to faithfully detect individual action potentials at some physiologically relevant firing rates [16]. While the vast majority of GCaMPs are green emitters, blue- and red-shifted GCaMPs are also available. Despite these many improvements, three core challenges remain in the application of these GECIs to the study of the fly brain. First, high signal-to-noise ratio GECIs tend to have low baseline fluorescence intensity, making identification of inactive cells difficult. Second, since these reporters were designed to facilitate spike detection, they generally respond strongly to increases in calcium concentration but relatively weakly to decreases. Third, in many parts of the fly brain, computation is believed to take place on time scales faster than the kinetics of the fastest calcium sensors. Nonetheless,

GCaMP imaging has become a standard for functional measurements, particularly in settings where electrophysiology is difficult or where subcellular spatial information is desired.

Other Measures of Activity

Measurements of intracellular calcium can be complemented by measurements of other physiologically distinct forms of neuronal signaling, for example, changes in membrane potential and synaptic vesicle release. These signals can also be measured using genetically encoded sensors. Neurons use membrane voltage to integrate synaptic inputs and rapidly send signals across long distances. Genetically encoded voltage indicators (GEVIs) have been used to track changes in membrane voltage across the complex arbor and cell body of clock neurons that underpin circadian rhythmicity [13]. Measures of synaptic vesicle fusion complement measures of calcium concentration and membrane voltage by providing a more direct measure of neurotransmitter release. The genetically encoded indicator synaptopHluorin is trafficked to synaptic vesicles and fluoresces when vesicles fuse with the cell membrane, exposing the reporter to a more neutral extracellular pH than the acidic environment inside vesicles [76]. This reporter has been used to quantify presynaptic inhibition of olfactory receptor neurons by a neuropeptide [49]. As genetically encoded activity sensors of neuronal activity become easier to use and more diverse, they will enable an increasingly nuanced view of neuronal computation at the subcellular and circuit levels.

2.3 Making Manipulations

To assess the role of a neuron within its circuit context, it is useful to cell-type specifically manipulate activity and then observe the consequences of this manipulation in the same neurons, in other neurons, or as manifested in behavior. Cell-type-specific genetic drivers can be used to express proteins that alter physiology in vivo. In the fly, neural plasticity, either morphological or physiological, appears to be the exception rather than the rule, with only sparse indications that changes in activity effect longer-lasting changes in brain function or behavior. Thus, current techniques used to manipulate neuron function, either brief or long lasting, can be informative.

Manipulation of Synaptic Release

One way to probe the function of a neuron is to abolish its synaptic output. This manipulation tests whether the synaptic output of a given neuron or class of neurons—and by inference the computation being performed by these neurons—is necessary for the normal activity of other neurons or for behavior. For example in the visual system, synaptic block of output from lamina monopolar cells L1 and L2 reveals their necessity in the detection of moving light and dark edges, respectively, as measured both by electrophysiological responses of downstream neurons and optomotor behavior [18, 52]. Two different genetic methods are commonly used to specifically block chemical synaptic transmission. First, synaptic output can be blocked by expressing tetanus toxin light chain [108]. Tetanus toxin disrupts synaptic vesicle fusion by proteolytically cleaving synaptobrevin, producing a permanent block of synaptic transmission. However, it is important to note that expressing tetanus toxin in cells of interest can also alter the trafficking of cell surface proteins and cause developmental defects [46]. One solution to this problem is to temporally restrict the expression of tetanus toxin using more refined genetic control, as discussed previously. A more temporally precise way to abolish synaptic output is to express a dominant negative temperature-sensitive form of dynamin, a GTPase that mediates vesicle endocytosis, including synaptic vesicle recycling [59]. When shifted from a permissive low temperature to a restrictive high temperature, this mutant protein, called shibire^{ts}, blocks synaptic vesicle recycling, causing vesicle depletion. As a result, synaptic transmission is severely impaired, if not abolished. However, as with tetanus toxin light chain, high levels of shibire^{ts} expression can also have nonspecific deleterious effects on cell function [34].

Manipulation of Membrane Potential

Another common strategy for testing neuron function is to manipulate membrane conductance by overexpressing an ion channel. This manipulation has the advantage of affecting both chemical and electrical synapses and can be made inducible by using channels with known temperature-, ligand-, or light-gated activity. Kir2.1 is an inward-rectifier potassium channel whose expression holds neurons at hyperpolarized potentials [6]. Other potassium channels and potassium channel mutants have also been used for this purpose [119] and for inducing the opposite effect, hyperexcitability [81]. Overexpression of sodium channels (i.e., NaChBac) can also achieve hyperexcitability [85].

A number of genetic tools allow for the rapid and reversible manipulation of membrane conductances. The cation channel TRPA1 depolarizes cell types of interest in response to temperature elevation [41]. The P2X receptor is a nonselective cation channel that is gated by ATP [65]. As ATP can be applied in an inactive, caged form and then uncaged by light, P2X receptors can be opened in a spatially and temporally precise manner. Finally, just as in many other model systems, other optogenetic manipulations based on channelrhodopsin and halorhodopsin are gated by different wavelengths of light and serve as additional tools for probing neuronal function [38, 50, 100]. Because light penetrates the fly cuticle, neurons throughout the brain can be optogenetically manipulated without surgery [65, 116].

Forward and Reverse Neuronal Screens

To identify the functional contributions of individual cell types to behavior, a common approach is to start with a driver that targets a large group of neurons whose activation or inactivation produces a specific behavioral phenotype of interest. Refinement of the driver line expression pattern, through intersectional or other strategies as discussed above, then tests the contribution of more specific subsets of neurons to the behavioral phenotype. For example, among neurons expressing *fruitless*, a master regulator of sexual dimorphism in the brain, olfactory receptor neurons were shown to be critical for male courtship behavior [104]. More recently, this approach has been used to identify other individual cell types critical for such diverse behaviors as post-mating female behavior, male song, feeding, aggression, and more [5, 72, 91, 123].

An alternative approach to identifying the functional roles of specific cell types takes inspiration from forward genetics. In a forward genetic screen, a large collection of genetic mutants are tested for a specific phenotypic change. Individual genes can then be inferred to contribute to specific traits. By analogy, large collections of transgenic flies can be used to drive effector expression in diverse, often sparse, sets of cells and screened for a specific behavioral change [35, 53, 102]. Such an approach allows functionally relevant neurons to be identified without a priori knowledge of their contribution to a particular phenotype.

3 Defining the Role of a Gene

Until now we have emphasized the methods used to elucidate relationships between brain function and behavior. What molecular mechanisms underlie brain function at the level of single neurons? The defining properties of a neuronal cell type—the identity and nature of its synaptic connections and its electric and biochemical properties—are all shaped by genes. Thus, to understand the molecular implementation of neuronal computation in cell types of interest, one can ask what role a gene of interest plays in these cell types. Some mutations affect specific behaviors even when the entire animal is mutant, for example, circadian rhythmicity, courtship, and learning [24, 40, 60]. These rare cases involve genes whose function is apparently restricted to a defined neural circuit. However, the development and function of most cells relies on the use of common subsets of genes. Because these genes play vital roles in many cell types, mutations affecting these genes are pleiotropic, making phenotypes difficult to interpret. In this section, we will review methods that describe how genes instruct brain function, specifically at the level of genetically accessible cell types.

A cornerstone in the rich history of *Drosophila* genetic manipulation is the use of clonal mosaic analysis to study the role of particular genes in specific tissues. In classic studies, flies were created bearing both a stable X chromosome containing a recessive mutation of interest and a mitotically unstable X chromosome that

Fig. 3 Schematic of mosaic gynandromorph flies that can be produced using a mitotically unstable X chromosome. XX tissue is shaded and XO tissue is unshaded to mark the clone region of the fly where the X chromosome recessive mutation of interest is uncovered (inspired by experiments in [48])

would be lost in mitosis with some frequency [48, 107]. Such animals were genetically mosaic, creating in each animal a stochastic distribution of XO tissue in an otherwise XX background and revealing the recessive phenotype when one X chromosome was lost (Fig. 3). Over multiple flies, the mutant phenotype could then be correlated with the varying distribution of mutant tissue. This method was used to map the functions of many genes onto neural tissue [48, 101]. Subsequent work used X-rays to extend these methods to other chromosomes in order to study neuronal development and fate. Improvements in imaging techniques allowed for analysis at the resolution of single cells. For example, clonal analysis of photoreceptors in the retina revealed the cell autonomous action of the gene *sevenless* in the developmental specification of a specific photoreceptor type [12].

Today, there are a number of tools for cell-type-specific gene manipulation. These tools combine binary systems, RNA interference, clonal analysis, and site-specific recombination. While many of these tools have been instrumental in defining the developmental roles of specific genes, they can also be used to understand how the functional properties of neurons emerge from their genetic programs.

3.1 Gene Manipulation Using Binary Systems

RNA Interference

Binary systems can be used to study gene function through the manipulation of endogenous gene activity within specific cells. One common goal is to disrupt the native function of a gene. This manipulation is most commonly achieved using genetically encoded RNA interference tools. RNA-mediated gene interference (RNAi) takes advantage of an endogenous biological process that targets specific mRNA sequences for degradation [29]. To implement RNAi under binary system control, inverted repeats of a fragment of a target gene are used to make a transgene that produces a double-stranded "hairpin" RNA molecule when the transgene is transcribed (Fig. 4a) [54]. Large public transgenic RNAi libraries now allow the targeting of almost any gene in the *Drosophila* genome [23, 84].

Whereas effectors like shibire^{ts} abolish all chemical synaptic output, RNAi has been used to investigate the contribution of various neurotransmitter receptors, and thus of specific kinds of neural input, to the response properties of particular cell types.

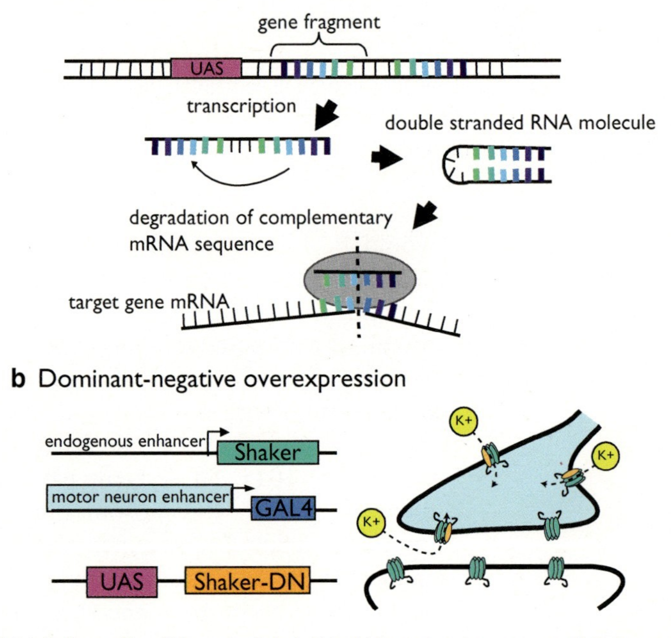

Fig. 4 (**a**) Schematic of the principle behind binary system compatible transgenic RNAi tools. A transgene is designed that produces a double-stranded hairpin RNA to knock down the target gene. (**b**) Binary system overexpression of a dominant negative form of Shaker allows for presynaptic specific manipulation of Shaker function and thus excitability [81]

For example, the RNAi knockdown of GABA-B receptors in olfactory receptor neurons reduced levels of presynaptic inhibition and impaired odor detection in a behavioral assay [96]. Other studies have taken similar approaches to assess the importance of other neurotransmitters or neuropeptides onto other cell types [49, 56, 121]. Because RNAi is relatively easy to use, RNAi screens for genes that play important roles in behavior have provided molecular entry points for subsequent circuit analysis [44, 125].

Because many cell-type-specific driver lines are expressed late in development at only modest levels, RNAi knockdown may be incomplete. As a consequence, it is important to validate the efficacy of gene knockdown. In cases where the manipulated cell type is anatomically prominent, immunohistochemical staining can provide a qualitative assessment of RNAi efficacy [96]. However, this approach works less well for manipulated cell types that are found along with many nontargeted types, and while it is possible to perform quantitative RT-PCR on single cell types in the brain [77, 110, 124], this technique has not yet been used to validate an RNAi knockdown. Another important caveat to RNAi is one of specificity: the hairpin RNAi molecule can affect the expression of multiple genes. The most common strategy for addressing this concern is to confirm results using multiple RNAi transgenes targeting different fragments of the same gene. When cell-type-specific validation of RNAi knockdown is inconclusive, alternative manipulations that affect the activity of the same gene, such as overexpression or traditional mutations in the locus, can help to increase confidence in a finding.

Expression of Dominant Negative Proteins

Dominant negative mutations provide an alternative means of reducing gene activity in a cell-type-specific manner. A dominant negative mutation results in a protein that no longer performs its wild-type function and that also compromises the activity of the wild-type protein. Effector lines can be designed to express dominant negative gene products under the control of a binary system, allowing them to "override" the function of wild-type endogenous protein in specific cell types [81](Fig. 4b). A good example of the use of dominant negative tools in the dissection of circuit function comes from the study of activity-dependent changes at the neuromuscular junction in *Drosophila*. Here a dominant negative allele of Shaker (Sh), a voltage-gated potassium channel, was used to increase excitability cell-type specifically (Fig. 4b) [81]. Presynaptic but not postsynaptic expression of the Shaker dominant negative drives expansion of axon arbors, providing strong evidence that activity-dependent retrograde signaling is dependent on postsynaptic receptor activation, as opposed to postsynaptic voltage [81]. While using binary systems to express dominant negative transgenes provides an elegant way to probe gene function in specific cell types, the application of this technique is limited to genes for which dominant negative alleles have been identified.

As a complement to gene disruption, binary system-mediated overexpression of wild-type genes can augment the native expression of genes in specific cells. Gene overexpression can probe whether a process of interest depends on the level of endogenous gene activity. For example, cell-type-specific overexpression of a neuropeptide receptor caused flies to behave as if they perceived lower levels of odorants, demonstrating a modulatory role for the corresponding neuropeptide in olfaction [49]. Binary system overexpression can also be used in combination with genomic mutations. Reintroducing a gene into a subpopulation of neurons within an otherwise genetically deficient background probes the sufficiency of the gene in a particular process in a given group of cells. This approach has been used extensively to study the circuit mechanisms underlying learning and memory [67, 127].

3.2 Recombinase-Mediated Genome Manipulation

Another class of cell-type-specific genome manipulation takes advantage of recombinases, specialized enzymes that mediate either inter- or intrachromosomal DNA rearrangements. Here we discuss the practical applications of these two kinds of recombinase-mediated reactions.

Interchromosomal Rearrangements: Mosaic Analysis

Interchromosomal DNA rearrangement requires the placement of two recombinase target sites on homologous chromosomes and expression of the recombinase before cell division (Fig. 5a). The exchange of homologous DNA strands prior to mitosis can create somatic mosaic animals, with cells homozygous for a mutation of interest within otherwise heterozygous tissue. As a result, the effects of mutations can be studied in very specific cell types. While many alternative strategies exist for identifying the clone [122], the most commonly used mosaic analysis technique in the brain is mosaic analysis with a repressible cell marker (MARCM) [63]. MARCM uses a repressible binary system to specifically label daughter cells that have become homozygous mutant for a gene following Flp-mediated mitotic recombination (Fig. 5b). In this scheme, all cells that have remained heterozygous express the binary system repressor, while those that have become homozygous mutant have lost the repressor, allowing the binary system driver to express effectors that mark the cell. MARCM and its variants [61–63, 94] have been used extensively to interrogate the cell autonomous and non-autonomous roles of genes in neuronal development. MARCM has also been used to examine cell-type-specific requirements for fruitless function in courtship behavior [57, 58]. However, MARCM's application to behavior is challenging. Many behaviors depend on the function of a given gene in many cells, so that a stochastic targeting method like MARCM will reach all of these cells together only rarely. Also, even when the number of cells that must be made homozygous to see a behavioral phenotype is relatively small, individual experimental animals must be

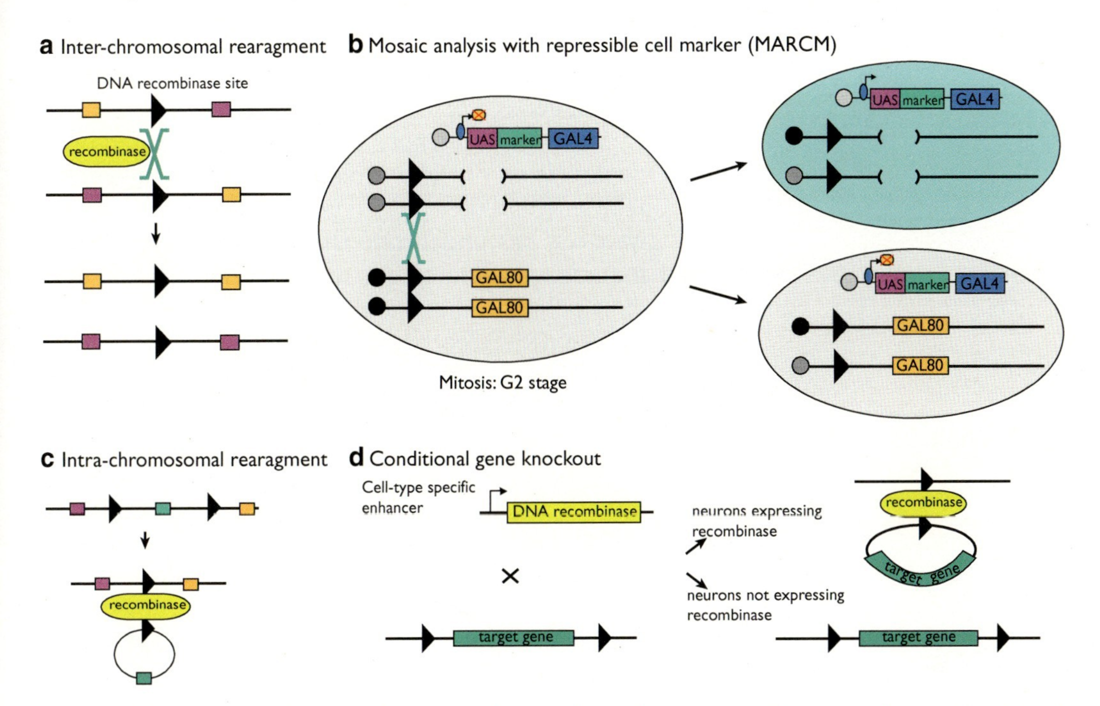

Fig. 5 (**a**) Interchromosomal interactions can be induced by placing corresponding recombinase sites on each chromosome and expressing recombinases prior to cell division. (**b**) Mosaic analysis with a repressible cell marker (MARCM). MARCM uses the loss of the Gal4 repressor Gal80 to specifically label daughter cells that have become homozygous mutant for a recessive gene of interest following mitotic recombination. All cells that have remained heterozygous express the binary system repressor (and are unlabeled), while those that have undergone the recombination event and become homozygous mutant have lost the repressor, allowing the cell to become marked by an effector. (**c**) Intrachromosomal rearrangement: by placing multiple recombination sites on the same chromosome DNA manipulations such as targeted deletions are possible. (**d**) Conditional gene knockout. By flanking a gene or exon with recombinase sites, a conditional allele is created in which the gene will only be disrupted in cell types that express the recombinase

characterized anatomically as well as behaviorally in order to correlate behavioral phenotype with cellular genotype. On the other hand, MARCM may be very useful for imaging or electrophysiological studies of neural activity. In such experiments, the responses of mutant cells can be compared to the responses of wild-type cells, potentially in the same individual.

Intrachromosomal Rearrangements to Create Gene Knockouts

Whereas the placement of recombinase sites on homologous chromosomes results in interchromosomal rearrangement, intrachromosomal rearrangements can be generated by placing recombinase sites at nearby locations on the same chromosome (Fig. 5c). When the recombinase target sites are in the same orientation, recombination results in removal of the intervening sequence. Thus, when these sites flank a gene of interest, this removal creates a gene knockout (Fig. 5d). This manipulation can be made conditional via spatial and temporal restriction of recombinase expression.

This form of genome editing has set the gold standard for creating conditional knockouts in mice for over a decade. Many transgenic mice exist in which genes or exons have been flanked by recombinase sites [64]. Such gene knockout tools have seen limited use in flies [39]. One limitation of these tools in flies is that recombination can be inefficient, and there are no general means to determine whether a given cell has undergone the rearrangement. Ideally, conditional mutations would target the same neurons in every animal, a level of precision that should be expected in *Drosophila* neurobiology given the stereotypy of the brain.

4 Outlook

An already extensive and still growing repertoire of genetic tools is enabling *Drosophila* neuroscientists to elucidate brain function and behavior at the level of single cell types and their connections. Indeed, there is hope that we may eventually understand the function of every neuron in the fly brain. As the pattern of connectivity is highly, if not completely, stereotyped in many fly brain areas, a description of the activity of every cell type, combined with a functional and anatomical connectome, may explain a great deal of animal behavior. However, across neuroscience, there is a growing awareness of the need to consider the coordinated, simultaneous activity of multiple cells distributed across a circuit. To achieve this end, new techniques to measure the activity of multiple cell types and brain regions at the same time would be immensely useful. Similarly, the ability to identify and genetically target neurons that are anatomically or functionally connected would enable experiments that target widely distributed circuitry. These approaches will be particularly useful in brain regions where neuronal connectivity appears to be less stereotyped. The tools used to unravel these complex circuits, and the insights gained from their analysis, will be instrumental in understanding brain function widely, in the fly and beyond.

References

1. Akerboom J et al (2009) Crystal structures of the GCaMP calcium sensor reveal the mechanism of fluorescence signal change and aid rational design. J Biol Chem 284(10): 6455–6464
2. Akerboom J et al (2013) Genetically encoded calcium indicators for multi-color neural activity imaging and combination with optogenetics. Front Mol Neurosci 6:2
3. Akerboom J et al (2012) Optimization of a GCaMP calcium indicator for neural activity imaging. J Neurosci 32(40):13819–13840
4. Alekseyenko OV et al (2013) Single dopaminergic neurons that modulate aggression in Drosophila. Proc Natl Acad Sci U S A 110(15): 6151–6156
5. Asahina K et al (2014) Tachykinin-expressing neurons control male-specific aggressive arousal in Drosophila. Cell 156(1–2): 221–235
6. Baines R et al (2001) Altered electrical properties in Drosophila neurons developing without synaptic transmission. J Neurosci 21(5):1523–1531

7. Bausenwein B, Fischbach K-F (1992) Activity labeling patterns in the medulla of Drosophila melanogaster caused by motion stimuli. Cell Tissue Res 270(1):25–35
8. Bausenwein B, Müller NR, Heisenberg M (1994) Behavior-dependent activity labeling in the central complex of Drosophila during controlled visual stimulation. J Comp Neurol 340(2):255–268
9. Bellen HJ et al (2011) The Drosophila gene disruption project: progress using transposons with distinctive site specificities. Genetics 188(3):731–743
10. Brand AH, Perrimon N (1993) Targeted gene expression as a means of altering cell fates and generating dominant phenotypes. Development 118(2):401–415
11. Cajal R (1909) Histologie du système nerveux de l'homme et des vertébrés
12. Campos-Ortega JA, Jfirgens G, Hofbauer A (1979) Cell Clones and Pattern Formation : Studies on sevenless, a Mutant of Drosophila melanogaster. Wilhelm Roux's Arch 50: 27–50
13. Cao G et al (2013) Genetically targeted optical electrophysiology in intact neural circuits. Cell 154(4):904–913
14. Chan C et al (2011) Article Systematic Discovery of Rab GTPases with Synaptic Functions in Drosophila. Curr Biol 21(20): 1704–1715
15. Chen C-C et al (2012) Visualizing long-term memory formation in two neurons of the Drosophila brain. Science (NY) 335(6069): 678–685
16. Chen T-W et al (2013) Ultrasensitive fluorescent proteins for imaging neuronal activity. Nature 499(7458):295–300
17. Chiang A-S et al (2011) Three-dimensional reconstruction of brain-wide wiring networks in Drosophila at single-cell resolution. Curr Biol 21(1):1–11
18. Clark D et al (2011) Defining the computational structure of the motion detector in Drosophila. Neuron 70(6):1165–1177
19. Cox RT, Spradling AC (2003) A Balbiani body and the fusome mediate mitochondrial inheritance during Drosophila oogenesis. Development 130(8):1579–1590
20. Datta SR et al (2008) The Drosophila pheromone cVA activates a sexually dimorphic neural circuit. Nature 452(7186):473–477
21. Davis I et al (1995) A Nuclear GFP That marks Nuclei in Living Drosophila Embryos: Maternal Supply Overcomes a Delay in the Appearance of Zygotic Fluorescence. Dev Biol 170(2):726–729
22. Denk W, Strickler J, Watt W (1990) Two-Photon Laser Scanning Fluorescence Microscopy. Science 248:73–76
23. Dietzl G et al (2007) ARTICLES A genome-wide transgenic RNAi library for conditional gene inactivation in Drosophila. Nature 448:151–156
24. Dudai Y, Jan Y (1976) dunce, a mutant of Drosophila deficient in learning. Proc Natl Acad Sci U S A 73(5):1684–1688
25. Estes PS et al (2000) Synaptic localization and restricted diffusion of a drosophila neuronal synaptobrevin - Green fluorescent protein Chimera in vivo. J Neurogenet 1(4): 233–255
26. Fan P et al (2013) Genetic and neural mechanisms that inhibit Drosophila from mating with other species. Cell 154(1): 89–102
27. Feinberg EH et al (2008) Neurotechnique GFP reconstitution across synaptic partners (GRASP) defines cell contacts and synapses in living nervous systems. Neuron 57:353–363
28. Fiala A et al (2002) Genetically Expressed Cameleon in Drosophila melanogaster Is Used to Visualize Olfactory Information in Projection Neurons. Curr Biol 12(02):1877–1884, http://www.sciencedirect.com/science/article/pii/S0960982202012393
29. Fire A et al (1998) Potent and specific genetic interference by double-stranded RNA in Caenorhabditis elegans. Nature 391(February):806–811
30. Fischbach K-F, Dittrich APM (1989) The optic lobe of Drosophila melanogaster. I. A Golgi analysis of wild-type structure. Cell Tissue Res 258:441–475
31. Fişek M, Wilson RI (2014) Stereotyped connectivity and computations in higher-order olfactory neurons. Nat Neurosci 17(2):280–288
32. Ghosh I et al (2000) Antiparallel leucine zipper-directed protein reassembly: application to the green fluorescent protein department of molecular biophysics and biochemistry the dissection and subsequent reassembly of a protein from peptidic fragments provides an avenue for. Mol Biol Cell 11:5658–5659
33. Gohl DM et al (2011) A versatile in vivo system for directed dissection of gene expression patterns. Nat Methods 8(3):231–237
34. Gonzalez-Bellido PT et al (2009) Overexpressing temperature-sensitive dynamin decelerates phototransduction and bundles microtubules in Drosophila photoreceptors. J Neurosci 29(45):14199–14210
35. Gordon MD, Scott K (2009) Motor control in a Drosophila taste circuit. Neuron 61(3): 373–384
36. Grueber WB et al (2007) Projections of Drosophila multidendritic neurons in the central nervous system: links with peripheral dendrite morphology. Development (Camb) 134(1):55–64

37. Hadjieconomou D et al (2011) Flybow: genetic multicolor cell labeling for neural circuit analysis in Drosophila melanogaster. Nat Methods 8(3)
38. Haikala V et al (2013) Optogenetic Control of Fly Optomotor Responses. J Neurosci 33(34):13927–13934
39. Hakeda-Suzuki S et al (2011) Golden Goal collaborates with Flamingo in conferring synaptic-layer specificity in the visual system. Nat Neurosci 14(3):314–323
40. Hall JC (1978) Courtship among males due to a male-sterile mutation in Drosophila melanogaster. Behav Genet 8(2):125–141
41. Hamada FN et al (2008) An internal thermal sensor controlling temperature preference in Drosophila. Nature 454(7201):217–220
42. Hampel S et al (2011) Drosophila Brainbow: a recombinase-based fluorescence labeling technique to subdivide neural expression patterns. Nat Methods 8(3):253–259
43. Han C, Jan LY, Jan Y-N (2011) Enhancer-driven membrane markers for analysis of non-autonomous mechanisms reveal neuron-glia interactions in Drosophila. Proc Natl Acad Sci U S A 108(23):9673–9678
44. Häsemeyer M et al (2009) Sensory neurons in the Drosophila genital tract regulate female reproductive behavior. Neuron 61(4): 511–518
45. Heim N, Griesbeck O (2004) Genetically encoded indicators of cellular calcium dynamics based on troponin C and green fluorescent protein. J Biol Chem 279(14):14280–14286
46. Hiesinger PR et al (1999) Neuropil pattern formation and regulation of cell adhesion molecules in Drosophila optic lobe development depend on synaptobrevin. J Neurosci 19(17):7548–7556
47. Horn C et al (2003) piggyBac-based insertional mutagensis and functional insect genomics. Genetics 163(2):647–661
48. Hotta Y, Benzer S (1970) Genetic Dissection of the Drosophila Nervous System by Means of Mosaics. Proc Natl Acad Sci U S A 67(3): 1156–1163
49. Ignell R et al (2009) Presynaptic peptidergic modulation of olfactory receptor neurons in Drosophila. Proc Natl Acad Sci U S A 106(31):13070–13075
50. Inada K et al (2011) Optical dissection of neural circuits responsible for Drosophila larval locomotion with halorhodopsin. PLoS One 6(12):29019
51. Jefferis G et al (2001) Target neuron prespecification in the olfactory map of Drosophila. Nature 414(November):204–208
52. Joesch M et al (2010) ON and OFF pathways in Drosophila motion vision. Nature 468(7321):300–304
53. Katsov AY, Clandinin TR (2008) Motion processing streams in Drosophila are behaviorally specialized. Neuron 59(2):322–335
54. Kennerdell JR, Carthew RW (2000) Heritable gene silencing in Drosophila using double-stranded RNA. Nat Biotechnol 18(8): 896–898
55. Kim J et al (2012) mGRASP enables mapping mammalian synaptic connectivity with light microscopy. Nat Methods 9(1):96–102
56. Kim WJ, Jan LY, Jan YN (2013) A PDF/NPF Neuropeptide Signaling Circuitry of Male Drosophila melanogaster Controls Rival-Induced Prolonged Mating. Neuron 80(5):1190–1205
57. Kimura K-I et al (2008) Fruitless and doublesex coordinate to generate male-specific neurons that can initiate courtship. Neuron 59(5):759–769
58. Kimura K-I et al (2005) Fruitless specifies sexually dimorphic neural circuitry in the Drosophila brain. Nature 438(7065):229–233
59. Kitamoto T (2000) Conditional modification of behavior in Drosophila by targeted expression of a temperature-sensitive shibire allele in defined neurons. Science (Abstract)
60. Konopka RJ, Benzer S (1971) Clock mutants of Drosophila melanogaster. Proc Natl Acad Sci U S A 68(9):2112–2116
61. Lai S-L, Lee T (2006) Genetic mosaic with dual binary transcriptional systems in Drosophila. Nat Neurosci 9(5):703–709
62. Lee T et al (2000) Essential roles of Drosophila RhoA in the regulation of neuroblast proliferation and dendritic but not axonal morphogenesis. Neuron 25(2):307–316
63. Lee T, Luo L (1999) Mosaic analysis with a repressible cell marker for studies of gene function in neuronal morphogenesis. Neuron 22(3):451–461
64. Lewandoski M (2001) Conditional control of gene expression in the mouse. Nat Rev Genet 2(10):743–755
65. Lima SQ, Miesenböck G (2005) Remote control of behavior through genetically targeted photostimulation of neurons. Cell 121(1): 141–152
66. Liu C et al (2012) A subset of dopamine neurons signals reward for odour memory in Drosophila. Nature 488(7412):512–516
67. Liu G et al (2006) Distinct memory traces for two visual features in the Drosophila brain. Nature 439(7076):551–556
68. Livet J et al (2007) Transgenic strategies for combinatorial expression of fluorescent proteins in the nervous system. Nature 450(7166):56–62
69. Maisak MS et al (2013) A directional tuning map of Drosophila elementary motion detectors. Nature 500(7461):212–216

70. Mank M et al (2008) A genetically encoded calcium indicator for chronic in vivo two-photon imaging. Nat Methods 5(9):805–811
71. Manoli DS et al (2005) Male-specific fruitless specifies the neural substrates of Drosophila courtship behaviour. Nature 436(7049): 395–400
72. Marella S, Mann K, Scott K (2012) Dopaminergic modulation of sucrose acceptance behavior in Drosophila. Neuron 73(5):941–950
73. Meinertzhagen I, O'Neil SD (1991) Synaptic organization of columnar elements in the lamina of the wild type in Drosophila melanogaster. J Comp Neurol 305(2):232–263
74. Mellert DJ, Truman JW (2012) Transvection is common throughout the Drosophila genome. Genetics 191(4):1129–1141
75. Melom JE, Littleton JT (2013) Mutation of a NCKX eliminates glial microdomain calcium oscillations and enhances seizure susceptibility. J Neurosci 33(3):1169–1178
76. Miesenböck G, Angelis DD, Rothman J (1998) Visualizing secretion and synaptic transmission with pH-sensitive green fluorescent proteins. Nature 394(July):192–195
77. Miller M et al (2009) TU-tagging: cell type specific RNA isolation from intact complex tissues. Nat Methods 6(6):439–441
78. Minsky M (1961) Microscopy Apparatus. , p.US Patent 3,013,467
79. Miyawaki A et al (1997) Letters to nature fluorescent indicators for Ca 2+ based on green fluorescent proteins and calmodulin. Nature 388:882–887
80. Morante J, Desplan C (2008) The color-vision circuit in the medulla of Drosophila. Curr Biol 18(8):553–565
81. Mosca TJ et al (2005) Dissection of synaptic excitability phenotypes by using a dominant-negative Shaker K+channel subunit. Proc Natl Acad Sci U S A 102(9):3477–3482
82. Muzumdar MD et al (2007) A global double-fluorescent Cre reporter mouse. Genesis 605(September):593–605
83. Nakai J, Ohkura M, Imoto K (2001) A high signal-to-noise Ca(2+) probe composed of a single green fluorescent protein. Nat Biotechnol 19(2):137–141
84. Ni J-Q et al (2009) A Drosophila resource of transgenic RNAi lines for neurogenetics. Genetics 182(4):1089–1100
85. Nitabach MN et al (2008) Electrical Hyperexcitation of Lateral Ventral Pacemaker Neurons Desynchronizes Downstream Circadian Oscillators in the fly circadian circuit and induces multiple behavioral periods. J Neurosci 26(2):479–489
86. O'Kane C, Gehring W (1987) Detection in situ of genomic regulatory elements in Drosophila. Proc Natl Acad Sci U S A 84(December):9123–9127
87. Otsuna H, Ito KEI (2006) Systematic analysis of the visual projection neurons of Drosophila melanogaster. I Lobula-Specific Pathways. J Comp Neurol 497:928–958
88. Peled ES, Isacoff EY (2011) Optical quantal analysis of synaptic transmission in wild-type and rab3-mutant Drosophila motor axons. Nat Neurosci 14(4):519–526
89. Pfeiffer BD et al (2008) Tools for neuroanatomy and neurogenetics in Drosophila. Proc Natl Acad Sci U S A 105(28):9715–9720
90. Pfeiffer BD, Truman JW, Rubin GM (2012) Using translational enhancers to increase transgene expression in Drosophila. Proc Natl Acad Sci U S A 109(17):6626–6631
91. von Philipsborn AC et al (2011) Article Neuronal Control of Drosophila Courtship Song. Neuron 69(3):509–522
92. Pielage J et al (2008) A presynaptic giant ankyrin stabilizes the NMJ through regulation of presynaptic microtubules and transsynaptic cell adhesion. Neuron 58(2):195–209
93. Pilling AD et al (2006) Kinesin-1 and dynein are the primary motors for fast transport of mitochondria in Drosophila motor axons. PLoS One 17(4):2057–2068
94. Potter C et al (2010) The Q System : A Repressible Binary System for Transgene Expression, Lineage Tracing, and Mosaic Analysis. Cell 141(3):536–548
95. Reiff DF, Thiel PR, Schuster CM (2002) Differential regulation of active zone density during long-term strengthening of Drosophila neuromuscular junctions. J Neurosci 22(21): 9399–9409
96. Root CM et al (2009) A Presynaptic Gain Control Mechanism Fine-Tunes Olfactory Behavior. Neuron 59(2):311–321
97. Rust MJ, Bates M, Zhuang X (2006) Sub-diffraction-limit imaging by stochastic optical reconstruction microscopy (STORM). Nat Methods 3(10):793–795
98. Ruta V et al (2010) sensory input to descending output. Nature 468(7324):686–690
99. Schnell B et al (2012) Columnar cells necessary for motion responses of wide-field visual interneurons in Drosophila. J Comp Physiol
100. Schroll C et al (2006) Light-induced activation of distinct modulatory neurons triggers appetitive or aversive learning in Drosophila larvae. Curr Biol 16(17):1741–1747
101. Siddiqi O, Seymour B (1976) Neurophysiological defects in temperature-sensitive paralytic mutants of Drosophila

melanogaster. Proc Natl Acad Sci U S A 73(9):3253–3257
102. Silies M et al (2013) Modular Use of Peripheral Input Channels Tunes Motion-Detecting Circuitry. Neuron 79(1):111–127
103. Stocker R et al (1990) Neuronal architecture of the antennal lobe in Drosophila melanogaster. Cell Tissue Res 262:9–34
104. Stockinger P, Kvitsiani D, Rotkopf S et al (2005) Neural circuitry that governs Drosophila male courtship behavior. Cell 121(5):795–807
105. Strausfeld NJ (1976) Atlas of an insect brain. Springer-Verlag, New York
106. Sun XR et al (2013) Fast GCaMPs for improved tracking of neuronal activity. Nat Commun 4:2170
107. Suzuki DT, Grigliatti T, Williamson R (1971) A mutation (parats) causing reversible adult paralysis. Proc Natl Acad Sci U S A 68(5): 890–893
108. Sweeney ST et al (1995) Targeted expression of tetanus toxin light chain in Drosophila specifically eliminates synaptic transmission and causes behavioral defects. Neuron 14(2):341–351
109. Takemura S et al (2013) A visual motion detection circuit suggested by Drosophila connectomics. Nature 500(7461):175–181
110. Takemura S et al (2011) Cholinergic circuits integrate neighboring visual signals in a Drosophila motion detection pathway. Curr Biol 21(24):2077–2084
111. Takemura S-Y, Lu Z, Meinertzhagen I (2008) Synaptic circuits of the Drosophila optic lobe: the input terminals to the medulla. J Comp Neurol 509(5):493–513
112. Thestrup T et al (2014) Optimized ratiometric calcium sensors for functional in vivo imaging of neurons and T lymphocytes. Nat Methods 11(2):175–182
113. Tian L et al (2009) Imaging neural activity in worms, flies and mice with improved GCaMP calcium indicators. Nat Methods 6(12): 875–881
114. Ting C-Y (2014) Photoreceptor-derived activin promotes dendritic termination and restricts the receptive fields of first-order interneurons in Drosophila. Neuron pp 1–17
115. Tuthill JC et al (2013) Contributions of the 12 neuron classes in the fly lamina to motion vision. Neuron 79(1):128–140
116. De Vries SEJ, Clandinin TR (2012) Loom-sensitive neurons link computation to action in the Drosophila visual system. Curr Biol 22(5):353–362
117. Wang JW et al (2003) Two-photon calcium imaging reveals an odor-evoked map of activity in the fly brain. Cell 112(2):271–282
118. Wang Q et al (2008) Structural basis for calcium sensing by GCaMP2. Structure (Lond 1993) 16(12):1817–1827
119. White BH et al (2001) Targeted attenuation of electrical activity in Drosophila using a genetically modified K(+) channel. Neuron 31(5):699–711
120. Wong AM, Wang JW, Axel R (2002) Spatial representation of the glomerular map in the Drosophila protocerebrum. Cell 109(2): 229–241
121. Wu C-L et al (2013) An octopamine-mushroom body circuit modulates the formation of anesthesia-resistant memory in Drosophila. Curr Biol 23(23):2346–2354
122. Xu T, Rubin GM (1993) Analysis of genetic mosaics in developing and adult Drosophila tissues. Development (Camb) 117(4): 1223–1237
123. Yang C-H et al (2009) Control of the post-mating behavioral switch in Drosophila females by internal sensory neurons. Neuron 61(4):519–526
124. Yang Z, Edenberg HJ, Davis RL (2005) Isolation of mRNA from specific tissues of Drosophila by mRNA tagging. Nucleic Acids Res 33(17):e148
125. Yapici N et al (2008) A receptor that mediates the post-mating switch in Drosophila reproductive behaviour. Nature 451(7174): 33–37
126. Yu JY et al (2010) Article Cellular Organization of the Neural Circuit that Drives Drosophila Courtship Behavior. Curr Biol 20(18):1602–1614
127. Zars T (2000) Localization of a Short-Term Memory in Drosophila. Science 288(5466): 672–675
128. Zhang YQ, Rodesch CK, Broadie K (2002) Living synaptic vesicle marker: synaptotagmin-GFP. Genesis (NY) 34(1–2):142–145
129. Zhao Y et al (2011) An expanded palette of genetically encoded Ca2+ indicators. Science 1888(2011)

Index

A

AAV. *See* Adeno-associated viruses (AAV)
Acetoxymethyl (AM)
 dye 118, 127
 ester 118
ACSF. *See* Artificial cerebral spinal fluid (ACSF)
Adeno-associated viruses (AAV)..................145, 150–151, 156, 237, 238
Agglutinin 31, 51–63
Alexa conjugated 58
Alexa Fluor..................11–13, 17, 26–28, 32, 34, 58, 103, 109, 111, 112, 121, 126, 127, 131, 133–135
Allatostatin..................149, 169–170
AMPA..................233
Amplifier..................62, 88, 89, 93, 94, 145, 154
Analog-to-digital..................145
Anesthesia 92, 114, 119, 121, 128
Anterograde..................2, 5, 6, 8, 11, 18–38, 52, 53, 56, 57, 67–71, 79, 259
Array tomography 236, 237
Artificial cerebral spinal fluid (ACSF)..................121, 127, 128, 130, 132, 135
Axon..................2, 4, 5, 13–18, 20, 22–26, 28, 30–32, 35–37, 39, 53, 55, 58, 59, 62, 67–69, 71, 111, 232, 233, 235, 242, 254, 259, 260, 267

B

Bacterial artificial chromosome (BAC)..................150, 177–213
BACTRAP..................202–203
BDA. *See* Biotinylated dextran amine (BDA)
Bicistronic..................240
Biosensor..................103, 110
Biotinylated dextran amine (BDA) 19–29, 32–35, 37, 39, 69–71, 79
Brainbow..................254
Bregma..................151, 153

C

CamKII..................238
cAMP. *See* Cyclic adenosine monophosphate (cAMP)
Cannula..................145, 152–158
Cell attached..................85
Channelrhodopsin (ChR)..................146–148, 228, 263
Channelrhodopsin-2 (ChR2)..................146–149, 153, 157, 178–180, 209, 228
ChARGe..................149
ChAT. *See* Choline acetyltransferase (ChAT)
Chemical genetics..................161–172
ChETA..................147
Cholera toxin..................17, 32, 68, 70
Choline acetyltransferase (ChAT)..................179, 182–184
ChR. *See* Channelrhodopsin (ChR)
ChR2. *See* Channelrhodopsin-2 (ChR2)
Circuit 12, 32–38, 51–54, 56, 58–61, 63, 71, 72, 85, 89, 101–103, 107, 109, 110, 112, 113, 118, 143–158, 161–163, 165, 166, 171, 172, 177–213, 217–229, 231, 235, 237, 243, 245, 249–270
Clozapine *N*-oxide (CNO)..................170, 171
Conditional knockout..................270
Confocal..................11, 12, 19, 22, 25–28, 30, 32, 34, 35, 37, 68, 69, 108, 119, 179, 233, 238, 239, 242, 244, 245, 258
Contour-clamped homogeneous electric field (CHEF), 186
Craniotomy 86, 92, 96, 121, 122, 130–135, 138, 139, 151, 154
Cre. *See* Cre/loxP
Cre/loxp..................203
Cryopreservation
Cyclic adenosine monophosphate (cAMP)..................169

D

DAB. *See* Diaminobenzidine (DAB)
Dendrite..................2, 4, 5, 10, 14, 15, 22, 23, 26–28, 33–35, 53, 67, 71, 110, 118, 149, 232, 241, 242, 244, 254, 259
Designer receptors exclusively activated by designer drugs (DREADD)..................149, 170–171
Dextran..................8, 11, 19–24, 28, 37, 59, 69, 103, 108, 109, 112
Diacylglycerol (DAG) 169
Diaminobenzidine (DAB)..................4, 6, 7, 21, 22, 79
Digoxigenin..................75
DMCM. *See* Methyl-6, 7-dimethoxy-4-ethyl-beta-carboline-3-carboxylate (DMCM)
DNA preparation 115, 196
DREADD. *See* Designer receptors exclusively activated by designer drugs (DREADD)
Drosophila..................54, 149, 169–171, 236, 249–270

E

Electrocompetent 192, 194, 210
Electrode 32, 88, 90, 92–93, 96, 104, 105, 112, 113, 119, 145, 154

Benjamin R. Arenkiel (ed.), *Neural Tracing Methods: Tracing Neurons and Their Connections*, Neuromethods, vol. 92, DOI 10.1007/978-1-4939-1963-5, © Springer Science+Business Media New York 2015

Electroporation....................101–115, 191–193, 206, 210, 213, 237, 238
Ensembl....................182–184
Epifluorescence....................13, 26, 35, 120, 121, 131–134
External solution....................105, 108

F

FITC....................17
FLEX
Flp/Frt....................190, 193, 195, 218, 228, 254–256
Flumazenil....................168
Fluorescent....................8–9, 11–17, 19, 22–27, 33, 35, 38, 56, 58–61, 68, 69, 73, 75, 77, 102, 103, 106–108, 110–112, 114, 115, 117–139, 150, 165, 166, 178, 219, 224, 233–235, 240, 241, 245, 246, 254, 258–260
Fluoro-gold....................8–15, 33, 34, 68, 70
Forster resonance energy transfer (FRET)....................260

G

GABA....................28, 29, 72, 150, 166, 167, 267
GABAergic....................12, 167, 178, 232, 234
GAL80....................255, 257, 269
GAL4/upstream activating sequence (UAS)....................252, 255, 257, 269
GCaMP....................228, 261, 262
Gene Expression Nervous System Atlas (GENSAT)....................178, 183, 202, 209
Genetically encoded calcium indicators (GECIs)....................260, 261
Genetically encoded voltage indicators (GEVI)....................262
Genotyping....................202
GENSAT. *See* Gene Expression Nervous System Atlas (GENSAT)
GFP. *See* Green fluorescent protein (GFP)
GFP reconstitution across synaptic partners (GRASP)....................234–236, 241
GIRK. *See* G protein-coupled inwardly rectifying potassium (GIRK)
Glutamate-gated chloride channel (GluCl)....................165, 166
Glutamatergic....................12, 178, 232–234
Golgi....................2, 22, 33, 36–38, 40, 51
G protein-coupled inwardly rectifying potassium (GIRK)....................169–171
G-protein-coupled receptors (GPCR)....................72–9, 149, 169, 171
GRASP. *See* GFP reconstitution across synaptic partners (GRASP)
Green fluorescent protein (GFP)....................38, 39, 103, 104, 106, 111, 115, 126, 127, 178, 227, 233–236, 238, 239, 241, 243, 246, 258–261

H

Halorhodopsin....................149, 263
Herpes simplex virus....................69, 218
hM4Di....................170, 171
hM3Dq....................170, 171
Horseradish peroxidase (HRP)....................3–8, 12, 15–17, 19, 24, 25, 31, 32, 36–38, 52, 54–56, 68, 69
Hyperpolarization....................149, 165, 169, 171

I

Immunohistochemical....................11, 12, 17, 18, 20, 21, 25, 26, 59, 68, 69, 71, 73, 239, 267
Inducible systems....................178
InSITE....................257
In situ hybridization....................16, 72–75, 79
In vivo....................11, 18, 39, 54, 56, 58, 59, 62, 85–115, 117–139, 144–146, 150–156, 163–165, 169, 171, 217, 228, 233–235, 237, 239–242, 257, 261, 262
Ionotropic....................162–169
Iontophoretic....................3, 14, 20, 24, 79, 104
Ivermectin....................165, 166

K

Knockin....................150, 164, 165, 239
Knockout....................269–270

L

Laser....................11, 12, 25–28, 30, 32, 34, 35, 37, 68, 86, 117, 120, 121, 124, 125, 136, 145, 146, 153–155, 245
Lectin....................8, 24, 31, 32, 38, 51–53, 69
LED. *See* Light-emitting diodes (LED)
Lentivirus (LV)....................145, 150–151, 156, 238
LexA-*lexAop*....................252
Lidocaine....................92
Light-emitting diodes (LED)....................111, 145, 146, 153
Light-sensitive....................144, 146, 148
Lipofectamine....................223, 224
Lipophilic....................39–40
Loose patch....................86, 90, 92, 93, 96
Loss of function....................148, 154, 220
Loxp. *See* Cre/loxp

M

MARCM. *See* Mosaic analysis with repressible cell marker (MARCM)
mEPSCs....................233
Messenger RNA (mRNA)....................72, 73, 76, 78, 79, 109, 206, 240, 266
Metabond....................89, 92, 145, 153
Metabotropic....................162, 163
Methyl-6, 7-dimethoxy-4-ethyl-beta-carboline-3-carboxylate (DMCM)....................167, 168
Microdrill....................89, 91
Microinjection....................182, 197, 199, 201, 212, 213
Micromanipulator....................88–89, 113

Microscopy........................... 3, 4, 6–9, 11, 12, 16, 17, 19, 20, 22, 24–26, 30, 32, 35–38, 57, 68, 69, 77, 90, 93, 101, 103, 105, 108–110, 114, 115, 117–139, 157, 231, 232, 236–238, 241–242, 244, 245, 258, 259
Microtome.. 7, 28, 33, 236
Monosynaptic.. 112, 218, 220, 235
Morpholino..109
Morphometrics....................................... 102, 106, 109–110
Mosaic...................... 202, 251, 254, 255, 264, 265, 268–269
Mosaic analysis with repressible cell marker (MARCM)... 251, 268, 269
Mouse..................................... 12, 28, 34, 39, 53–58, 60, 61, 77, 86, 87, 91, 92, 94, 95, 121–124, 128–130, 165, 166, 170, 178, 179, 182, 183, 186, 191, 197, 203, 204, 207, 209, 212, 232, 236, 239–241, 246, 254
mRNA. *See* Messenger RNA (mRNA)

N

NaChBac...263
nAChR. *See* Nicotinic acetylcholine receptor (nAChR)
Nanoparticles.. 15–16, 57, 165
Natronomonas pharaonis, halorhodopsin (NpHR).............149
Neuroanatomical............................. 1–40, 54, 67–69, 79–81
Neurobiotin.. 36–37, 39, 90, 92–94
Nicotinic acetylcholine receptor (nAChR)......................168
N-methyl-d-aspartate (NMDA)...............................31, 234
NpHR. *See Natronomonas pharaonis*, halorhodopsin (NpHR)

O

Oncoretrovirus..238
Opsin............................... 144, 146–151, 155, 157, 162, 178
Optogenetics..144, 146, 153, 154, 156, 157, 161, 172, 178
Oregon Green BAPTA (OGB).......................................106
Oscilloscope...89, 94, 105, 107, 114

P

PFGE. *See* Pulsed-field gel electrophoresis (PFGE)
PHA-L. *See Phaseolus vulgaris*-leucoagglutinin (PHA-L)
Pharmacologically selective actuator modules (PSAM)...168
Pharmacologically selective effector molecules (PSEM)...168
Phaseolus vulgaris-leucoagglutinin (PHA-L)...................... 21, 23–27, 32, 35, 69, 71, 79
Phenotype...............................202, 203, 264, 265, 268, 269
PhiC31...252, 254
Photoactivateable green fluorescent protein (GFP)..........259
Photoconversion
Photomultiplier tubes (PMTs)..........121, 124, 125, 136, 245
PLA. *See* Protein-protein interactions (PLA)
PolyA Signal
Postsynaptic..22, 28, 37, 70, 106, 167, 232–234, 237, 258, 260, 267
Presynaptic... 28, 218, 219, 228, 229, 233–235, 243, 259, 262, 266, 267
Pronuclear........................179, 196, 197, 201, 202, 212, 213
Propidium iodide...8, 68
Protein-protein interactions (PLA)......................76–79, 109
PSAM. *See* Pharmacologically selective actuator modules (PSAM)
PSD-95.............106, 107, 109, 110, 233, 234, 238, 239, 243
PSEM. *See* Pharmacologically selective effector molecules (PSEM)
Pulsed-field gel electrophoresis (PFGE)......... 181, 186–188, 192, 195–197, 201, 204–206, 208, 212

Q

QF-QUAS..252, 255

R

Rabies virus..................................69, 70, 112, 217–229, 235
Recombination............. 57, 58, 150, 189–191, 193–195, 203, 205–207, 209, 210, 240, 251, 253, 254, 265, 268–270
Recombineering........................181, 188, 190–196, 202–206
Retrograde.......................................2–17, 19, 20, 22, 23, 25, 31–38, 52, 53, 56, 67–81, 112, 218–220, 235, 259, 267
Riboprobe...72–75
RNA-mediated gene interference (RNAi)....... 109, 266, 267
Rosa 26 reporter...228, 229

S

SCE. *See* Single-cell electroporation (SCE)
shRNA...109, 220
Single cell.. 33, 101–115, 223, 224, 250, 251, 253, 254, 258, 265, 267, 270
Single-cell electroporation (SCE)...........................101–115
Split GFP...236, 259–261
Step-function opsins (SFO)..148
Stereotaxic...20, 28, 33, 91, 119, 121–124, 126, 128, 130, 132, 243
Synapse.......................................2, 5, 23, 26, 32, 37, 53–55, 69, 106, 107, 109, 110, 231–247, 250, 258, 263
Synapsin...238

T

T2A..240
Time-lapse...............................102, 104, 109, 110, 112, 242
Tracing.. 1–40, 51–63, 68–70, 72–81, 85, 101, 103, 217–229, 235, 243, 259
Tract-tracing.. 3, 9, 31, 35, 67–81
Transactivator...163
Transfection....... 102–104, 106–109, 112, 147, 163, 223, 233
Transformation...210
Transgene..150, 177, 178, 180, 182, 189, 191, 194, 202, 204–206, 209, 212, 220, 222, 238, 239, 241, 244, 251–253, 258, 260, 266, 267

Transgenesis 202, 238, 239, 258, 260
Transient receptor potential cation channel subfamily V member1 (TrpV1)................................ 149, 164, 165
Transneuronal..17, 54, 57, 241, 243
Transposon 252, 253
Transsynaptic..................................54, 70, 71, 217–229, 235
TrpV1. *See* Transient receptor potential cation channel subfamily V member1 (TrpV1)
Two-photon (2-photon)................................ 103, 104, 108, 109, 111, 241–242, 259

U

Ultrastructural 3, 7, 23, 25, 56, 68, 69, 71, 232, 233, 235, 237
Untranslated regions (UTRs)..258

V

Vectors.................................... 16, 22, 36, 79, 145, 150, 157, 163, 170, 179, 180, 183, 184, 186, 187, 189–191, 193, 194, 199, 203, 204, 206, 207, 209, 210, 213, 217–223, 228, 236, 238, 240, 241, 243, 244
Vesicular glutamate transporter 2 (VGlut2)... 28–30, 72, 178
Virus..................................... 58, 69, 70, 102, 112, 145, 146, 150, 151, 153–157, 165, 167, 217–229, 235, 240

W

Wheat germ agglutinin (WGA) 16, 17, 19, 25, 31, 32, 51–63
Wheat germ agglutinin – horseradish peroxidase (WGA–HRP) 17, 25, 31, 32, 53–56, 59
Wheel velocity sensor...88, 91, 95
Whole cell 121
WPRE

X

Xenopus.. 103, 108, 109, 113–114
XFP 240, 246

Z

Zolpidem..166–168

Printed by Printforce, the Netherlands